D1473438

EMERGING COMMUNICATIONS TECHNOLOGIES

Other Books by Uyless Black:

Computer Networks: Protocols, Standards, and Interfaces, 2/E, © 1993, 450 pp., cloth, ISBN: 0-13-175605-2

OSI: A Model for Computer Communications Standards, © 1991, 640 pp., cloth, ISBN: 0-13-637133-7

Data Networks: Concepts, Theory, and Practice, © 1990, 650 pp., cloth, ISBN: 0-13-198466-7

Data Communications and Distributed Networks, 3/E, © 1993, 448 pp., cloth, ISBN: 0-13-203464-6

Data Link Protocols, © 1993, 271 pp., cloth, ISBN: 0-13-204918-X

EMERGING COMMUNICATIONS TECHNOLOGIES

PRENTICE HALL SERIES IN
ADVANCED COMMUNICATIONS TECHNOLOGIES

UYLESS BLACK

PTR Prentice Hall
Englewood Cliffs, New Jersey 07632

CARROLL COLLEGE LIBRARY
Waukesha. Wisconsin 53186

Library of Congress Cataloging-in-Publication Data

Black, Uyless D.
    Emerging communications technologies / Uyless Black.
        p. cm.
    Includes bibliographical references and index.
    ISBN 0-13-051500-0
    1. Telecommunication. I. Title.
    TK5101.B553 1994
    621.382—dc20   94-20007
                                                                CIP

Series Editor: Uyless Black
Acquisition Editor: Mary Franz
Cover Design: Bruce Kenselaar
Cover Photo: Reginald Wickham
Cover Design Director: Eloise Starkweather Muller
Manufacturing Manager: Alexis R. Heydt
Compositor/Production Services: Pine Tree Composition, Inc.

© 1994 PTR Prentice Hall
Prentice-Hall, Inc.
A Paramount Communications Company
Englewood Cliffs, New Jersey 07632

The publisher offers discounts on this book when ordered in bulk
quantities. For more information, contact Corporate Sales Department,
PTR Prentice Hall, 113 Sylvan Avenue, Englewood Cliffs NJ 07632.
Phone: 201–592–2863; FAX: 201–592–2249.

All rights reserved. No part of this book may be
reproduced, in any form or by any means,
without permission in writing from the publisher.

Printed in the United States of America

10 9 8 7 6 5 4 3 2

ISBN: 0-13-051500-0

Prentice-Hall International (UK) Limited, *London*
Prentice-Hall of Australia Pty. Limited, *Sydney*
Prentice-Hall Canada Inc., *Toronto*
Prentice-Hall Hispanoamericana, S.A., *Mexico*
Prentice-Hall of India Private Limited, *New Delhi*
Prentice-Hall of Japan, Inc., *Tokyo*
Simon & Schuster Asia Pte. Ltd., *Singapore*
Editora Prentice-Hall do Brasil, Ltda., *Rio de Janeiro*

With no disrespect intended for the citizens of Maine and Texas, this book is dedicated to Henry David Thoreau, who, in 1845, penned these words in his book *Walden:*

> We are in great haste to construct a magnetic telegraph from Maine to Texas; but Maine and Texas, it may be, have nothing important to communicate.

It appears now, after some 150 years, that not only do the citizens of Maine and Texas have something to communicate, but much of the world as well. This book is a modest attempt to explain the technical underpinnings of the new "magnetic telegraphs" that will carry our information societies into the twenty-first century.

At the same time it is a good idea to remember another Thoreau thought, also from *Walden:*

> After all, the man whose horse trots a mile in a minute does not [necessarily] carry the most important message.

Contents

CHAPTER 2   **Foundations for the Emerging
Technologies**   **39**

CHAPTER 5   **Fiber Distributed Data Interface
                  (FDDI)**   **134**

CHAPTER 6   **Metropolitan Area Networks (MANs) and Switched Multi-Megabit Data Service (SMDS)**   **162**

Preface

Like most of the more recent books I have written, I decided to write this book as part of my ongoing work with my clients because of the absence of a systematic analysis of the subject matter. In this case, my clients could not find a book that provided a description and comparison of the new technologies that are appearing in the telecommunications industry. Generally, some books are available that are accurate, and well-written, but they treat only one, two, or three of the technologies. Obviously, the problem was not specific to my clients alone, so I decided to prepare this book for other interested readers as well.

The book is the culmination of a series of lectures I have been conducting in North America and Europe, the fruition of my lecture notes, and a reflection of ongoing consulting work with my clients. I thank those people who attended these lectures; in effect, they were my "referees" of the technical content of this book. Additionally, I wish to thank three other referees, furnished by my publisher, who provided very valuable input to the finished manuscript.

The book is intended for readers who are interested in the fields of telecommunications and computer-based networks. It can also serve as supplemental reading for advanced networking classes in colleges and universities.

My goals in writing this book are threefold. First, I wish to provide the reader a general description of how the emerging communications technologies operate and where they fit in a computer/communications network and in customer equipment. Second, I wish to provide the reader with a general comparison of the technologies, their pros and cons, where they do or do not compete with each other, and their targeted applications. Third, I wish to provide the reader with a general explanation of the international standards that are published for these emerging communications technologies. I have attempted to achieve a balance between a detailed and general treatment of the subject matter.

To aid the reader in understanding the material, chapters devoted to a specific technology have the same (or nearly the same) format to allow comparison of the technologies. With some minor exceptions, the format is as follows (where XXX is the name of the technology):

Introduction
The Purpose of XXX
Pertinent Standards
Typical XXX Topology (or Topologies)
The XXX Layers
The XXX Protocol Data Unit (PDU) (or units)
XXX Operations in More Detail
Network-to-Network (NNI) Operations (if appropriate)
Other Notable Aspects of XXX
The XXX MIB
XXX Worksheet
Summary

Each technology-specific chapter contains a "Worksheet" which the reader can fill in after reading the chapter. I have included a completed worksheet with an explanation of the reasons for my answers.

Tutorial information is also provided for the reader—Appendix A is a basic tutorial on communications systems and networks; Appendix B is a tutorial on layered protocols; Appendix C is a tutorial on management information bases (MIBs); and Chapter 2 is a tutorial on how to analyze a communications protocol. Even though Chapter 2 can be considered as a tutorial, I think that it will be useful to all readers, because it provides a systematic approach to the analysis of communications protocols.

The emphasis of this book is on the wire-based emerging communications technologies, and the principal focus is on the role of optical fiber in these technologies. The emerging communications technologies using wireless media, such as cellular and cordless systems, are of such scope to warrant another book. Notwithstanding, for purposes of comparisons and completeness, I have included an overview of this subject in Chapter 9.

The reader will notice that I have included some topics that could be considered as "old," not emerging technologies—as example, frame relay and FDDI. I chose to include frame relay because, while the underlying technology is not new, the service offerings are new, and the ideas of committed information rate and traffic tagging, for example, are also new. Additionally, FDDI is not new, but some vendors are positioning FDDI to compete with some of the emerging technologies. Therefore, frame relay and FDDI are appropriate subjects for this book.

The very nature of the subject matter of this book, *emerging* communications technologies, makes it difficult to write about the topic as if these systems and protocols were cast in stone. Nonetheless, this book reflects the most up-to-date information available at the time the book went to press. The latest information from the ATM Forum, The Frame Relay Forum, The Network Management Forum, the SMDS Interest Groups, the standards groups, and my ongoing work are included in this book.

Notes for the reader:

The International Telegraph and Telephone Consultative Committee (CCITT) has changed its name to the International Telecommunication Union-Telecommunication Standardization Sector (ITU-T).

Unless identified otherwise, the term bandwidth is used in this book to describe a channel's capacity in bits per second, and not a frequency spectrum.

The term NNI is used in this book to mean the network-to-network interface. NNI is also known as the node-to-node interface. With either term, NNI is supposed to describe a switch-to-switch interface, which could operate within a network or between networks. The major goal of an NNI is to allow switches from different vendors to interwork with each other.

The initials ICI means intercarrier interface, and is used in this book to convey the same meaning as NNI.

The actual node-to-node interface specifications (say, for operations within a network) for several of the technologies discussed in this book are being developed, and are not yet available to the public.

CHAPTER 1

Introduction

INTRODUCTION

This chapter examines the present communications system infrastructure and focuses on some of the problems that are associated with using the current technology. In so doing, a case is made for the need to implement new communications technologies to overcome deficiencies of the current technologies. An analysis is made of the communications requirements of upcoming (and present) user applications. Also, and as a prelude to subsequent chapters, a general overview is provided of the emerging communications technologies that are being developed to meet the needs of these applications.

THE NEED FOR ENHANCED SERVICES

The Past

The computer/communications industry is undergoing one of the most intense and dramatic changes in its brief history. In the past few years, the progress made in the development of high-speed, inexpensive computers has opened a vista to new and powerful applications. As examples, interactive real-time simulations, three-dimensional modeling, and color images are redefining how a person interacts with the computer. The positive responses from the users of these systems and their demand for more functions in their applications are paving the way toward even more powerful and productive applications.

With the rapid evolution and deployment of automated, complex, and large information systems (ISs), there has been a concomitant need for powerful computers, databases, and networks to support the IS applications. At the time of this book's publication, the telecommunications infrastructure is being revamped to adequately support upcoming information systems. It is now recognized that what has been sufficient in the past will not be sufficient for the future.

The T1/E1 and X.25 legacies. During the 1970s and early 1980s, many people expected that the evolving digital networks centering on the T1/E1 carrier systems would continue to be sufficient to handle most user applications. After all, these systems provide transfer rates of 1.5 to 2.048 million bits per second (Mbit/s), respectively, and have proven adequate to meet the requirements of most user applications (which typically need communications networks that support a transfer rate of a few kilobits per second [kbit/s]).

Notwithstanding, some people in that "distant past" recognized that future applications would require much greater transmission capacities. The increased use and experimentation with integrated voice, video, and graphics applications revealed that future applications would require networks that operate in the multi-megabit per second range. Of course, the communications requirements of these applications would overwhelm the communications facilities of that time.

During the 1970s and 1980s, X.25 was also considered to be an effective solution for transporting end user data traffic across a network to another end user, and X.25 has served the industry well for two decades. However, the X.25 technology was targeted for low-capacity wide area networks (WANs) and was not designed for high-speed requirements— for example, the megabit rates of local area networks (LANs).

LAN internetworking. Nonetheless, many user applications that were developed in the early 1980s began to operate on high-capacity LANs, and the revolution in the personal computer (PC) industry spawned a wide variety of new applications that worked well with the rapidly expanding LAN technology.

As these LANs interconnected to each other to share resources and exchange data, telephone companies were also improving their wide area technology by installing optical fiber in their trunk lines. The idea was to enhance the quality and speed of their networks to ease the task of inter-networking LANs and, of course, to encourage users to stay with their services. The telephone companies also installed faster equipment and brought 45 Mbit/s (in North America) and 140 Mbit/s (in Europe) services to the user.

Unfortunately, some of the LAN interconnection efforts and most of the development of the high-speed carrier equipment took place without regard for the need to use standardized protocols between different vendors' hardware and software. Consequently, many systems in the marketplace today do not interface gracefully with each other, even though the systems provide similar services. The result is an overly complex and expensive environment which entails extensive and time-consuming conversion operations to enable the heterogeneous systems to interwork with each other.

While compatibility problems remained a problem, the end user was not affected directly, and the 45 Mbit/s and 140 Mbit/s services provided adequate bandwidth for most large enterprises. But for smaller enterprises, they were too costly. The leasing of a high-speed circuit to connect two computers at a price of several thousand dollars a month was simply beyond the budget of a great number of potential network users.

For a growing number of users, and telecommunications providers, the systems in place in the 1980s were perceived to be inadequate to meet the needs of both users' and vendors' applications. Consequently, efforts began in the mid-1980s to address the needs of future applications.

The Future Has Become the Present

In just a few short years, the future applications are now the present applications. The cliché "the future is now" is silly, but it certainly captures the milieu of the dynamic computer/communications industry, because the environment is changing; and it is changing rapidly.

The communications industry is moving toward powerful gigabit networks that support any type of application. Moreover, for the first time in the history of the computer/communications industry, network administrations throughout the world are embracing a worldwide set of computer/communications standards for use on high-capacity, multimedia networks. Without question, this trend will lead to enhanced services to the end user, and will result in significant gains in productivity with a decrease in the cost of doing business.

The underlying foundations for these networks—the emerging communications technologies—are the subject of this book.

Goals of the Emerging Communications Technologies

The emerging communications technologies support integrated, high-capacity networks that contain multi-vendor computers and software packages that operate with different carriers. Presently, the vast majority of communications packages differ across vendor networks. One key goal of the emerging technologies is to develop a set of procedures and standards that apply equally well across different vendors and networks.

Of course, the creation of integrated networks using common interfaces and protocols, while certainly beneficial, is not the only goal of these emerging technologies. Another important goal is to provide more throughput capacity (in bit/s) for user applications and to perform network operations faster on behalf of the applications.

This means that the emerging technologies are designed to provide for high throughput, with very fast transmission speeds, while incurring little delay. Indeed, integrated standards, with high throughput and low delay, are the cornerstones of these technologies.

Another important goal of these technologies is to support any type of application, such as voice, video, music, FAX, telemetry, etc. (actually, we shall see that most of the technologies are not quite this ambitious). An appropriate term for this service is multi-application networks, although most people use the term multimedia networks. The cost-effective implementation of multimedia networks is proving to be one of the biggest challenges facing the industry, and we shall have much to say about this subject throughout the book.

From the standpoint of the network provider, another important goal of the emerging communications technologies (at least, some of them) is to provide for more and better network management tools. At first glance, this factor may not mean much to an end user, but these tools allow the network provider to monitor carefully the network facilities and to provide a robust and relatively error-free service for the user applications. In contrast to the T1/E1-based systems, which have very limited network management features, the new carrier technologies use about 5 percent of the network's bandwidth for management. Since the communications channels are comprised of optical fiber, sufficient bandwidth is available to support this important operation.

Finally, another of the principal goals of several of the emerging technologies is the provision for "seamless" interconnections between network hardware/software and "seamless" internetworking between networks. The term *seamless* is used here to connote that an end user (or even a network administrator) is not aware that user traffic is transported through equipment that is manufactured by different vendors over different networks. The networks may be local or remote; they may consist of one vendor's equipment and software, or many. Yet, operations are transparent to the user (and ideally, to a network administrator).

Without question, these are lofty goals, but they can be achieved.

LAN Interconnectivity

As we have just learned, one of the driving forces behind several of the emerging communications technologies is the need of many enterprises to connect two or more LANs together across WANs. Of course, the network administrator would like to be provided with seamless interconnectivity between these networks.

This requirement is becoming quite pervasive throughout the computer/communications industry. The need for LAN interconnectivity is a result of the rapid growth in the number of LANs installed and the number of PCs (and other devices) attached to these LANs. The installed LAN base in the world is continuing to grow, as is the attachment of PCs and workstations to these LANs.

We can summarize some of the thoughts of the previous discussions by stating that the increasing power of computers continues to allow us to develop more powerful and productive applications. But if these applications are to communicate effectively with each other, it becomes essential that the intervening networks have the capacity to support their interactions in a seamless manner. We have made the case that, until recently, these intervening networks were somewhat adequate for the task. Let us now take a look at what the upcoming requirements will be for these new technologies.

Need for Greater Communications Capacity

In the early days of computing, a network or communications link that operated in the range of 300 to 600 bit/s was considered adequate to handle most user applications. (These "early" days were less than 20 years ago.) Consider the industry only five years ago: a 9.6 kbit/s channel with attached modems was considered to be a rather esoteric and high-speed technology. For most users, the 9.6 kbit/s channel was certainly considered to be a "high-capacity" communications support function for a personal computer or a workstation.

Well, how fast things do change in our industry. Today, a 9.6 kbit/s speed is woefully inadequate for many applications. As an example, the reader might logon to a network and retrieve a weather map. At a transfer speed 9.6 kbit/s, it takes about 40 seconds to fill the screen with a low-quality black and white map. One must wait several minutes for a high-resolution color map to be completely transmitted. This delay may be acceptable for some users and some applications, but it is clearly not a very productive way to work.

The weather map retrieval and display is actually a modest requirement for bandwidth. The need for more capacity, beyond this simple black-and-white image transmission, is growing as more applications are developed that provide data, voice, and video services. With the increased interest and need for applications that support color images, voice, and video, a multi-megabit communications infrastructure becomes essential.

Figure 1–1 shows the cumulative bandwidth needs for these new applications. In the 1980s, most applications concerned themselves with text and numbers, and there was little need for high-capacity networks to support the transmissions of this traffic between user applications. In just a few short years, we have witnessed the rise of applications that support bit-mapped graphics, color, sound, and even video. The cumulative bandwidth requirements for *one* of these applications to send and receive all these images to another similar application now approach the multi-megabit range.

One simple example of a current application requirement will demonstrate the need for increased bandwidth—the FAX. Consider an application that transmits several pages of a bank account document between two computers. If this document is transmitted through a FAX machine that does not perform data compression, approximately 40–50 million bits are needed to represent the information!

Of course, one can say this supposition is faulty, because FAX machines perform compression operations on the traffic before it is transmitted. Yet, compression techniques are limited in their ability to decrease the number of bits representing an image. As effective as they are, we cannot continue to rely on compression techniques to mask the limited capacity of the communications channel. Nor should we; optical fiber provides the bandwidth we need.

Eventually, as optical fiber finds its way into local loops and the interfaces on workstations, one may not be too concerned about the

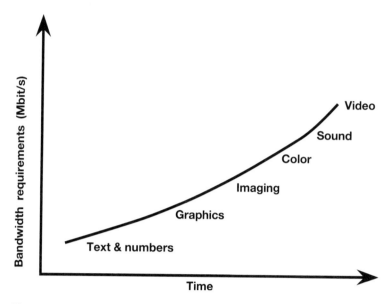

Figure 1–1
The increased need for more capacity. Note that increased capabilities over time create a cumulative increase in bandwidth. [BNR 92]

amount of data that needs to be transmitted vis-à-vis the transmission speeds available on the media. For the foreseeable future, limited bandwidth will remain a problem. For example, when using a 64 kbit/s circuit, several to many seconds are required to send a relatively small amount of traffic (such as spreadsheets, source programs, and graphical images).

What the applications will need. What are the requirements for these powerful applications? A study conducted by the Xerox Palo Alto Research Center (PARC) shows that a wide array of applications need a large communications capacity (large bandwidth). Table 1–1 lists several types of applications that require large bandwidths. They are typically video transmission applications or high-quality full-color still images. Until recently, cost-effective network solutions to support these types of applications appeared somewhat dim (no pun intended). However, significant advances have been made in compression techniques, which allow some current technologies to handle these applications.

Table 1–1 also shows the requirements for the various types of applications, both for uncompressed and compressed traffic. The high-

Table 1–1   Various Applications Needs (Image Transmission) and Bandwidth Requirements [LYLE 92]

Description of Media	Uncompressed Speed Range (Mbit/s)	Compressed Speed Range (Mbit/s)
Broadcast-quality NTSC video	120	3–6
Studio-quality NTSC video	216	10–30
High-definition video broadcast quality	1500	20–30
Visualization imagery, full motion @ 60 fps, 1000 × 1000 pixels	1500–2500	50–200
Monochrome binary still* images, 600 × 600 spots/inch, 135 pp/min	120	5–50
Full-color still images** 400 × 400 pixels/inch, 60 pp/min	500	45

* Derived from characters of the Xerox Docutech Electronic Copier System

** Authors' estimates of characteristics of hypothetical high-performance distributed color copiers

definition TV (HDTV) technology (which is expected to become a significant consumer-driven technology over the next decade) requires 1500 Mbit/s for uncompressed traffic and even 20–30 Mbit/s for compressed traffic.

Interestingly, still images have a very large bandwidth requirement vis-à-vis moving images. The reason for this seeming anomaly is that still images require very high resolution for display. Resolution for still images must be higher than for moving display images, because the human eye tends to "study" this image more closely than a moving image. In addition, the very high-speed printers today print at such a fast rate (about 2 pages per second) that these printers have a frame rate which is only about 15 times slower than NTSC video.

The bottom line is that for high-quality video to reside at a user workstation, PARC's study shows that the industry will need at least a 100 Mbit/s link at a user desktop workstation.

While it might appear that television and color images represent the extent of the need for gigabit networks, several other applications are surfacing that need this capacity; multimedia conferencing is one of them. Teleconferencing can save an enterprise a considerable amount of money in travel, lodging, and lost time, and some of my clients have reduced their budgets dramatically in this area. Yet, *each* multimedia conference must operate in the megabit range, and individual conferencing between two people requires the same amount of bandwidth. In some enterprises, several of these conferences are conducted *continuously each day*.

Another bandwidth-thirsty application that is just coming into being is a concept called *virtual reality* (which is discussed in more detail in Chapter 2). While the initial implementations of virtual reality are somewhat frivolous, the technology does hold tremendous potential. Visualization systems can be used by explorers, geologists, medical examiners, etc., to facilitate their work and activities. It is clear that virtual reality systems will be operating in the megabit range.

Beyond the raw bandwidth needs of these applications, it is also becoming clear that some applications require more control over a variety of operations that, heretofore, have been relegated to the control of the network administrator. Some users, for example, need to know what the network delay will be between the two communicating applications. Equally important to some applications is the ability to have control over the loss of traffic, as well as failure recovery. Even the control of network/user set-up times is becoming important for certain applications, as is the ability to predict the set-up time vis-à-vis the amount of traffic and delay. For example, a name server query may require numerous transactions through an internet to/from various name servers. For one query, the service set-up time may be quite significant, but for a longer session, the set-up time may not be significant. Therefore, some applications are emerging that will need these types of quality of service features.

LAN AND WAN INTERNETWORKING

As mentioned earlier, one of the issues that has gained considerable attention in the past few years is the need to interconnect LANs and WANs. Before discussing this issue, it should prove helpful to compare and contrast these two types of networks (Table 1–2).

Until recently, it has been relatively easy to define WANs and LANs and to point out their differences. It is not so easy today, because the terms *wide area* and *local area* do not have the meaning they once had. For example, a LAN in the 1980s was generally confined to a building or a campus where the components were no more than a few hundred or a

Table 1–2   LANs and WANs Compared and Contrasted

Wide Area Networks (WANs):

- Multiple user computers connected together

- Machines are spread over a "wide" geographical region

- Communications channels between the machines are usually furnished by a *third party* (for example, the telephone company, a public data network, a satellite carrier)

- Channels are of relatively low capacity (measuring throughput in kilobits per second [kbit/s])

- Channels are relatively *error-prone*
  (for example, a bit error rate of 1 in 100,000 bits transmitted)

Local Area Networks (LANs):

- Multiple user computers connected together

- Machines are spread over a "small" geographical region

- Communications channels between the machines are usually *privately* owned

- Channels are of relatively high capacity (measuring throughput in megabits per second [Mbit/s])

- Channels are relatively *error-free* (for example, a bit error rate of 1 in 10^9 bits transmitted)

few thousand feet from each other. Today, LANs may span scores of miles.

Nonetheless, certain characteristics are unique to each of these networks. A WAN is usually furnished by a third party. For example, many WANs are termed public networks because the telephone company or a public data network (PDN) vendor owns and manages the resources and rents these services to users. In contrast, a LAN is usually privately owned. The cables and components are purchased and managed by an enterprise.

LANs and WANs can also be contrasted by their transmission capacity (in bit/s): Most WANs generally operate in the kbit/s range, whereas LANs operate in the Mbit/s range.

Another principal feature that distinguishes these networks is the error rate. WANs are generally more error prone than LANs due to the wide geographical and relatively hostile (weather, etc.) terrain over which the media must be laid. In contrast, LANs operate in a relatively benign environment, because the data communications components are housed in buildings where humidity, heat, and electricity ranges are controlled.

Costs of Connecting Dispersed LANs

In the past few years, LANs and WANs have been interconnected with bridges, routers, gateways, and packet-switched networks. These internetworking units are interfaced into the LANs and WANs through dedicated communications channels (leased lines).

As a general practice, leased lines are "nailed up" end-to-end between the user's customer premises equipment (CPE). The user is provided with the leased line on a dedicated basis, and the full transmission capacity is available 24 hours a day (with some exceptions). Therefore, the user pays for the circuit regardless of its utilization. Moreover, if a connection is needed to yet another location (say, another city in the country), another leased line must be rented from the common carriers, once again on an end-to-end, continuous basis.

The use of these lines to connect internetworking units and LANs and WANs is a very expensive process. Moreover, reliability problems occur because individual point-to-point leased lines have no backup capability. In addition, the extensive processing of the traffic (edits, error checks) has created unacceptable time delays for certain applications.

A better approach is to develop a LAN/WAN-carrier network which provides efficient switching technologies for backup purposes as well as high-speed circuits—a network that will allow users to share the expensive leased lines. This concept is called a *virtual private network* (VPN).

The Virtual Private Network (VPN)

The VPN is so named because an individual user shares communications channels with other users. Switches are placed on these channels to allow an end user to have access to multiple end sites. Ideally, users do not perceive that they are sharing a network with each other, thus the term virtual private network—you think you have it, but you don't.

Figure 1–2 illustrates the concepts of VPNs and their advantage over fully meshed private point-to-point systems. In Figure 1–2a, four customer sites are connected to the customer premises equiment (CPE, which is typically a router or some other type of internetworking unit) through local loops to interexchange carriers (IXC). The local exchange

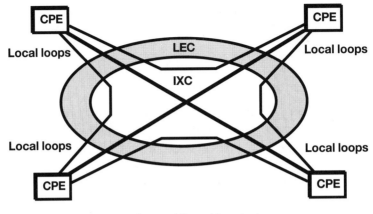

a. Leased lines (4 nodes)

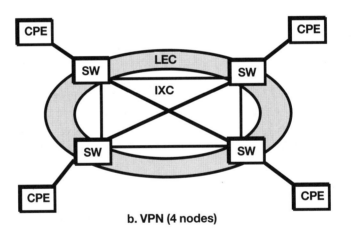

b. VPN (4 nodes)

Figure 1–2
Virtual private networks vs. leased lines. [LISO91], [BNR92a]

carriers (LECs) terminate their local loops to the CPE and the IXC. In this example, a fully meshed leased line network requires six IXC leased lines, 12 dedicated lines to the IXC (local loops), and 12 router ports (3 at each router).

In contrast, through the use of a VPN (Figure 1–2b), the CPE only requires one dedicated local loop to the VPN switch (SW). Thereafter, the

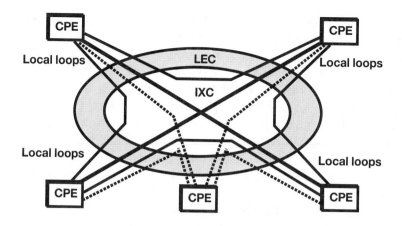

a. Leased lines (5 nodes)
Dashed lines indicate new facilities

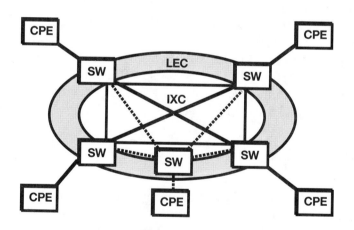

b. VPN (5 nodes)
Dashed lines indicate new facilities

Figure 1–3
Virtual private networks vs. leased lines. [LISO91], [BNR92a]

VPN provider creates a fully meshed network. Well, maybe; the network provider determines how the backbone network is constructed. Nonetheless, this approach allows the traffic to be routed to various endpoints and does not require the end-user devices to "nail up" private leased lines. Figure 1–2b shows that the VPN requires six IXC lines, but only four router ports, and only four dedicated local loops are needed.

The difference between fully meshed leased lines and VPNs is even more dramatic when another CPE is added to the network. In Figure 1–3a, a fifth CPE has been added to the network. If the user wishes full connectivity to all sites, this approach has rather frightening consequences in that ten interexchange private lines are now needed as well as 20 local loops, and 20 router ports. In contrast, as illustrated in Figure 1–3b, the addition of this fifth CPE only requires the addition of another router port and another private local line—at least from the perspective of the end user.

Of course with a fully meshed network, the same number of private lines are required between the switches. However, it is important to note that the switches are relaying traffic from multiple users. If another router CPE were added, then it would only require the leasing of one dedicated local loop to the most convenient switch. The number of interexchange lines would remain the same as long as additional switches were not added to the network.

VPN is a relatively new term in the computer/communications industry. Yet, this new term describes an old concept. The ideas behind the VPN are not new at all. Public X.25 networks have offered VPN services for years, and switched T1 services also offer VPN-like features. However, we shall see that several of the emerging technologies offer more powerful VPNs than these older technologies.

PROPOSED SOLUTIONS

To ameliorate and/or solve the problems cited thus far in this chapter, a number of solutions have been proposed, most of which involve VPNs. These proposed solutions are the subject of this book. Some of these solutions are complementary to each other and others are competitive. This section provides a terse and general overview of these newer technologies in order to set the stage for the other chapters in the book. Table 1–3 provides a summary of the types of emerging communications technologies discussed in this section.

The *frame relay* technology proposes to solve the WAN bottleneck problem by offering a "fast packet service" (actually a fast frame service). The approach is to provide minimal network support by relying on the user to recover from problems. In addition, frame relay offers bandwidth

Table 1–3.   Types of Emerging Communications Technologies

- Frame Relay

    A fast relay service, with minimal network support offering bandwidth on demand for data applications

- Metropolitan Area Networks (MANs)

    A fast distributed queue, dual bus technology to support multimedia applications

- Switched Multi-megabit Data Service (SMDS)

    A public high-speed transport system, based on the MAN technology

- The Fiber Distributed Data Interface (FDDI)

    A high-speed dual ring LAN using token passing technology (now supports multimedia applications)

- Broadband ISDN (BISDN)

    A combination of high-speed switching and carrier technologies, supporting a wide range of multimedia applications

- Asynchronous Transfer Mode (ATM)

    A high-speed virtual circuit, cell relay technology for voice, video, and data transmissions

- The Synchronous Optical Network (SONET) or Synchronous Digital Hierarchy (SDH)

    A high-speed synchronous carrier system, based on the use of optical fiber technology, and a defined digital multiplexing hierarchy

- Cellular and Cordless Systems

    A wide array of wireless systems designed to support mobile communications traffic

on demand services. Currently, frame relay is targeted for data (not voice) applications.

The *metropolitan area network* (MAN) is an IEEE standard (IEEE 802.6) that offers a very fast distributed queue dual bus (DQDB) protocol to support integrated networks for multimedia applications. This standard is so named because it is targeted to interconnect LANs across a metropolitan area.

The *switched multi-megabit data service* (SMDS) relies on the MAN technology and provides a public high-speed transport system. SMDS is being offered by the telephone companies in the United States and several public carriers in Europe as a service for high-speed data applications that require bursts of high bandwidth transmissions for applications such as file transfer, CAD/CAM, imaging, etc.

The *fiber distributed data interface* (FDDI) is a LAN technology, offering high-speed support for data applications. Recent enhancements to FDDI (FDDI II) allow it to support multimedia applications as well.

The *broadband ISDN* (BISDN) technology is actually a combination of several technologies. The technologies are combined to offer high-speed switching, multiplexing, and carrier services to support any type of user application. From the perspective of the International Telecommunication Union-Telecommunication Standardization Sector (ITU-T), BISDN encompasses ATM and SONET/SDH.

The *asynchronous transfer mode* (ATM) is part of the BISDN solution. It is a cell relay technology that includes high-speed multiplexing and switching services for voice, data, and video applications.

The *synchronous optical network* (SONET)/ *synchronous digital hierarchy* (SDH) is a high-speed carrier system (in telephone terms, a transport system), based on synchronous timing in which network devices are synchronized from a master clock (or clocks). Using optical fiber technology, it is designed to support current carrier systems such as T1, T3, E1, E3, as well as LAN traffic, and frame and cell relay traffic. Eventually, it is targeted to replace the T1/T3 and E1/E3 carrier systems.

The *cellular* and *cordless* technologies are (as their name implies) systems that provide communications facilities using high-frequency radio waves. Presently, they are designed as an enhancement to the current communications infrastructure, such as local loops and local telephone services. Obviously, they provide a very flexible means for handling mobile users. Eventually, as the technology improves, they will certainly be competitors to some of the other technologies discussed in this book.

FAST RELAY SYSTEMS

Much of the emerging technology is based on the idea of relaying traffic as quickly as possible. This idea is often called *fast packet relay* or *fast packet switching*. These names are considered generic terms in this book and are used in a variety of ways in the industry. We will use the term fast relay systems. Currently, fast relay comes in two forms: frame relay and cell relay. Figure 1–4 shows the relationships of these two forms of fast relay systems.

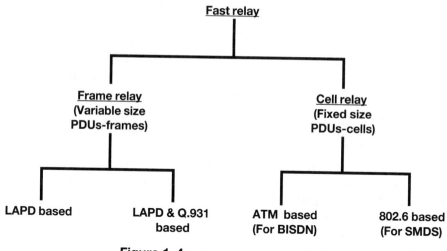

Figure 1–4
Types of relay systems. [DAVIS 91]

Frame relay uses variable-sized protocol data units (PDUs) which are called *frames*. The technology is based on the link access procedure for the D channel (LAPD), which has long been used in integrated services digital network (ISDN) systems. Most frame-based implementations are using LAPD as the basic frame format for the relaying of the traffic across permanent virtual circuits (PVCs). In the near future, it is anticipated that a modified version of ISDN's Q.931 will also be used for control signaling and to set up connections between the user and the network. Although not used much in frame relay today, Q.931 can form the basis for public switched offerings (switched virtual calls [SVCs]). The reader may refer to Chapter 3 for a tutorial on LAPD and Q.931.

In contrast to frame relay, cell relay uses a fixed length PDU which is called a *cell*. The cell usually consists of a 48-octet payload with a 5-octet header, although some implementations use different cell sizes. This cell (with slight variations) is being used on both the asynchronous transfer mode (ATM) and the IEEE 802.6 standard, more commonly known as the Metropolitan Area Network (MAN) specification. In turn, the 802.6 standard is being used as a basis for the switched multimegabit data service (SMDS).

The industry is still divided on the advantages and disadvantages of the use of cell relay versus frame relay. While there are arguments for both, the trend is toward the use of the cell relay technology. The reasons are many, but before they are explained, a brief summary of cell relay technology is in order.

At the customer premises equipment (CPE), such as a computer, a PBX, etc., a customer's traffic, which could be variable in length, is segmented into a smaller fixed length units called cells. As stated earlier, in most cell relay systems, the cells are only 53 bytes in length with 5 bytes devoted to a header and the remaining 48 bytes consisting of user information (the payload). The term cell is used to distinguish this PDU from variable length PDUs (frames and packets).

Cell relay is an integrated approach to networking in that it supports the transmission and reception of voice, video, data, and other applications. This capability is of particular interest to large companies that have developed multiple networks to handle multiple transmission schemes. As examples, common carriers, telephone companies, and Postal Telephone and Telegraph ministries (PTT) have to support many types of applications, and historically have implemented a variety of networks to support them.

Why do many people prefer the cell technique over the frame technique? First, the use of fixed length cells provides for a more predictable performance in the network than with variable length frames. Transmission delay is more predictable as is queuing delay inside the switches. In addition, fixed length buffers (with cell relay technology) are easier to manage than variable length buffers. In essence, a cell relay system is more deterministic than the use of a technology with variable length data units. It should also be emphasized that cell relay is easier to implement in hardware than variable length technology.

Some people have expressed concern about the high overhead of cell relay—because of the ratio of 5 bytes of header to every 48 bytes of user payload. But many people in the industry believe that the constant concern with efficient utilization of raw bandwidth is not a sound approach for the future. With the capacity of (a) optical fiber and (b) high-speed processors that are entering the marketplace, the approach of cell relay is to concentrate on superior quality-of-service features to the user.

The philosophy is straightforward: Let the fast optical channels and the fast computers handle the transmission and processing of the overhead traffic. Is this philosophy realistic? Only time will tell. Some people believe the small cell will not work and will not scale well with large networks. Others are equally adamant that the opposite is true.

One of the most challenging problems is how to use *one* network to support voice, data and video services. As of this writing, it is far from clear that one network (in the near future) will indeed be the best answer. Some researchers hold that it may make sense to use cell technology to support two networks, one for voice/video, and one for data.

We will return to these issues later in the book. For the present, let us examine how the communications and computer industries are evolving with regard to speed, capacity, and costs.

TRENDS IN TECHNOLOGY

Hardware and Software

Since the widespread commercial inception of computers and communications networks in the 1960s, the trend in public and private networks has been toward providing faster services to the user. One reason is the increased use of hardware to provide these services.

In the 1970s, a computer's communications architecture was almost all software (see Figure 1–5). Typically, the physical layer (labeled as 1 in Figure 1–5) (using the OSI reference model concept) consisted of hardware, and the data link and network layers were comprised of software. The software was written in machine language or assembler code in order to operate as efficiently and fast as possible.

As standards began to emerge in the late 1970s and early 1980s, it became economical to implement into hardware some of the data link layer functions (labeled as 2 in Figure 1–5). This move was fostered by the success of the High Level Data Link Control (HDLC) standard published by the ISO. It led to off-the-shelf hardware products, such as link access procedure, balanced (LAPB) and link access procedure for the D Channel (LAPD), that could be offered in chip sets along with firmware.

In the late 1980s and early 1990s, some designers and members of standards bodies recognized that (in some systems and standards) several redundant operations existed at the data link layer and network layer (labeled as 3 in Figure 1–5). As examples, error checking, flow control, and sequencing services sometimes existed (and still exist in many systems) in both of these layers. Therefore, new standards were developed to eliminate some of this redundancy.

Several new technologies, covered in this book, have focused on merging these two layers, and some now treat them as one. This practice was also fostered by the maturation of the standards at these layers, which once again permitted the use of off-the-shelf hardware. Of course,

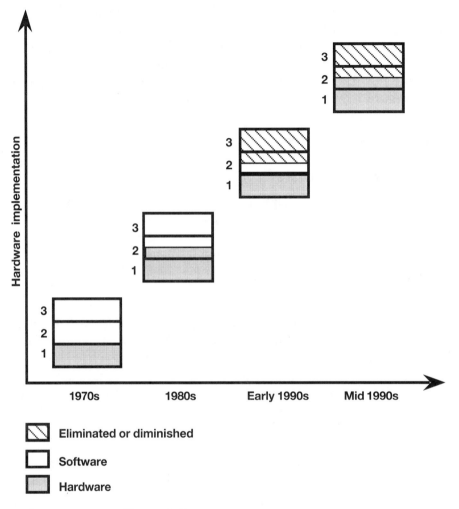

Figure 1–5
Trends in software and hardware.

executing the data link and network layer protocols in hardware made things work much faster.

Transmission channels. The foundation for much of the emerging technology is optical fiber. A series of rather extraordinary break-throughs in the last 15 years have lowered costs and increased the capacity of optical fiber. To gain an appreciation of how rapidly the optical fiber industry has progressed in such a brief period of time, Figure 1–6

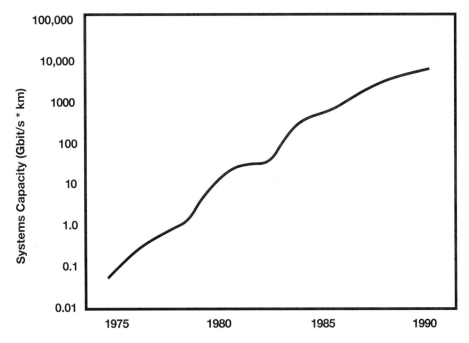

Figure 1–6
Transmission capacity of optical fiber. [CHEU92]

shows the trend in the capacity of lightwave systems since the early 1970s. The figure shows system capacity in gigabit/s times kilometer (km) spanning several years. The increase in the capacity of optical fiber is the result of technological breakthroughs occurring every 4 to 5 years.

Optical-based systems have increased dramatically in capacity, and at the same time, have decreased dramatically in cost. As the technology has matured and as mass production of systems has increased, the cost has dropped exponentially since 1975. Figure 1–7 illustrates this trend. Several years ago, 40 Mbit/s was considered to be a very high-speed system. Today, 2.5 Gbit/s commercial systems are commonplace, and 10 Gbit/s systems are now being installed.

The progress in lightwave carrier technology comes at a fortuitous time, because the industry has been making rapid progress in computer-generated video graphics. For a true multimedia environment to work efficiently and effectively, high-quality video graphics are essential. The use of an integrated workstation for voice, data, and video images is a certainty. However, the trends cited in these figures must continue for

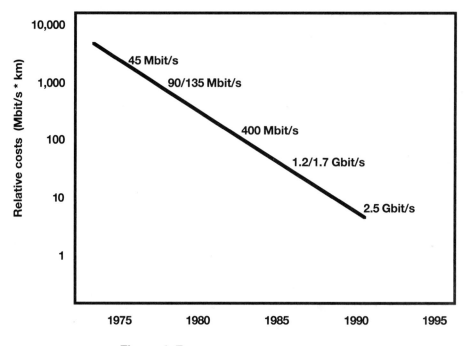

Figure 1–7
Cost trends in fiber optics. [CHEU92]

the next few years, if this integrated workstation is to become a practical, cost-effective tool in the workplace.

Computers. Nowhere is the advance of technology more evident than in the increased power of the central processing unit (CPU) and the memory speed of computers. In just a few short years, computing power has increased exponentially several times.

As Figure 1–8 illustrates, the world's first recognized computer, the ENIAC, developed in 1946, processed 100,000 instructions per second (IPS). Progress from 1945 through 1955 was relatively slow. However, with the extraordinary gains made in transistor technology, by 1975 an IBM 370 model 168 could process 2 million instructions per second.

From that time, the computing power of machines has increased at a fantastic rate. IBM's 1988 PS/2 model 70 processes 5 million instructions per second, and Dell's 1992 model 486-based desktop computer processes 19.4 million instructions per second. Consequently, many of

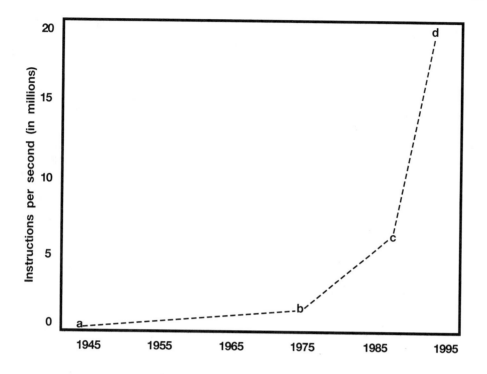

a = 1946, ENIAC, 100,000 IPS (30 tons)
b = 1975, IBM 370/168, 2 million IPS
c = 1988, PS2/70, 5 million IPS
d = 1992, Dell 486P50, 19.4 million IPS

Figure 1–8
Trends in computing power. [DELL92]

the concepts of the new, advanced technologies are being made possible through the speed of these new machines.

The speed of the processors is increasing at an extraordinary rate. Indeed, machines such as personal computers are experiencing obsolescence approximately every six months (from the standpoint of processor speed). Today, laptop computers operate with a 33 MHz CPU, and this book is being typed on a laptop that has the capacity of several 1980 mainframes (it's a Mac, so it's just about as expensive as a mainframe).

By the time this book is published, it is likely that CPUs will be available to operate laptops and notebooks well above the 60–100 MHz range, and this writer buys each new Mac that provides more "horse-power". It is that vital to my productivity.

In addition to the rather extraordinary gain in computing speeds made during the last two decades, the industry has made noticeable gains in reducing the costs of computers as well. As an example, the ENIAC cost approximately 3 million dollars.

Between the years of 1946 and 1975, the cost of computing dropped exponentially. Figure 1–9 may lead the reader to think that computing actually went up with the large mainframe computers of the 1960s and 1970s. However such is not the case. For example, the cost of the IBM 370 model 168 was approximately 8 million dollars, but this cost included peripheral devices (disks, tape drives, etc.) and services such as

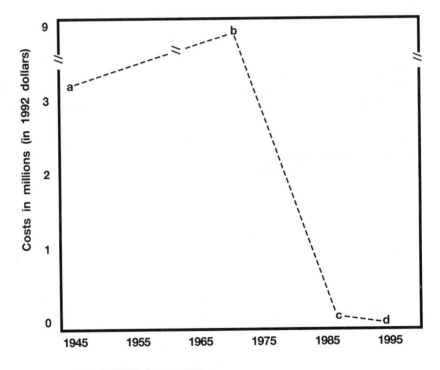

a = 1946, ENIAC, $3,223,846
b = 1975, IBM 370/168, $8,828,625
c = 1988, PS2/70, $13,840
d = 1992, Dell 486P50, $2,188

Figure 1–9
Trends in computing costs. [DELL92]

software for databases and operating systems, which did not exist with ENIAC.

The point of this discussion is to examine Figure 1–9 from the year 1975 to the present. During this period, costs have dropped an extraordinary rate where a 486-based desktop computer costs slightly more than $2,000.

New Technologies—To Use Them or Not to Use Them?

As with any complex and expensive undertaking, there is often a reluctance to implement a technology and develop applications that use it if the success of the technology is not assured. This situation often exists in the computer and communications industry, and is sometimes compared to the age-old chicken and egg question—except the question in this industry is not which came first, but which *should* come first: the applications that use the technology or the technology that supports the applications? Lest the reader think this is not an important issue, just consider the initial experience (the first ten years) with ISDN. Here was a technology that outpaced the applications that were supposed to use it, because most enterprises were quite content with the V.24/V.28/EIA-232-E technology, operating up to 20 kbit/s; and V.35 and T1/E1 systems, for higher-speed requirements. In essence, few enterprises had any motivation to migrate to ISDN because (a) it was not prevalent in the marketplace, (b) it did not fulfill an end user organization's requirements, and (c) it was not cost-effective.

Therefore, a number of companies that were trailblazers for ISDN found in the late 1980s and early 1990s that they had gone to the expense of giving a party (ISDN) and few showed up for the fun. It does not take many ISDNs to put a chill on trailblazing and party-giving. ISDN is becoming more accepted now, a topic that will be discussed in Chapter 3.

Given the chicken (communications technology) and egg (applications) problem, how will the communications industry evolve in the latter part of this century? I can answer this question easily: No significant communications vendor can afford to sit on the sidelines and not become involved. If the vendor does not continue to press forward and develop new technologies, the competitive marketplace will dig that vendor's grave and cover it up. Indeed, today's technology is moving so fast and technical innovations are occurring so rapidly that successful companies must adopt and implement the "self-cannibalization" rule: develop new technologies and products to compete with the company's ongoing prod-

These conclusions are providing the impetus and driving force for the new technologies:

Conclusion: More powerful applications have a direct and positive effect on human productivity.

Conclusion: Bandwidth is becoming available to support these applications that need gigabit pipes.

Conclusion: Decreased costs to use this bandwidth will allow expanded use of high-speed circuits and switches to support applications.

Conclusion: Increased computing power will support new applications.

Conclusion: Decreasing costs for computing will make gigabit-based applications cost-effective to implement.

Figure 1–10
Trends and conclusions.

ucts as soon as possible. If company A does not do it to itself, company B will.

So, if there is no issue about building new technologies, the question remains: Which technology should I (the network provider) build? The network provider does not have this answer, nor does anyone else. Some of the emerging technologies are competitive, and some vendors are developing more than one offering in hopes of selecting a winner (or winners).

Given the above discussions, it seems reasonable to make the conclusions that are summarized in Figure 1–10. The issue is not if new technologies are to be implemented, but what they will be. After all, we know that increased productivity can result from more powerful applications; we also know that productivity can be increased by allowing computers to interconnect with each other. Therefore, it can be concluded that more powerful networks are essential to our continuing technological progress and our productivity.

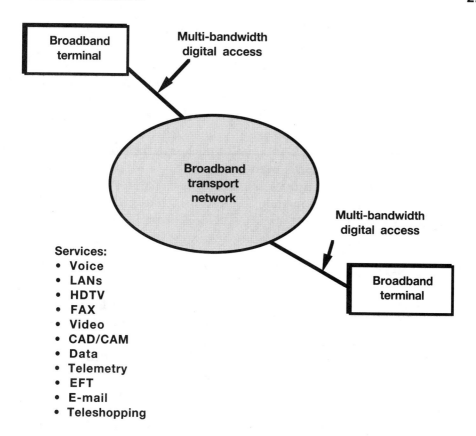

Figure 1–11
Applications support with broadband networks.

BROADBAND NETWORKS

Many of the emerging technologies are called *broadband multime-dia networks*. The purpose of a broadband multimedia network is to provide a transport service for any type of application, thus the term, multimedia network.[1] Telephone service, video service, data service, tele-

---

[1]This book uses term multimedia network in accordance with common industry practice. I prefer the term multi-application network, because it conveys more accurately the nature of the network. Mr. Webster and I view the term media to be (a) a means of communications (newspapers, radios, etc.), or (b) a physical channel.

shopping, CAD/CAM, FAX, etc. are all supported by a broadband multimedia network as suggested in Figure 1–11.

These emerging technologies are generally referred to as broadband networks. The term broadband has been in use for several years and has several definitions applied to it. The earliest definition is: a network that uses high-frequency analog technology as the transport mechanism for user traffic. Another definition is: any network that operates above the voice frequency range (0–4 kHz). Still another definition is: any network that operates above the ISDN primary rate (1.544 Mbit/s in North America and 2.048 Mbit/s in Europe). This book uses the term to describe any high-capacity system (a megabit/s system) that offers a multimedia transport system.

BROADBAND SIGNALING HIERARCHIES

At the beginning of this chapter, it was stated that for the first time in the history of the computer/communications industry, networks throughout the world are embracing a worldwide set of computer/communications standards. One of these standards deals with multiplexing and signaling hierarchies for carrier transport networks. Because of its importance, we shall introduce the subject here, and examine it more closely in later chapters.

During the past 30 years, three different digital multiplexing and signaling hierarchies have evolved throughout the world. These hierarchies were developed in Europe, Japan, and North America. Fortunately, all are based on the same pulse code modulation (PCM) signaling rate of 8,000 times a second, yielding 125 microsecond sampling slots (1 second / 8000 samples = 0.000125). Therefore, the basic architectures interwork reasonably well.

But the multiplexing hierarchies differ considerably, which results in extensive and expensive conversion operations, if traffic is exchanged between them (See Figure 1–12). Moreover, the analog-to-digital conversion schemes also differ between North America and Europe, which further complicates interworking the disparate systems.

Japan and North America base their multiplexing hierarchies on the DS1 rate of 1.544 Mbit/s. Europe uses a 2.048 Mbit/s multiplexing scheme. Thereafter, the three approaches multiplex these payloads into larger multiplexed packages at higher bit rates, and use different values for the multiplexing integer, n.

As depicted in Figure 1–12, the synchronous digital hierarchy

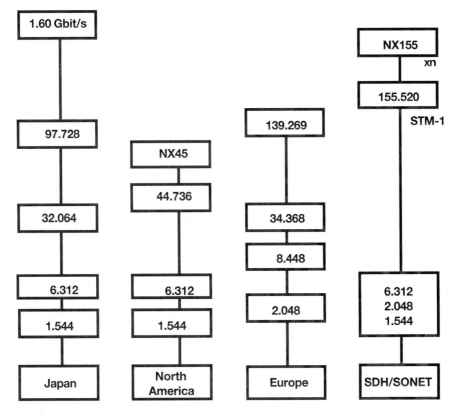

Note: Unless noted otherwise, speeds in Mbit/s

Figure 1–12
SONET and SDH hierarchies.

(SDH, European term) and the synchronous optical network (SONET, North American term) support the schemes that have been in existence for many years but specify a different multiplexing hierarchy.

The basic SDH/SONET rate is 155.52 Mbit/s. SDH/SONET then uses an x 155.52 multiplexing scheme, with x as the multiplexing factor (e.g., 155.52 * x, where x is 3, 6, etc.). Rates smaller than 155.52 Mbit/s are available at 51.840 Mbit/s. The electrical signals in the hierarchy are called the synchronous transport signal (STS-n). The optical counterpart of STS-1 is called the optical carrier-level 1 signal (OC-1). We shall have more to say about Figure 1–12 in later chapters and the other terms in this figure will be explained at that time.

At long last, worldwide agreement has been reached (with minor exceptions) on a common transport scheme. This agreement can only foster new technologies and decrease their implementation costs.

APPLICATIONS SUPPORTED BY THE NEW TECHNOLOGIES

Given that the communications industry has finally agreed upon a standard transport system for the emerging technologies, what applications are these technologies supposed to support? First, some of them are designed to provide a seamless connection between LANs and WANs. This aspect really does not support applications, but it does support the networks.

Several of the new technologies are targeted for specific applications that require a cost-effective megabit pipe. Examples are data, voice, video, CAD/CAM, x-ray, and most any type of imagery. There are exceptions to this statement. ATM and SONET/SDH are positioned to be application independent (although we shall see that ATM employs another protocols to accommodate to diverse applications).

Other technologies, such as frame relay and SMDS, are positioned to support most applications except interactive voice and video. Figure 1–13 provides a summary of the principal targeted applications.

NEW TECHNOLOGIES: COMPETITIVE OR COMPLEMENTARY?

As of this writing, it is not certain how all these technologies will interwork, but things are moving rapidly and by the time of this book's publication, a clearer picture will surely have emerged. But it is clear now, as we shall see in this book, that a few of the technologies are competing and others are complementary. Figure 1–14 summarizes some of the more likely competitive pairings. The figure is self-explanatory, so I shall not embellish it further.

Not all carrier companies are investing in all technologies. While some companies, such as large computer/communications carriers (AT&T, British Telecom, PTTs, etc.), may be able to afford to develop products for all these technologies, the vast majority of vendors have decided to "partition" and invest in those technologies that they think will be the most successful. Thus, as we shall see, certain organizations have decided not to implement all of these technologies and have gambled on others.

- *Frame relay and SMDS*
    Seamless LANs and WANs
    Graphics
    X-ray
    Large database transfers
    CAD/CAM
    LAN interconnections
    Medium quality, still-image video
    Connectionless data transfer
    Connection-oriented data transfer
    Almost everything except video and voice

- *MAN*
    All of the above as well as high-quality video and voice

- *ATM*
    All of the above

- *SONET/SDH*
    Anything, it is simply a physical carrier technology—a transport
        service for ATM, SMDS, frame relay, T1, E1, etc.

- *FDDI*
    Higher speed LANs, to (a) combat predicted obsolescence of
        1980s technology, and (b) support data *and* voice (with FDDI II)

- *Cellular and Cordless*
    Any low-speed data application, voice, and low-to-medium qual-
        ity video

Figure 1–13
Targeted applications?

If several of these technologies are targeted for the same applications, why are more than one of them being offered? The answer is that
no one knows which solution will be better from the standpoint of performance and cost. But we do know that some of the technologies compete
with each other to support the same types of user applications. In addi-

Technology	Competes with?
Frame relay	1. Leased lines 2. X.25 based networks 3. SMDS 4. ATM—probably, too soon to know
SMDS	1. Leased lines 2. X.25 based networks 3. Frame relay 4. ATM—probably, too soon to know 5. High-speed bypass
ATM	1. Too soon to know, but probably all of the above and FDDI
SONET/SDH	1. Private leased lines networks (T1 or E1 based) 2. Otherwise, complements and supports all of the above *and* T1/E1
FDDI	1. Eventually, the lower speed LANs (Ethernet, token ring) 2. The new 100 Mbit/s Ethernet 3. Presently, complements LANs as a backbone
Cellular and cordless	1. Local loops 2. Wired LANs 3. If capacity increased, frame relay, SMDS, ATM

Figure 1–14
Competition analysis.

tion, most certainly do compete with many of the existing technologies—the "emerged technologies" (the subject of Chapter 3).

PERFORMANCE AND DISTANCE CONSIDERATIONS

Figure 1–15 represents a view of the bandwidth and distance capabilities of the major emerging technologies. As the figure suggests, the current LAN IEEE 802.3, 802.4, and 802.5 technologies are distance-

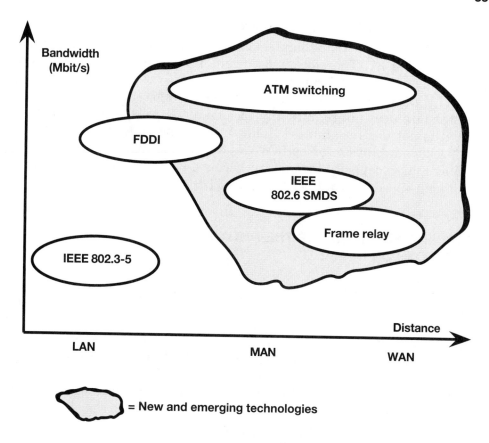

Figure 1–15
Network technologies.

limited to a few hundred meters and capacity-restricted up to 16 Mbit/s. FDDI provides a higher capacity than the 802 LANs and also supports greater geographical distances. However, all these networks are restricted in relation to their transmission capacity and their ability to span large geographical areas.

The metropolitan area network (MAN), published as IEEE 802.6 (and implemented with the switched multi-megabit data service [SMDS]), offers an interesting combination of distance and capacity. It can operate in the multi-megabit range and can span a larger geographical area in comparison to conventional LANs. Moreover, while no technical reason exists for it not to span a large geographical area, it is being implemented as a network encompassing a limited geographical area, such as a city (thus the name, metropolitan).

Frame relay is targeted for WANs; it is not constrained by distance. The physical layer is being implemented on the current carrier technologies of T1 and E1. Therefore, most frame relay implementations do not operate above the T1/E1 rates.

ATM technology is not constrained to any particular line speed and the standard does not require any specific physical layer solution. Moreover, it is not distance-limited. Indeed, it can be viewed as a switching and backbone solution for both WANs and LANs. It is shown in Figure 1–15 as a high capacity technology, because it is intended to operate on top of high-speed optical fiber and SONET/SDH.

OBTAINING SERVICES FOR THE NETWORKS: BANDWIDTH ON DEMAND

A key factor in the use of these emerging technologies is a concept called *bandwidth on demand*. The term means that a user can request and receive network capacity dynamically (i.e., bandwidth). The tool for obtaining bandwidth on demand is called the committed information rate (CIR) or the sustained information rate (SIR) (see Figure 1–16). It is a tool that is used by SMDS, frame relay, and ATM networks to (a) regulate the flow of user traffic, (b) allow the user some choice in a throughput rate, and (c) determine certain pricing structures for the service.

Several of the new networks provide the user with a guaranteed service (relating to throughput), if the user's input rate is below some predefined CIR. In Figure 1–16, as long as the arrow on the gauge remains below the CIR, the user should be guaranteed the services of the network. If the user exceeds the CIR for some period of time, the network should not discard traffic beyond the CIR, *if* the user traffic only exceeds CIR for a brief duration. However, any traffic sent that exceeds CIR may be discarded. The word "may" means that the network will most likely not discard traffic if it has sufficient resources to transport the user traffic during the time the CIR is exceeded.

In most implementations, the current rate of a user's input might exceed not only a short "burst rate" beyond CIR, but also a permitted maximum rate—an "excess rate". If this situation occurs, the network may discard all excess traffic beyond this rate.

Regardless of which technology is chosen as a mechanism for the user-to-network interface, a user should examine carefully the exact method in which bandwidth on demand is provided. These key questions should be posed to the network provider:

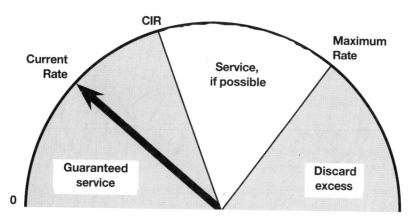

Figure 1–16
Using the CIR/SIR for service decisions. (Bandwidth on demand)

- What should the carrier charge for services up to CIR?
- What should the carrier charge for services beyond CIR?
- How should customer be informed about discarded traffic?
- How should prices be apportioned for discarded traffic?

WHERE SERVICES ARE PROVIDED

The user obtains the network services (and bandwidth on demand) as illustrated in Figure 1–17. The operations between the user and the network consists of a set of conventions and protocols through which the user and network send and receive traffic to and from each other.

This traffic is in the form of control packets and user payload. In this book, the term *payload* is used to describe user traffic. It is so named because the user payload (the vendor hopes) will provide the revenue to defray the costs of operating the network and pay a dividend to the network owners and stockholders.

This interface also provides conventions for the user and network to provision/negotiate certain quality of service (QOS) features. As examples, the negotiation of a delay, CIR, throughput, etc., is implemented at this interface.

Several of the newer technologies also define the operations between networks. The interface between networks is very important,

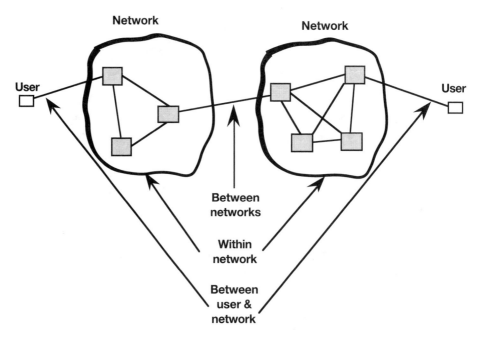

Note: Shaded boxes represent switches.

Figure 1–17
Where the services are provided.

since most organizations allow their users to log on to one network and then communicate with another organization in yet another network. While the operations, conventions, and protocols between networks may not be of direct concern to the end user, they are of great concern to the network administrators.

One might ask: Why is another set of conventions needed? Why not just use the conventions that exist between the user and the network? The answer is that different operations are required between different networks. For example, networks do not need to provide certain types of flow control mechanisms between network boundaries that are needed at the boundary between the user and the network.

Finally, a set of conventions and protocols are also needed within the network. While many networks have been implemented with proprietary protocols inside the network *cloud,* this approach is changing and, increasingly, standardized conventions and protocols even define how an internal network communicates between its switches. Thus, emerging technologies may consist of three major sets of protocols and conven-

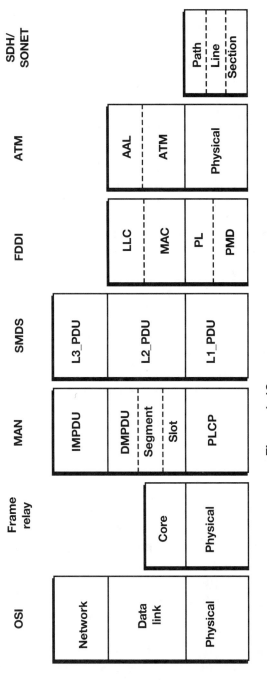

Figure 1-18
Layers of the emerging technologies.

tions: those between the user and the network, those within the network, and those between networks. We shall have more to say about Figure 1–17, and we will embellish it with other information in Chapter 2 in the section titled "Network Interfaces".

LAYERED ARCHITECTURES OF THE EMERGING TECHNOLOGIES

Figure 1–18 shows the layers of the major emerging technologies. It is obvious that these technologies rest on the lower three layers of the OSI Model. Indeed, it is the intent of the OSI Model that communications services reside in these layers. Interestingly, some technologies do not use the network layer. As discussed earlier, the network and data link layers increasingly are being merged together.

Three of the OSI layers are shown on the lefthand side of Figure 1–18. The bottom three layers of the OSI Model are known as *bearer services* in the sense that they support (and bear) the user traffic emanating from the upper four layers of the Model. None of the emerging technologies describe operations at the upper four layers of the OSI Model. Some are restricted to layers 1, 2, and 3; while others are restricted to layers 1 and 2; and SONET/SDH restricts its operations to layer 1 (the physical layer) only. We will use this figure extensively throughout the book. For the uninitiated reader, a tutorial on the OSI Model layers is provided in Appendix B.

SUMMARY

It is recognized that upcoming user applications are demanding greater throughput and lower delay from communications systems. The emerging technologies are designed to meet this need. In addition, some of the emerging technologies transport voice, video, and data images. With some exceptions, the emerging technologies are designed to support multimedia application requirements.

Optical fiber systems are providing the foundation for the media support of these technologies, and high-speed processors are providing the speed to process the traffic.

Finally, the movement toward worldwide standards for carrier transport technology, multiplexing techniques, and switching methodologies is at last providing the groundwork for a comprehensive and homogeneous approach to computer/communications networks.

Foundations for the Emerging Technologies

INTRODUCTION

This chapter examines the concepts on which voice, data, and video networks are founded. An understanding of these concepts is essential for comprehending the functions of the emerging technologies. This chapter can also be considered as a tutorial on how to analyze a computer/communications system. The reader may also find this material helpful in understanding everyday computer-communications operations. A number of my clients use this material to brief their staffs on how their products use these concepts.

The following concepts are explained in this order:

1. Virtual circuits
2. Connection-oriented and connectionless systems
3. Variable bit rate (VBR) and constant bit rate (CBR) applications
4. Flow control and congestion management
5. User payload integrity management
6. Layered protocols and protocol data units (PDUs)
7. Addressing and identification schemes

8. Multiplexing methodologies
9. Switching, routing, and relaying
10. Network interfaces
11. Convergence, segmentation, and reassembly operations

The reader can pick and choose the topics that require further edification before delving into subsequent chapters. Also, some of these topics are explained in Appendix A, which is a tutorial on communications networks. To avoid redundancy, I refer the reader to the relevant part of Appendix A, when appropriate.

VIRTUAL CIRCUITS

End users operating terminals and computers communicate with each other through a communications channel called a *physical circuit*. These physical circuits are known by other names such as channels, links, lines, and trunks. Physical circuits can be configured wherein two users communicate directly with each other through one circuit, and no one uses this circuit except these two users.

In more complex systems, such as networks, circuits are shared with more than one user. Within a network, the physical circuits are terminated at intermediate points at machines that provide relay services on to another circuit. These machines are known by such names as switches, routers, bridges, gateways, etc. They are responsible for relaying the traffic between the communicating users.

This arrangement can result in more than one user sharing a physical circuit. Since many communications channels have the capacity to support more than one user session, the network device (such as the switch or multiplexer) is responsible for sending and receiving multiple user traffic to/from a circuit.

In an ideal arrangement, a user is not aware that the physical circuit is being shared by other users. Indeed, the circuit provider attempts to make this sharing operation transparent to all users. Moreover, in this ideal situation, the user thinks that the circuit directly connects only the two communicating parties. However, it is likely that the physical circuit is being shared by other users.

The term *virtual circuit* is used to describe a shared circuit (or circuits) wherein the sharing is not known to the circuit users. The term was derived from computer architecture in which an end user perceives that a computer has more memory than actually exists. This other "virtual" memory is actually storage on hard disk. Therefore, a virtual circuit (and virtual memory) exists in the mind of the beholder (the user).

A Brief Digression

This section can be skipped and the reader will not suffer any ill effects. As an aside, the term "virtual" has entered the mainstream of our consumer society with another "emerging technology" called *virtual reality*. Virtual reality is obtained by a person slipping on a headset (containing sound and vision) and watching reality go by (the reality is created by a computer).

And then there are the virtual companies—they that have no physical location, no permanent staff, but rely on mailboxes and answering services to provide a front to their customers. At the rate that many companies are shedding full-time staff and taking on part-time contractors, we may be seeing the coining of a new term—the virtual employee. Virtual companies, with virtual employees, providing (one hopes) non-virtual services to customers.

As often as the term *virtual* is being introduced into our vocabulary, it will eventually give rise to a plethora of virtual things. I thought it would be interesting to build a small matrix to contrast some of these terms, one of which is virtual. Figure 2–1 depicts these concepts. As just

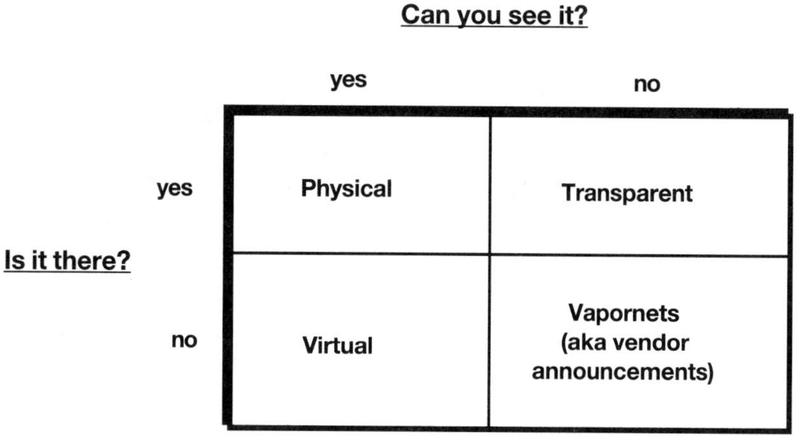

Can you see it?

		yes	no
Is it there?	**yes**	Physical	Transparent
	no	Virtual	Vapornets (aka vendor announcements)

Figure 2–1
Virtual, transparent, and physical things.

mentioned, the term virtual is applied to an entity in which a person perceives it can be seen, but it does not exist. Consequently, answering the questions: "Can you see it? Yes. Is it there? No." refers to a virtual entity, like a virtual circuit. On the other hand, answering yes to questions "Can you see it? Is it there?" references a physical entity, such as a physical circuit. Yet, another possibility is when questions: "Can you see it? No. Is it there? Yes." refer to transparent operations. A good example of a transparent operation is a protocol converter. It operates transparently to the user but is performing the useful operation of converting one protocol's transmissions to a form that is acceptable to another protocol that is receiving the transmission.

Finally, to complete the matrix, a answer of no to: "Can you see it? Is it there?" refers to another buzz word called *vapornet*. An example of this part of the matrix is many vendor announcements: Not only can you not see them but the products do not exist.

We could superimpose upon this matrix, in place of *yes* or *no,* the answer of *somewhat* to the questions: "Can you see it? Is it there?" Then this matrix probably would accurately describe most things in life (as well as the upcoming technology called fuzzy logic).

On a more serious note, let us return to the subject at hand and focus on three types of virtual circuits: (a) permanent virtual circuit (PVC), (b) switched virtual circuit (SVC), and (c) semi-permanent virtual circuit (SPVC).

Permanent Virtual Circuit (PVC)

A virtual circuit may be provided to a user (provisioned) on a continuous basis. With this approach, the user has the service of the network at any time. This concept is called a PVC. A PVC is established by creating entries in tables in the network nodes' databases. These entries contain a unique identifier of the user payload which is known by various names such as a logical channel number (LCN), virtual channel identifier (VCI), or virtual path identifier (VPI). Entries are also stored in the databases to identify the destination user, and the types of services that are to be provided to the two users. These entries are created as part of a contract agreement between the user and the network, and they must be created before the user can use the network.

Once the contract is signed, and the network has stored all this information, the user can obtain the services of a network by transmitting a packet to the network that contains data and the reserved logical channel number or virtual circuit identifier. Upon receiving the packet, the network node compares the LCN/VCI with its database to discern what kind of quality of services (QOS) the user wants, and with whom the user wishes to communicate. This latter statement is important because PVCs are provisioned on a user-to-user (end-to-end) basis. Consequently, a PVC is not only a permanent relationship between a user and a network, but also between a user and another user.

Network features such as throughput, delay, security, etc. (also known as QOS) are also provisioned before the user ever submits traffic to the network. Consequently, a pure PVC does not allow a user to negotiate services. If different types of services are desired, and if different destination endpoints must be reached, then the user must submit a different PVC identifier with the appropriate user payload to the network. This PVC is provisioned to the different endpoint, and perhaps with different services.

Switched Virtual Circuit (SVC)

In contrast to a PVC, a switched virtual circuit (SVC) is not "preprovisioned". When a user wishes to obtain network services to communicate with another user, it must submit a connection request packet to the network. This packet usually identifies the originator in the form of a network address. It *must* provide the address of the receiver, and it must also contain the virtual circuit number that is to be used during the session. This virtual circuit value is used during this communications session to associate the destination and sources addresses of a label. Once the session is over, this value is made available to any other user that wishes to "pick it out" of a table.

SVCs entail some delay during the call set-up process (and a connection may be denied), but they are quite flexible in that they allow the user to select dynamically the receiving party simply by inserting the appropriate destination address in the connection request packet. In addition, networks that support SVCs usually allow the negotiation of network features, such as delay and throughput, on a call-by-call basis.

Semi-permanent Virtual Circuits (SPVC)

Many networks support another virtual circuit service which is called by various names; we will use the term semi-permanent virtual circuit (SPVC). With this approach, a user is preprovisioned, as in a regular PVC. A reserved label is also stored and associated with the user. The identity of the user's destination node is also stored. Like a PVC, the network node contains information about the communicating parties and the type of services desired.

However, this type of virtual circuit does not guarantee that the users will obtain their level of requested service. For example, if the network is congested and the SPVC user wishes a session that requires a large amount of network resources, the user could be denied the service.

A more likely occurrence would be that the user initially is granted the services of the network, but is denied continuous services because of some action that violates the user contract with the network, such as sending data at a rate greater than was agreed upon by the network and the user.

CONNECTION-ORIENTED AND CONNECTIONLESS SYSTEMS

Connection-Oriented Systems

Systems that employ the concepts of virtual circuits (discussed in the previous section) are said to be *connection-oriented*. As we just learned, the network maintains information about the users, such as their addresses and their ongoing QOS needs. Often, this type of system uses state tables that contain rules governing the manner in which the user interacts with the network. While these state tables clarify the procedures between the user and the network, they do add overhead to the process.

Connectionless Systems

In contrast, systems that do not employ virtual circuits are said to be *connectionless* systems. They are also known as datagram networks and are widely used throughout the industry. The principal difference between a connection-oriented and a connectionless operation is that connectionless protocols (as the name implies) do not establish a virtual circuit for the end user communications process. Instead, traffic is presented to the service provider in a somewhat ad hoc fashion. Handshaking arrangements (such as the negotiation of services) are minimal and perhaps nonexistent. The network service points and the network switches (if employed) maintain no ongoing knowledge about the traffic between the two end users (such as state tables). Therefore, datagram services provide no a priori knowledge of user traffic (such as a PVC) and they provide no ongoing, current knowledge of the user traffic.

The Pros and Cons of Connection-Oriented Systems and Connectionless Systems

In the past, a connectionless system was designed to make things happen faster than in a connection-oriented system because connectionless service entails less processing overhead. For example, state tables in connectionless systems are not maintained; therefore, the network nodes can execute their operations more quickly.

A connection-oriented system is an attractive option from the network point of view because it gives the network more control over the traffic and in servicing the users' QOS needs. In the past, connection-oriented systems have usually entailed additional processing overhead.

In recent systems, connection-oriented networks are more efficient because they maintain a limited amount of information about the user, and state tables have been reduced or eliminated. More will be said about this feature later in this chapter and in later chapters. For the present, it is sufficient to state that the idea of a connection-oriented system is to speed things up yet maintain a precise awareness of the user's traffic in relation to the overall network operations.

The Coexistence of Connection-Oriented Systems and Connectionless Systems

Many networks today use a combination of connection-oriented and connectionless protocols. The most common implementation of this hybrid operation is found in wide area data networks. The operations at the boundary of the network (between the user device and the network node) are connection-oriented and the operations inside the network (between the network nodes) are connectionless. This is sometimes referred to as a *connectionless subnet with connection-oriented interfaces*. An example of this approach is the use of the X.25 interface at the user-to-network interface and a proprietary connectionless protocol within the network itself.

This example is not meant to imply that a connectionless subnet with connection-oriented interfaces is the only approach used in the industry. Another widely used combination is the provision for connection-oriented operations within and outside the network. A common example of this approach is a telephone network that is connection-oriented between the user and the network as well as connection-oriented through-out the network.

Some networks are completely connectionless within the network, at the user-to-network interfaces, and even between networks. Many data networks fall into this category, especially LANs. Of course, while the use of the connectionless approach keeps things quite simple and may speed up the operations, the end user might opt for some type of connection-oriented system at a layer above the network-bearer services. A common approach is to provide for connectionless operations at the lower layers and implement a connection-oriented service between the end-user applications themselves.

VARIABLE BIT RATE (VBR) AND CONSTANT BIT RATE (CBR) APPLICATIONS

A useful method to describe the nature of applications traffic is through two concepts known as variable bit rate (VBR) and constant bit rate (CBR).

VBR Applications

An application using VBR schemes does not require a constant and continuous allocation of bandwidth. These applications are said to be *bursty,* which means that they transmit and receive traffic asynchronously (at any time, and possibly, with periods in which nothing is sent or received). Examples of VBR applications are most any type of *data* communications process. These applications permit the queuing of traffic in a variable manner from the standpoint of time, and they do not require a fixed timing relationship between the sender and the receiver. Therefore, if traffic is sent from the sender and is buffered (queued) for variable periods, the receiver is not disturbed. Typical applications using VBR techniques are interactive terminal-to-terminal dialogues, inquiry-response operations, client-server systems, and bulk data transfer operations.

While VBR schemes permit loose timing and asynchronous operations between the sender and the receiver, most VBR applications do require some type of timing constraint. For example, an interactive operation between the user at an automated teller machine and the bank's central database can vary, even on the order of seconds. But a variance in minutes will certainly lead to unhappy (and eventually) "vapor" customers.

CBR Applications

In contrast, an application using CBR schemes requires a constant and continuous (or nearly so) allocation of bandwidth and a precise timing relationship between the sender and receiver. These applications are said to be *non-bursty*. The term non-bursty has to be used carefully because some of the applications will tolerate a certain amount of burstyness.

Typical CBR-based applications are voice and video transmission. These applications require a guaranteed bandwidth and a constant and continuous timing relationship between the sending and receiving devices. They also require a predictable and fixed delay of transmission time between the sender and the receiver.

To illustrate, let us imagine that this writer is having a telephone conversation from Virginia with his mother, who resides in New Mexico. It is certainly reasonable to expect that the technology carrying this conversation will provide continuous service for the conversation. When I say, "Hello mother," the manner in which I state these words to the sending device (the telephone) should be exactly the same at the receiving device. If the service provider does not provide a guaranteed minimal delay and queues portions of this greeting (say, in a buffer), my mother may receive not "Hello, mother" but "Hello, ma . . ." (or some other type of *fuzzy* greeting).

Like most aspects of life, and as in each section of this chapter, exceptions are inevitable. In today's technology, systems exist which permit bursty voice schemes. The prevalent approach is to take advantage of the periods of silence in voice conversations (which are quite few in some) and eliminate the signals that would represent these periods of silence. This approach is implemented widely today and allows voice networks to achieve a somewhat bursty nature.

Other CBR applications are not tolerant of bursty schemes. A principal example is a high-quality voice-video application such as television or quality video-conferencing. These systems require non-bursty support and precise timing relationships between the sender and the receiver.

Some of the emerging communications technologies are confronted with the requirement to provide an integrated network for VBR and CBR applications. In today's environment, it is quite easy to convert any type of application image to digital syntax—some people consider this to be an integrated network. However, the transmission of all digital images is the tip of the iceberg. Floating below the surface of this technology is the challenge of managing the traffic and providing a fair and equitable service for the wide range of applications that require different variations of VBR and CBR schemes.

Therefore, the challenge of an integrated, multimedia network is not just to digitize all traffic—that is a rather trivial task. The major challenge is to manage the user traffic in such a manner that the user does not realize that only one network is supporting multiple combinations of CBR and VBR traffic. This concept surely warrants a new buzzword—let's call it *virtual multiple networks*.

FLOW CONTROL AND CONGESTION MANAGEMENT

Another important aspect of communications systems is the nature in which the traffic is controlled at the ingress and egress points of the network. The operation by which user traffic is controlled by the network is called, logically enough, *flow control*. Flow control assures that traffic does not saturate a network or exceed the network's capacity. So, flow control is used to manage congestion.

The three methods used for the control of traffic are called *explicit flow control, implicit flow control,* and *no flow control*.

Explicit Flow Control

As the name suggests, explicit flow control is a specific limitation on how much user traffic can enter the network. If a network issues an explicit flow control message to the user, the user has no choice but to stop sending traffic (or reduce transmissions in accordance with some direction from the network), and the user may not send any more traffic until the network releases this "throttle". The reader may be familiar with these forms of explicit flow control operations with the familiar terms receive not ready (establishing flow control) and receive ready (releasing flow control).

Implicit Flow Control

Implicit flow control does not restrict flow absolutely. Rather, it suggests that the user either stop sending traffic or at least reduce the amount of traffic it is sending to the network. Typically, the implicit flow control message is a warning to the user that (a) the user is violating its service contract with the network, and/or (b) the network is congested. In any case, if the user continues to send traffic, it risks having traffic discarded by the network.

This idea is rather new to the communications industry. In the past, most networks issued explicit flow control signals to users, but several of the emerging communications technologies are incorporating implicit flow control into their operations.

No Flow Control

Ironically, networks may establish flow control by performing no flow control at all. Generally, an absence of flow control means that the network can discard any traffic that is creating problems. While this approach certainly provides superior congestion management from the standpoint of the network, it may not perform the type of flow control and congestion management that the user has in mind. Notwithstanding, most connectionless networks (such as LANs) will regulate the flow of traffic through the use of a carrier-sense collision detection mechanism, the issuance of tokens, etc.

USER PAYLOAD INTEGRITY MANAGEMENT

Some people in the industry confuse connection management with payload management. They are separate from each other in how they operate in a network. However, this confusion is understandable because, in the past, most connection-oriented protocols also provided a variety of payload integrity and data management services with features such as the acknowledgment (ACK) of traffic for correctly received traffic, the negative acknowledgment (NAK) for incorrectly received traffic, and other features such as explicit or implicit flow control. Notwithstanding, the reader should understand that even though a system may offer connection-oriented services with PVCs and SVCs, these services do not imply that the network will be responsible for the integrity and delivery of user payload.

In order to account for user traffic, rather extensive operations must be performed by the network on behalf of the user. For example, error checking of the received user payload is performed at each node in the network. As a result of this error check, the sender receives either an ACK or NAK from the receiver that performed the error check on the traffic. If a NAK is received, the originator must resend the affected traffic to the receiver. In effect, a system that provides end-to-end traffic integrity requires that the user payload be send from the source to the destination with absolute assurance that it will arrive safely and correctly at that destination.

Payload integrity operations consume resources in a network. Consequently, the network provider and the network user should understand clearly the tradeoffs, and whether or not payload integrity management is worth the cost. Certainly, one can pose a scenario where this type of support is absolutely essential, and an application that comes to mind is electronic funds transfer (EFT). Surely, one would not want to process EFTs from one bank to another if the data were distorted en route and arrived with the decimal place misaligned in the transfer. (Of course, where one stands on an issue depends on where one sits. The movement of the decimal place might have a serendipitous effect on the bank customer, but not on the bank manager.)

In addition, errors are tolerable in certain types of applications. For most voice applications and many video applications, the occasional errored frame and subsequent loss of this traffic does not affect substantially the perceived quality of the sound or image once it is reconstructed back to an analog signal for the human ear or eye. Therefore, and once again, we face the challenge of managing different types of data traffic, user traffic, and voice traffic in a different manner, depending on the nature of the traffic.

LAYERED PROTOCOLS AND PROTOCOL DATA UNITS

Most communications systems today are designed around the concept of layered protocols. These protocols transmit and receive units of traffic called protocol data units (PDUs). These concepts are explained in considerable detail in Appendix B and should be reviewed if ideas associated with layered protocols and PDUs are not familiar to the reader.

The term PDU, while useful and applicable to all layers, is often replaced by more common terms. The use of these terms vis-à-vis the layers is depicted in Figure 2–2. The PDUs created and used at the application layer (APDU) are known by various names. The terms *message* and *envelope* are used usually for PDUs created by electronic mail (E-mail) systems. The terms *file* and *record* are generally associated with traffic created and used by file transfer systems and data management sys-

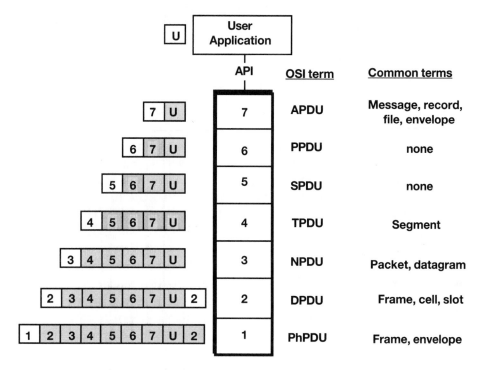

Figure 2–2
Protocol data units (PDUs).

tems. This traffic is sent to the lower layers, and (as discussed in Appendix B) encapsulated into headers at these lower layers which create larger PDUs.

Some systems associate the term *segment* with PDUs created and used at the transport layer (TPDU). This term is commonly used with the widely used transmission control protocol (TCP). In turn, the segment is sent to the network layer where it is further encapsulated.

The term *packet* is normally associated with connection-oriented network layer protocols, such as X.25. The term *datagram* is also a PDU term normally associated with connectionless protocols, such as OSI's connectionless network protocol (CLNP) and the widely used Internet Protocol (IP).

Finally, this traffic is encapsulated into the *frame* at the data link layer (DPDU). Some vendors and designers also associate the frame with the physical layer.

As explained in Chapter 1, another term to describe a PDU is called a *cell*. This term was developed (coined) to explain the concept of a small and fixed length PDU (the term cell usually connotes something small). Cells are normally associated with a PDU at the data link layer because systems that use this concept, such as asynchronous transfer mode (ATM), operate by dividing the user traffic into cells at this layer. Some systems use the term *slot* in place of the term cell. The slot has the same characteristics as the cell.

One other term used in these emerging technologies is the term *envelope* which, like many of these terms, can be associated with different protocols and layers. But, increasingly, it is used to describe traffic in E-mail at layer 7 (the application layer), in which traffic is placed inside an electronic envelope or at layer 1 (the physical layer[PhPDU]) in which traffic is placed in a payload envelope such as a SONET/SDH envelope.

One hopes this discussion has clarified the terms—at least as best as they can be clarified. The reader should be aware that many of these terms are used to describe the same thing. Also, please be aware that these terms are not made up by this writer!

In using these terms, the reader should be aware that there is no industry-wide consensus on which layers create and use these particular types of PDUs. In this book, we will use these terms as depicted in Figure 2–2, and also use the terms *unit, traffic,* or *payload* when providing a generic description of user information (regardless of whether the traffic is voice, data, or video).

ADDRESSING AND IDENTIFICATION SCHEMES

In order for user traffic to be sent to the proper destination, it must be associated with an identifier of the destination. Most systems use one or two kinds of identifiers which are called (a) explicit addresses, or (b) labels.

An *explicit address* has a location associated with it. It may not refer to a specific geographic location but rather a name of a network or a device attached to a network. As an example, the Internet Protocol (IP) address has a structure that permits the identification of a network, a subnetwork attached to the network, and a host device (a computer) attached to the subnetwork. As another example, the ITU-T X.121 address has a structure which identifies a country, a network within that country, and a device within that network. Other values are used with these addresses to identify protocols and applications running within the host such as file servers and message servers.

Explicit addresses are used by switches, routers, and bridges as an index into routing tables. These routing tables contain information about how to route the traffic to the destination.

Another identifying scheme is known by the term *label,* although the reader may be more familiar with other terms, such as logical channel number (LCN) or virtual circuit identifier (VCI). A label contains no information about network identifiers or physical locations. It is simply a value that is assigned to a user's traffic that identifies each data unit of that user's traffic.

In an earlier discussion in this chapter, PVCs and SVCs were explained. A common practice is to assign a label (such as a VCI) for PVCs. This means that the user of the PVC need only submit to the network the label in the header associated with the user's traffic. The network then uses this label to examine tables to determine the explicit location information. For SVC traffic, the typical practice is to submit a label and an explicit address during the connection set-up. This information is mapped into tables for the later management of the traffic. Once this mapping has occurred, the user need only submit to the network a label such as a VCI. The VCI is used as an index into the table to find an explicit address.

Almost all connectionless systems use explicit addresses, and the destination and source addresses must be provided with every PDU in order for it to be routed to the proper destination.

Most of the emerging technologies discussed in this book use labels, such as virtual circuit identifiers. The reason for this approach is that the labels are shorter than explicit addresses (which can be quite long). These short labels can be examined rapidly which results in shorter processing time at the switches. Indeed, the concept of cells implies very short headers with the use of labels. This approach allows for much of the routing logic to be implemented in hardware, yet provides another way to speed up routing operations.

MULTIPLEXING METHODOLOGIES

All the emerging technologies discussed in this book use some form of multiplexing. Multiplexing operations accept lower-speed voice, video, or data signals from terminals, telephones, and user applications and combine them into a high-speed stream for transmission onto a link. A receiving device demultiplexes and converts the combined stream into the original multiple lower-speed signals. Since several separate transmissions are sent over the same line, the efficiency of the path is improved.

Appendix A contains more detailed information on multiplexing and compares frequency division, time division, and statistical time division multiplexing. The interested reader should refer to this appendix.

SWITCHING, ROUTING, AND RELAYING

The method in which user traffic is routed within and between networks is performed with one of two methods: *source routing* and *non-source routing*. The majority of the emerging technologies use non-source routing.

Source routing derives its name from the fact that the transmitting device (the source) dictates the route of the PDU through a network or networks. The source (host) machine places the addresses of the "hops" (the intermediate networks or routers) in the PDU. Such an approach means that the internetworking units need not perform address maintenance, but they simply use an address in the PDU to determine where to route the frame.

In contrast, non-source routing requires that the switches make decisions about the route. The switch does not rely on the PDU to contain information about the route. Non-source routing is usually associated with bridges and is quite prevalent in LANs. Most of the emerging technologies implement this approach with the use of a virtual circuit identifier (VCI). This label is used by the network nodes to determine where to route the traffic.

Appendix A contains more detailed information on source and non-source routing, as well as information on adaptive and nonadaptive routing. The interested reader should refer to this appendix.

NETWORK INTERFACES

Several of the emerging technologies are organized around two basic sets of protocols: (a) The protocol that governs the interface between the user and the network, and (b) the protocol that governs the in-

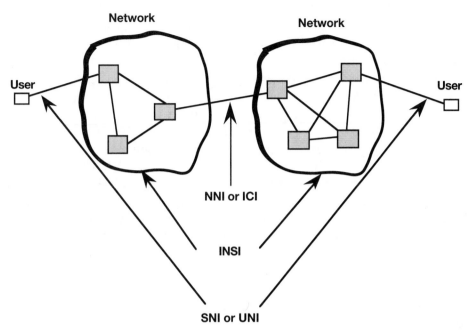

Where:

NNI = Network to network interface
ICI = Inter-exchange carrier interface
INSI = Intra-network switching interface
SNI = Subscriber network interface
UNI = User network interface

Figure 2–3
Network interfaces.

terface between networks. The former is described as the user-to-network interface (UNI) or the subscriber network interface (SNI) and the later is described as the network-to-network interface (NNI) or the inter-exchange carrier interface (ICI). These interfaces are illustrated in Figure 2–3.

I have coined another term to define the protocols that operate between the switches, multiplexers, and other components inside the network. It is the intra-network switching interface (INSI). This term often describes proprietary protocols, although in the past few years increasing pressure has been placed on vendors to adapt common conventions for operations within their network cloud.

Convergence, Segmentation, and Reassembly Operations

Some networks (especially those in the emerging communications technologies) support convergence, segmentation, and reassembly operations on the user traffic.

The term *convergence operations* refers to a protocol entity that accepts user traffic and provides services that allow the user to gain access to a network. The operations usually reside in part of a protocol layer; thus, they are also called the convergence sublayer (CS).

Convergence services are performed *before* the user traffic enters the network. The services are defined usually at the UNI. The most common convergence operations entail synchronization functions between a sender and receiver, and accommodating different transfer rates between the user and the network (in bit/s). Convergence operations may also perform error detection and error correction of the user traffic. CS operations are quite application-specific. Thus, CS services differ between voice, video, connectionless data, and connection-oriented data applications.

In some systems, the CS operates with another part of a protocol layer that performs segmentation and reassembly operations on the user payload. This service is called the segmentation and reassembly sublayer (SAR). *Segmentation* is performed on the user traffic *before* it enters the network; *reassembly* is performed *after* the traffic leaves the network—that is to say, at the UNI.

The idea of using CS and SAR is to make the traffic appear "uniform" to the network. Therefore, the network is not concerned with application timing requirements, variable size messages, etc. This approach allows the network to perform its transport functions very quickly and efficiently.

SUMMARY

The emerging communications technologies employ a number of common techniques to support the end user's traffic. The manner in which these techniques are used varies, and in some instances, a particular technology may not use the technique. Subsequent chapters will focus in more detail about how these techniques are employed by each technology.

Emerged Technologies

INTRODUCTION

This chapter provides an overview of the communications technologies that have been in use in the industry for the past two to three decades—thus the term, emerged technologies. These technologies are covered:

- T1 and E1 carrier systems
- X.25
- Integrated Services Digital Network (ISDN)
- Signaling system number 7 (SS7)

This chapter is not designed to provide a detailed treatise on the emerged technologies. Its goal is to acquaint the reader with the major terms and concepts as they relate to the subject matter of this book. The more experienced reader can skip much of this material, but be aware that subsequent chapters assume the reader has a firm grasp of these current systems and technologies. The reader should also review the section in Appendix A titled "Analog-to-Digital Conversion" if the subject of how analog signals are converted into digital signals is not familiar, and/or if the significance of the 125 microsecond (μsec.) time division multiplexed (TDM) slot is not understood.

Please note that the terms user-to-network interface (UNI) and subscriber-to-network interface (SNI) are used in this chapter to describe the same interface.

T1/E1 CARRIER SYSTEMS

The Purpose of T1 and E1

The T1/E1 systems are high-capacity networks designed for the digital transmission of voice, video, and data. The original implementations of T1/E1 digitized voice signals in order to take advantage of the superior aspects of digital technology. Shortly after the inception of T1 in North America, the ITU-T published the E1 standards, which were implemented in Europe. E1 is similar to (but not compatible with) T1.

T1 is based on multiplexing 24 users onto one physical TDM circuit. T1 operates at 1,544,000 bit/s, which was (in the 1960s) about the highest rate that could be supported across twisted wire-pair for a distance of approximately one mile. Interestingly, the distance of one mile (actually about 6000 feet) represented the spacing between manholes in large cities. They were so spaced to permit maintenance work such as splicing cables and the placing of amplifiers. This physical layout provided a convenient means to replace the analog amplifiers with digital repeaters.

The term T1 was devised by the telephone company to describe a specific type of carrier equipment. Today, it is used to describe a general carrier system, a data rate, and various multiplexing and framing conventions. A more concise term is DS1, which describes a multiplexed digital signal which is carried by the T carrier. To keep matters simple, this book uses the term T1 synonymously with the term DS1, and the term T3 synonymously with DS3. Just be aware that the T designator stipu-

Table 3-1　Carrier Systems Multiplexing Hierarchy

North America	Japan	Europe
64 kbit/s	64 kbit/s	64 kbit/s
1.544 Mbit/s 24 voice channels	1.544 Mbit/s 24 voice channels	2.048 Mbit/s 30 voice channels
6.312 Mbit/s 96 voice channels	6.312 Mbit/s 96 voice channels	8.448 Mbit/s 120 voice channels
44.736 Mbit/s 672 voice channels	32.064 Mbit/s 480 voice channels	34.368 Mbit/s 480 voice channels
274.176 Mbit/s 4032 voice channels	97.728 Mbit/s 1440 voice channels	139.264 Mbit/s 1920 voice channels

lates the carrier system, but the digital transmission hierarchy schemes are designated as DS-n, where n represents a multiplexing level of DS1. Table 3–1 lists the more common digital multiplexing schemes used in Europe, North America, and Japan.

Today, the majority of T1/E1 offerings digitize the voice signal through pulse code modulation (PCM), or adaptive differential pulse code modulation (ADPCM). Whatever the encoding technique, once the analog images are translated to digital bit streams, then many T1 systems are able to time division multiplex voice and data together in 24 user slots within each frame.

Typical Topology

Figure 3–1 shows a T1 topology (the same type of topology is permitted with the E1 technology). Actually, there is no typical topology for these systems. They can range from a simple point-to-point topology shown here, wherein two T1 multiplexers operate on one 1.544 Mbit/s link, or they can employ with digital cross connect systems (DCS) that add, drop, and/or switch payload as necessary across multiple links.

Voice, data, and video images can use one digital "pipe". Data transmissions are terminated through a statistical time division multiplexer (STDM), which then uses the TDM to groom the traffic across the transmission line through a T1 channel service unit (CSU); or other equip-

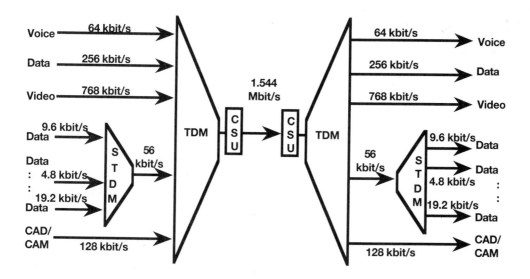

Figure 3–1
A typical T1 topology.

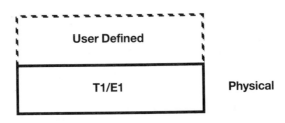

Figure 3–2
T1 and E1 layers.

ment, such as a data service unit (DSU), or the DSU and CSU may be combined as well. The purpose of the CSU is to convert signals at the user device to signals acceptable to the digital line, and vice versa at the receiver. The CSU performs clocking and signal regeneration on the channels. It also performs functions such as line conditioning (equalization), which keeps the signal's performance consistent across the channel bandwidth; signal reshaping, which reconstitutes the binary pulse stream; and loop-back testing, which entails the transmission of test signals between the DSU and the network carrier's equipment.

The bandwidth of a line can be divided into various T1 subrates. For example, a video system could utilize a 768 kbit/s band, the STDM in turn could multiplex various data rates up to a 56 kbit/s rate, and perhaps a CAD/CAM operation could utilize 128 kbit/s of the bandwidth.

T1 and E1 Layers

In relation to the OSI Model, (see Appendix B), the T1 and E1 layers reside in only one layer—the physical layer (see Figure 3–2). This layer defines the connectors, signaling conventions, framing formats, etc., that are found in most physical layers.

T1/E1 PDUs

The T1 frame (or the OSI term, PDU) consists of 24 8-bit slots and a framing bit (see Figure 3–3). To decode the incoming data stream, a receiver must be able to associate each sample with the proper TDM channel. At a minimum, the beginning and ending of the frame must be recognized. The function of the framing bit is to provide this delineation. The framing bit is in the 193rd bit of each frame. It is not part of the user's information, but added by the system for framing. The use of this bit varies, depending on the type of the T1 system and the age of the technology.

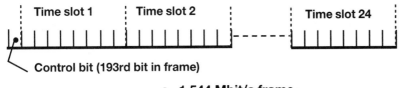

Control bit (193rd bit in frame)

a. 1.544 Mbit/s frame

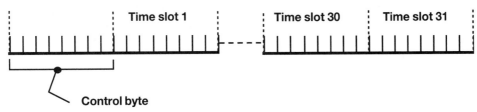

Control byte

b. 2.048 Mbit/s frame

Figure 3–3
T1 and E1 frames.

Some T1 channel banks use the 8th bit in every 8-bit slot for control signaling. Examples of control signaling are off-hook, on-hook, ringing, busy signals, battery reversal, etc. During the development of the later T1 equipment, the designers recognized that every 8th bit was not needed for signaling. Consequently, later equipment, known as D2 channel banks, uses the 8th bit of every 6th and 12th frame to provide signaling information. The least significant bit in these frames is overwritten with a signaling bit. This concept is called *bit robbing* and the respective 6th and 12th robbed frame bits are called the *A and B bits*.

For the transmission of data, the 8th bit is unreliable, so most vendors have chosen to ignore this bit for data signaling. As a result, the majority of T1 and related systems use a 56 kbit/s transmission rate (8000 slots/second * 7 bits per slot = 56000) instead of the 64 kbit/s rate (8000 slots/second * 8 bits per slot = 64000)

The rate for the 2.048 Mbit rate is similar to its counterpart T1 in that they both use a 125 μsec. timeframe (see Figure 3–3). However, the E1 frame is divided into 31 TDM slots and preceded by a 8-bit time slot that is used for control purposes. E1 reserves this slot in the frame for its "out-of-band" signaling, and thereby obviates the cumbersome bit robbing scheme found in T1. Robbing is not a good idea—nor is bit robbing; therefore, the emerging technologies discussed in this book (with some minor exceptions like T1), use out-of-band signaling instead. That is to say, bandwidth is reserved for control signaling rather than robbing bits to perform the function.

Conclusions on T1/E1

The T1/E1 systems have served the industry well. However, they are quite limited in their management operations, and they provide very little support for end-user control for the provisioning of services. In the old days, the use of bandwidth (control headers) for network management was not encouraged due to the limited transmission capacity of the facilities to accommodate this overhead traffic. Today, the prevailing idea is to exploit the high capacity of optical fibers and the processors, and allocate a greater amount of bandwidth (larger control headers) to support more network management services.

Moreover, these older technologies use awkward multiplexing schemes. Due to their asynchronous timing structure (each machine runs its own clock, instead of using a central clock in the network for all machines), timing differences between machines are accommodated by stuffing extra bits (bit stuffing) periodically in the traffic streams. These bits cannot be unstuffed when the traffic is demultiplexed from the higher rates (see Table 3–1) to the lower rates. Indeed, the traffic must be completely demultiplexed at the multiplexers and/or switches to make the payload accessible for further processing.

One should be careful about criticizing a technology that was conceived and implemented over thirty years ago. In retrospect, T1 and E1 were significant steps forward in the progress of the telecommunications industry. But like most everything else in this industry (thankfully, not life), things do not improve with age, and must be replaced. As we shall see, SONET/SDH are those replacements.

X.25

The Purpose of X.25

X.25 was designed to perform a function similar to that of the ISDN (discussed later): to provide an interface between an end user device and a network. However, for X.25, the end user device is a *data* terminal and the network is a packet-switched data network; whereas, for ISDN, the end user device was (originally) a voice terminal (telephone) and the network was (originally) a circuit-switched voice network. The comparison is apt, because both protocols have similar architectures, as will be explained in this section.

The idea of the X.25 interface, as conceived by the ITU-T study groups in the early 1970s, was to define unambiguous rules about how a public packet data network would handle a user's payload and accommodate to various quality of service (QOS) features (called X.25 facilities) that were requested by the user. X.25 was also designed to provide strict flow control on user payloads and to provide substantial management services for the user payload, such as the sequencing and acknowledgment of traffic.

The ITU-T issued the X.25 Recommendation in 1974. It was revised in 1976, 1978, 1980, 1984, and the last major revision was made in 1988. Since 1974, the standard has been expanded to include many options, services, and facilities, and several of the newer OSI protocols and service definitions operate with X.25. X.25 is now the predominant interface standard for wide area packet networks. Unlike, T1/E1, X.25 uses STDM techniques and is designed as a transport system for data—not voice.

Typical Topology

The placement of X.25 in packet networks is widely misunderstood. X.25 is *not* a packet switching specification. It is a packet network *interface* specification (see Figure 3–4). X.25 says nothing about operations

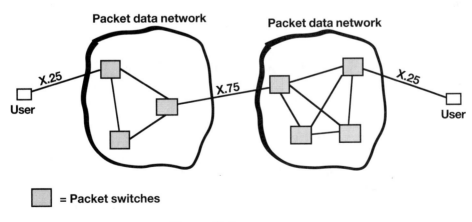

Figure 3–4
Typical X.25 topology.

within the network. Hence, from the perspective of X.25, the internal network operations are not known. For example, X.25 is not aware if the network uses adaptive or fixed directory routing, or if the internal operations of the network are connection-oriented or connectionless. The reader may have heard of the term "network cloud". Its origin is derived from these concepts.

It is obvious from an examination of Figure 3–4 that X.25 is classified as a subscriber network interface (SNI). It defines the procedures for the exchange of data between a user device (DTE) and the network (DCE). Its formal title is "Interface between Data Terminal Equipment and Data Circuit Terminating Equipment for Terminals Operating in the Packet Node on Public Data Networks." In X.25, the DCE is the "agent" for the packet network to the DTE.

X.25 establishes the procedures for two packet-mode DTEs to communicate with each other through a network. It defines the two DTE's sessions with their respective DCEs. The idea of X.25 is to provide common procedures between a user station and a packet network (DCE) for establishing a session and exchanging data. The procedures include functions such as identifying the packets of specific user terminals or computers, acknowledging packets, rejecting packets, initiating error recovery, flow control, and other services. X.25 also provides for a number of QOS functions such as reverse charge, call redirect, transit delay selection, etc., which are called X.25 facilities.

X.75 is a complementary protocol to X.25. X.75 is a network-to-network interface (NNI), although it is used often today for other configurations, such as amplifying and enhancing an interface on ISDN systems to support X.25-based applications.

X.25 Layers

The X.25 Recommendation encompasses the lower three layers of the OSI Model. Like ISDN, the lower two layers exist to support the third layer. Figure 3–5 shows the relationships of the X.25 layers. The physical layer (first layer) is the physical interface between the DTE and DCE, and is either a V-Series, X.21, or X.21*bis* interface. Of course, X.25 networks can operate with other physical layer interfaces (as examples, V.35, the EIA 232-E standard from the Electronic Industries Association, and even high-speed 2.048 Mbit/s interfaces). X.25 assumes the data link layer (second layer) to be link access procedure, balanced (LAPB). The LAPB protocol is a subset of HDLC.

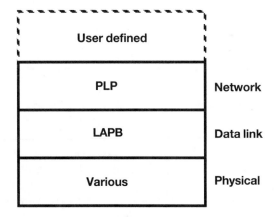

Various: **V Series, X.21, X.21bis, etc.**
LAPB: **Link access procedure, balanced**
PLP: **Packet layer procedures**

Figure 3–5
The X.25 layers.

X.25 PDUs

The X.25 packet is carried within the LAPB frame as the I (information) field (see Figure 3–6). LAPB ensures that the X.25 packets are transmitted across the link, after which the frame fields are discarded (stripped) and the packet is presented to the network layer. The principal function of the link layer is to deliver the packet error-free, despite the error-prone nature of the communications link. In this regard, it is quite similar to LAPD in ISDN. In X.25, a *packet* is created at the network layer and inserted into a *frame* which is created at the data link layer.

The network layer, also called packet layer procedures, or PLP, is responsible for establishing, managing, and tearing down the connections between the communicating users and the network.

Figure 3–6 illustrates that sequence numbers, ACKS, NAKs, and flow control techniques are implemented in two different fields residing in the X.25 PDU. First, the control field in the link header is used at layer 2 (LAPB) to control operations on the link between the user device and network node. Second, the packet header is used at layer 3 (PLP) and contains sequence numbers, ACKs, etc., to control *each user session* running on the link between the user device and network node. As we

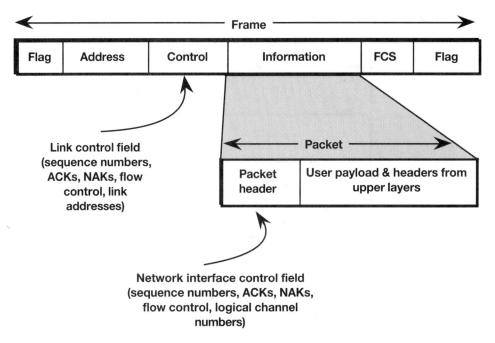

Figure 3–6
X.25 PDUs (packet and frame).

shall see in this book, some of the emerging technologies take the stand that these operations are either redundant or unnecessary, and so they are either eliminated or decreased substantially.

Other Noteworthy Aspects of X.25

X.25 uses logical channel numbers (LCNs) to identify the DTE connections to the network. An LCN is really nothing more than a virtual circuit identifier (VCI). As many as 4095 logical channels (i.e., user sessions) can be assigned to a physical channel; although, in practice, not all numbers are assigned at one time due to performance considerations. The LCN serves as an identifier (a label) for each user's packets that are transmitted through the physical circuit to and from the network. Typically, the virtual circuit is identified with two different LCNs—one for the user at the local side of the network and one for the user at the remote side of the network.

X.25 defines quite specifically how logical channels are established, but it allows the network administration considerable leeway in how the virtual circuit is created. Notwithstanding, the network administration must "map" the two LCNs together from each end of the virtual circuit through the network so they can communicate with each other. How this is accomplished is left to the network administration, but it must be done if X.25 is to be used as specified.

X.25 interface options. X.25 provides two mechanisms to establish and maintain communications between the user devices and the network: (a) Permanent virtual circuit (PVC), and (b) switched virtual call (SVC).

A sending PVC user is assured of obtaining a connection to the receiving user and of obtaining the required services of the network to support the user-to-user session. X.25 requires that a PVC be established before a session can begin. Consequently, an agreement must be reached by the two users and the network administration before the PVC is allocated. Among other things, agreement must be made about the reservation of LCNs for the PVC session and the establishment of facilities.

An SVC requires that the originating user device must transmit a call request packet to the network to start the connection operation. In turn, the network node relays this packet to the remote network node, which sends an incoming call packet to the called user device. If this receiving DTE chooses to acknowledge and accept the call, it transmits to the network a call accepted packet. The network then transports this packet to the requesting DTE in the form of a call connected packet. To terminate the session, a clear request packet is sent by either DTE. It is received as a clear indication packet, and confirmed by the clear confirm packet.

X.25 also provides extensive QOS options to the user. These options are called *facilities*. These facilities are preprovisioned for PVCs or are provided during the call establishment phase of the SVC hook up. The user is allowed to obtain features such as call redirections, security features (closed user groups), throughput, delay, reverse charge, reverse charge prevention, and a wide variety of other useful application support operations.

Conclusions on X.25

One could surmise that any data communications technology developed during the 1970s cannot be appropriate to fulfill the requirements of modern applications. Additionally, a substantial segment of the telecommunications industry believes that X.25 is ineffective as an SNI, because of its "overly connection-oriented nature".

It should be remembered that X.25 is old. It was designed to support user traffic on error-prone networks, with the supposition that most user devices were relatively unintelligent. Moreover, X.25 was designed to operate on physical interfaces that are also old (and therefore inherently slow), such as EIA-232-E, V.28, etc.

Nonetheless, X.25 usage continues to grow throughout the world because (a) it is well understood, (b) it is available in off-the-shelf products, (c) extensive conformance tests are available for the product, and (d) it is a cost-effective service for bursty, slow-speed applications, of which there are many.

This writer is neither a connection-oriented or connectionless proponent. Some applications benefit from a connectionless system and others from a connection-oriented system. It seems wiser to choose an appropriate protocol to support the application, be it connection-oriented or connectionless, than to decree that everything should be connectionless or connection-oriented. Fortunately, the data communications industry is coming around to this premise.

But, let there be no misunderstanding: If one cares for one's data and wishes to maintain some type of control over it, then there must be a connection-oriented data management protocol residing somewhere in the protocol suite—but it need not exist at the network layer. This idea is central to the behavior of several of the emerging technologies explained in this book: Yes, one needs to take care of the user's data (payload), but the network is not the place to do it—it is more appropriate for the user to perform this function. Later chapters will examine this premise.

X.25 will likely see decreasing usage as the emerging technologies and physical interfaces mature, become available in off-the-shelf products, and prove to be more efficient than X.25 and current interfaces. So, without burying X.25 prematurely, it can be stated that X.25 has served the industry well. X.25 pioneered the concepts of PVCs, SVCs, and VCIs; and in so doing, X.25 has left a efficacious legacy to the telecommunications industry.

ISDN

The Purpose of ISDN

The initial purpose of the Integrated Services Digital Network (ISDN) was to provide a digital interface between a user and a network node for the transport of digitized voice and (later) data images. It is now designed to support a wide range of services. In essence, all images (voice, data, television, fax, etc.), can be transmitted with ISDN technology.

ISDN has been implemented as an evolutionary technology, and the committees, common carriers, and trade associations who developed the standards wisely recognized that ISDN had to be developed from the long-existing telephone-based integrated digital network (IDN). Consequently, many of the digital techniques developed for T1 and E1 are used in ISDN. This includes signaling rates (e.g., 32 or 64 kbit/s), transmission codes (e.g., bipolar), and even physical plugs (e.g., the jacks to the telephone). Thus, the foundations for ISDN have been in development since the mid-1970s.

Typical Topology

The user interface to ISDN is a very similar topology to that of X.25. An end user device connects to an ISDN node through a UNI protocol. Of course, the ISDN and X.25 interfaces are used for two different functions. The X.25 UNI provides a connection to a packet-switched data network; while ISDN provides a connection to an ISDN node, which can then connect to a voice, video, or data network.

Before we begin an analysis of ISDN, two terms must be defined: *functional groupings* and *reference points*. First, functional groupings are a set of capabilities needed in an ISDN user-access interface. Specific functions within a functional grouping may be performed by multiple pieces of equipment or software. Second, reference points are the interfaces dividing the functional groupings. Usually, a reference point corresponds to a physical interface between pieces of equipment. With these thoughts in mind, please examine Figure 3–7, which shows several possibilities for setting up the ISDN components at the SNI (others are permitted).

The reference points labeled R, S, T, and U are logical interfaces between the functional groupings, which can be either a terminal type 1 (TE1), a terminal type 2 (TE2), or a network termination (NT1, NT2) grouping. The purpose of the reference points is to delineate where the responsibility of the network operator ends. If the network operator responsibility ends at reference point S, the operator is responsible for NT1 and NT2, and LT/ET.

The U reference point is the reference point for the 2-wire side of the NT1 equipment. It separates a NT1 from the line termination (LT) equipment. The U interface is a national standard, while interfaces implemented at reference points S and T are international standards. The R reference point represents non-ISDN interfaces, such as EIA-232-E, V.35, etc.

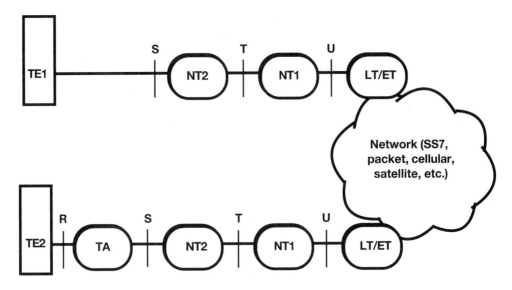

Figure 3–7
Typical ISDN topology.

The end-user ISDN terminal is identified by the ISDN term TE1. The TE1 connects to the ISDN through a twisted pair 4-wire digital link. Figure 3–7 illustrates other ISDN options—one is a user station called a TE2 device, which represents the current equipment in use, such as IBM 3270 terminals, Hewlett-Packard, and Sun workstations, telex devices, etc.

The TE2 connects to a terminal adapter (TA), which is a device that allows non-ISDN terminals to operate over ISDN lines. The user side of the TA typically uses a conventional physical layer interface such as EIA-232-E or the V-series specifications, and it is not aware that it is connected into an ISDN-based interface. The TA is responsible for the communications between the non-ISDN operations and the ISDN operations.

The TA and TE2 devices are connected to either an ISDN NT1 or NT2 device. The NT1 is a device which connects the 4-wire subscriber wiring to the conventional 2-wire local loop. ISDN allows up to eight terminal devices to be addressed by NT1. The NT1 is responsible for the physical layer functions, such as signaling synchronization and timing. NT1 provides a user with a standardized interface.

The NT2 is a more intelligent piece of equipment. It is typically

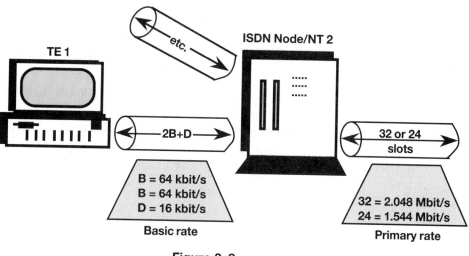

Figure 3–8
ISDN configuration.

found in a digital PBX and contains the layer 2 and 3 protocol functions. The NT2 device is capable of performing concentration services. It multiplexes 23 B+D channels onto the line at a combined rate of 1.544 Mbit/s or 31B+D channels at a combined rate of 2.048 Mbit/s. The NT1 and NT2 devices may be combined into a single device called NT12. This device handles the physical, data link, and network layer functions.

In summary, the TE equipment is responsible for user communications and the NT equipment is responsible for network communications.

As illustrated in Figure 3–8, the TE1 connects to the ISDN through a twisted pair 4-wire digital link. This link uses TDM to provide three channels, designated as the B, B, and D channels (or 2 B+D). The B channels operate at a speed of 64 kbit/s; the D channel operates at 16 kbit/s. The 2 B+D is designated as the basic rate interface (BRI). ISDN also allows up to eight TE1s to share one 2 B+D link. The purpose of the B channels is to carry the user payload in the form of voice, compressed video, and data. The purpose of the D channel is to act as an out-of-band control channel for setting up, managing, and clearing the B channel sessions.

In other scenarios, the user DTE is called a TE2 device. As explained earlier, it is the current equipment in use such as IBM 3270 terminals, telex devices, etc. The TE2 connects to a TA, which is a device that allows non-ISDN terminals to operate over ISDN lines. The user side of the TA typically uses a conventional physical layer interface such as EIA-232-E or the V-series specifications. It is packaged like an external modem or as a board that plugs into an expansion slot on the TE2 devices.

As explained earlier, ISDN supports yet another type of interface called the primary rate interface (PRI). It consists of the multiplexing of multiple B and D channels on to a higher-speed interface of either 1.544 Mbit/s (used in North America and Japan) or 2.048 Mbit/s (used in Europe). The 1.544 Mbit/s interface is designated as 23 B+D, and the 2.048 Mbit/s interface is designated as 31 B+D to describe how many B and D channels are carried in the PRI frame.

ISDN Layers

The ISDN approach is to provide an end-user with full support through the seven layers of the OSI Model, although ISDN confines itself to defining the operations at layers 1, 2, and 3 of this model. In so doing, ISDN is divided into two kinds of services: (a) the bearer services, responsible for providing support for the lower 3 layers of the seven-layer standard; and (b) teleservices (for example, telephone, Teletex, Videotex message handling), responsible for providing support through all 7 layers of the model and generally making use of the underlying lower-layer capabilities of the bearer services. The services are referred to as low-layer and high-layer functions, respectively. The ISDN functions are allocated according to the layering principles of the OSI Model.

Figure 3–9 shows the ISDN layers. Layer 1 (the physical layer) uses either the basic rate interface (BRI) or 2 B+D, or the PRI, which is either 23 B+D or 31 B+D. These standards are published in ITU-T's I Series as I.430 and I.431, respectively. Layer 2 (the data link layer) consists of LAPD and is published in the ITU-T Recommendation Q.921. Layer 3 (the network layer), is defined in the ITU-T Recommendation Q.931.

The reader might wish to compare Figure 3–9 with the X.25 layer structure in Figure 3–5. The basic architectures are quite similar in that the data link layer is responsible for the conveyance of the layer 3 (network layer) traffic across the SNI, and the network layer is responsible for establishing and releasing connections. However, with X.25, the network layer sessions are virtual circuits used for the transmission and reception of data and with ISDN the circuits are B channels used for the conveyance of (initially) voice and other types of traffic.

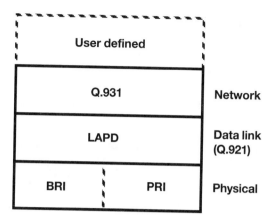

BRI: Basic rate interface (I.430)

PRI: Primary rate interface (I.431)

LAPD: Link access procedure for D channel

Figure 3–9
The ISDN layers.

ISDN PDUs

This section provides a more detailed view of the ISDN PDUs than previous discussions on the PDUs of the other technologies, because the ISDN layer 3 messages are used (with some modifications) in several of the emerging communications technologies (i.e., frame relay and ATM).

Data link layer frame (LAPD). The ISDN provides a data link protocol for devices to communicate with each other across the D channel. This protocol is LAPD, which is a subset of HDLC. The protocol is independent of a transmission bit rate and it requires a full duplex, bit transparent, synchronous channel. Figure 3–10 depicts the LAPD frame and its relationship to the ISDN layer 3, Q.931 specification. The Q.931 message is carried within the LAPD frame in the I (information) field. LAPD ensures that the Q.931 messages are transmitted across the link, after which the frame fields are stripped and the message is presented to the network layer. The principal function of the link layer is to deliver the Q.931 message error-free despite the error-prone nature of the communications link. In this regard, it is quite similar to LAPB in X.25.

LAPD has a frame format similar to HDLC, but LAPD provides for octets for the address field. This is necessary for multiplexing multiple sessions and user stations onto the BRI channel. The address field con-

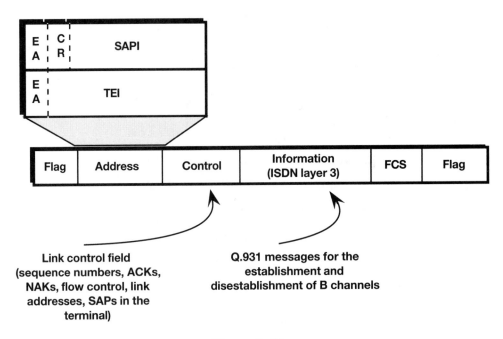

Figure 3–10
LAPD frame.

tains several control bits (which are not relevant to this discussion), a
service access point identifier (SAPI), and a terminal endpoint identifier
(TEI). The SAPI and TEI fields are known collectively as the data link
connection identifier (DLCI). These entities are discussed in the following paragraphs.

The SAPI identifies the entity where the data link layer services are
provided to the layer above (that is, layer 3). (Appendix B provides information on OSI SAPs.) At the present time, these SAPIs are defined in
ISDN:

SAPI Value	*Frame Carries*
0	signaling information
16	user traffic
63	management information

The TEI identifies either a single terminal (TE) or multiple terminals that are operating on the BRI link. The TEI is assigned automati-

cally by a separate assignment procedure. A TEI value of all 1s identifies a broadcast connection.

ISDN layer 3 Q.931. The ISDN layer 3 messages are used to manage ISDN connections on the B channels. These messages are also used (with modifications) by frame relay and ATM for setting up calls on demand at a SNI, and for provisioning services between networks at a NNI. Table 3–2 lists these messages and a short explanation is provided later in this section about the functions of the more significant messages.

The Q.931 messages all use a similar format. Figure 3–11 illus-

Table 3–2 ISDN Layer 3 Messages

Call Establishment Messages	Call Disestablishment Messages
ALERTING	DETACH
CALL PROCEEDING	DETACH ACKNOWLEDGE
CONNECT	DISCONNECT
CONNECT ACKNOWLEDGE	RELEASE
SETUP	RELEASE COMPLETE
SETUP ACKNOWLEDGE	
Call Information Phase Messages	*Miscellaneous Messages*
RESUME	CANCEL
RESUME ACKNOWLEDGE	CANCEL ACKNOWLEDGE
RESUME REJECT	CANCEL REJECT
SUSPEND	CONGESTION CONTROL
SUSPEND ACKNOWLEDGE	FACILITY
SUSPEND REJECT	FACILITY ACKNOWLEDGE
USER INFORMATION	FACILITY REJECT
	INFORMATION
	REGISTER
	REGISTER ACKNOWLEDGE
	REGISTER REJECT
	STATUS
	STATUS ENQUIRY

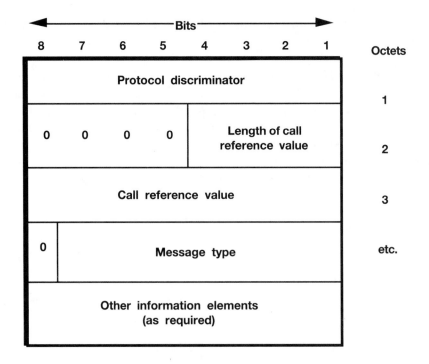

Figure 3–11
The Q.931 message.

trates this format. The message contains several parameters to define the circuit connection, and it must contain these three parameters:

- *Protocol discriminator:* Distinguishes between user-network call control messages and others, such as other layer 3 protocols (X.25, for example).
- *Call reference:* Identifies the specific ISDN call at the local UNI. It does not have end-to-end significance.
- *Message type:* Identifies the message function, such as a SETUP, DISCONNECT, etc.

The *other information elements* field may consist of many entries, and its contents depend on the message type.

The SETUP message is sent by the user or the network to indicate a call establishment. In addition to the fields in the message just described, as options, other parameters include the specific ISDN channel identification, originating and destination address, an address for a redirected call, the designation for a transit network, etc.

The SETUP ACKNOWLEDGE message is sent by the user or the network to indicate the call establishment has been initiated. The parameters for this message are similar to the SETUP message.

The CALL PROCEEDING message is sent by the network or the user to indicate the call is being processed. The message also indicates the network has all the information it needs to process the call.

The CONNECT message and the CONNECT ACKNOWLEDGE messages are exchanged between the network and the network user to indicate the call is accepted between either the network or the user. These messages contain parameters to identify the session as well as the facilities and services associated with the connection.

To clear a call, the user or the network can send a RELEASE or DISCONNECT message. Typically, the RELEASE COMPLETE is returned, but the network may maintain the call reference for later use, in which case, the network sends a DETACH message to the user.

A call may be temporarily suspended. The SUSPEND message is used to create this action. The network can respond to this message with either a SUSPEND ACKNOWLEDGE or a SUSPEND REJECT.

During an ongoing ISDN connection, the user or network may issue CONGESTION CONTROL messages to flow-control USER INFORMATION messages. The message simply indicates if the receiver is ready or not ready to accept messages.

The USER INFORMATION message is sent by the user or the network to transmit information to a (another) user.

If a call is suspended, the RESUME message is sent by the user to request the resumption of the call. This message can invoke a RESUME ACKNOWLEDGE or a RESUME REJECT.

The STATUS message is sent by the user or the network to report on the conditions of the call, or other administrative matters. The STATUS ENQUIRY is sent by the user or the network (but usually the user) to inquire about a state or operation.

ISDN also allows other message formats to accommodate equipment needs and different information elements. This feature provides considerable flexibility in choosing other options and ISDN services.

Conclusions on ISDN

ISDN cannot be considered successful based on its performance since inception in the early 1980s. In North America, the progress of ISDN has been much slower than in Europe because of the lack of a cohesive nationwide implementation policy. This situation has changed in the past few years with the Regional Bell Operating Companies (RBOC) and Bellcore aggressively implementing "National ISDN" throughout the United States. This is leading to extensive central office implementations in both primary and basic rate offerings. Nonetheless, progress has still been slow.

Notwithstanding, ISDN can be judged successful in another way. It has served the industry well with its specifications of LAPD and the Q.931 messaging protocol. Indeed, these two protocols are found throughout the communications industry. For example, LAPD has been one of the foundation technologies for frame relay as well as the link access procedure for modems (LAPM), and Q.931 is used extensively in other signaling systems such as mobile radio, frame relay, and ATM.

SIGNALING SYSTEM NUMBER 7 (SS7)

The Purpose of SS7

We now turn our attention to Signaling System Number 7 (SS7), a clear channel signaling specification published by the ITU-T. SS7 is the prevalent signaling system for telephone networks for setting up and clearing calls and furnishing services such as 800 operations. It is designed also to operate with the ISDN UNI.

SS7 defines the procedures for the set-up, ongoing management, and clearing of a call between telephone users. It performs these functions by exchanging telephone control messages between the SS7 components that support the end users' connection. Table 3–3 provides a summary of the major functions of SS7.

The SS7 signaling data link is a full duplex, digital transmission channel operating at 64 kbit/s. Optionally, an analog link can be used with either 4 or 3 kHz spacing. The SS7 link operates on both terrestrial and satellite links. The actual digital signals on the link are derived from pulse code modulation multiplexing equipment, or from equipment that employs a frame structure. The link must be dedicated to SS7. In accordance with the idea of clear channel signaling, no other transmission can be transferred with these signaling messages and extraneous equipment must be disabled or removed from an SS7 link.

Table 3–3 Examples of the SS7 Functions.

- Setup and clear down a telephone call
- Provide the called party's number (caller id)
- Indication that a called party's line is out of service
- Indication of national, international, or other subscriber
- Indication that called party has cleared
- Nature of circuit (satellite/terrestrial)
- Indication that called party cleared, then went off-hook again
- Use of echo-suppression
- Notification to reset a faulty circuit
- Language of assistance operators
- Status identifiers (calling line identity incomplete; all addresses complete; use of coin station; network congestion; no digital path available; number not in use; blocking signals for certain conditions)
- Circuit continuity check
- Call forwarding (and previous routes of the call)
- Provision for an all digital path
- Security access calls (called closed user group [CUG])
- Malicious call identification
- Request to hold the connection
- Charging information
- Indication that a called party's line is free
- Call set-up failure
- Subscriber busy signal
- Identifiers of: circuits signaling points, called and calling parities, incoming trunks, and transit exchanges

Typical Topology

Figure 3–12 depicts a typical SS7 topology. The subscriber lines are connected to the SS7 network through the Service Switching Points (SSPs). The purpose of the SSPs is to receive the signals from the CPE and perform call processing on behalf of the user. SSPs are implemented

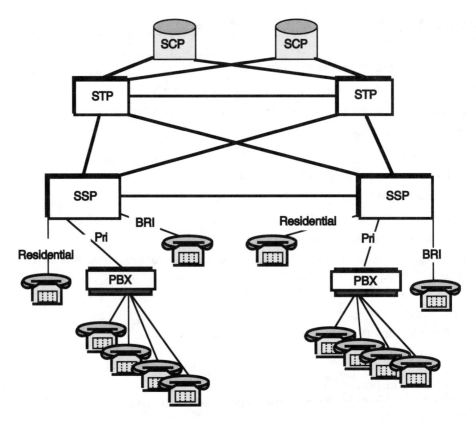

Figure 3–12
Typical SS7 topology.

at end offices or access tandem devices. They serve as the source and destination for SS7 messages. In so doing, SSP initiates SS7 messages either to another SSP or to a signaling transfer point (STP).

The STP is tasked with the translation of the SS7 messages and the routing of those messages between network nodes and databases. The STPs are switches that relay messages between SSPs, STPs, and service control points (SCPs). Their principal functions are similar to the layer 3 operations of the OSI Model.

The SCPs contain software and databases for the management of the call. For example, 800 services and routing are provided by the SCP. They receive traffic (typically requests for information) from SSPs via STPs and return responses (via STPs) based on the query.

Although Figure 3–12 shows the SS7 components as discrete entities, they are often implemented in an integrated fashion by a vendor's equipment. For example, a central office can be configured with a SSP, a STP, and a SCP or any combination of these elements. These SS7 components are explained in more detail later in this section.

SS7 Layers

Figure 3–13 shows the layers of SS7. The right part of the figure shows the approximate mapping of these layers to the OSI Model. Beginning from the lowest layers, the message transfer part (MTP) layer 1 defines the procedures for the signaling data link. It specifies the functional characteristics of the signaling links, the electrical attributes, and the connectors. Layer 1 provides for both digital and analog links, although vast majority of SS7 physical layers are digital. The second layer is labeled MTP layer 2. It is responsible for the transfer of traffic between SS7 components. It is quite similar to an HDLC-type frame and indeed was derived from the HDLC specification. The MTP layer 3 is somewhat related to layers 3 of ISDN and X.25 in the sense that this layer provides the functions for network management, and the establishment of message routing as well as the provisions for passing the traffic to the SS7 components within the SS7 network. Many of the operations at this layer pertain to routing, such as route discovery and routing around problem areas in an SS7 network.

Layer 3 of SS7 is organized into functional modules. Their functions are:

- *Message routing:* Selects the link to be used for each message.
- *Message distribution:* Selects the user part at the destination point.
- *Message discrimination:* Determines at each signaling point if the message is to be forwarded to message routing or to message distribution.
- *Signaling traffic management:* Controls the message routing functions of flow control, rerouting, changeover to a less faulty link, and recovery from link failure.
- *Signaling link management:* Manages the activity of the layer 2 function and provides a logical interface between layer 2 and layer 3.
- *Signaling route management:* Transfers status information about signaling routes to remote signaling points.

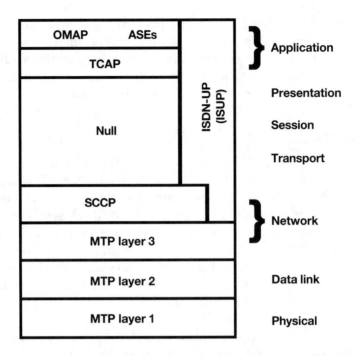

OMAP: Operations, maintenance and administration part

ASEs: Application service elements

TCAP: Transaction capabilities part

IDSN-UP: ISDN user part

SSCP: Signaling connection part

MTP: Message transfer part

Figure 3–13
The SS7 layers.

The signaling connection control point (SCCP) is also part of the network layer, and provides for both connectionless and connection-oriented services. The main function of SCCP is to provide for translation of addresses, such as ISDN and telephone numbers to identifiers used in the SS7 network.

The ISDN user part (ISUP) is responsible for transmitting call control information between SS7 network nodes. In essence, this is the call control protocol, in that ISUP sets up, coordinates, and takes down trunks within the SS7 network. It also provides features such call status checking, trunk management, trunk release, calling party number information, privacy indicators, detection of application of tones for busy conditions, etc. ISUP works in conjunction with ISDN Q.931. Thus, ISUP translates Q.931 messages and maps them into appropriate ISUP messages for use in the SS7 network.

The transaction capabilities application part (TCAP) is an application layer running in layer 7 of the OSI Model. It can be used for a variety of purposes. One use of TCAP is the support of 800 numbers transferred between SCP databases. It is also used to define the syntax between the various communicating components, and it uses a standard closely aligned with OSI transfer syntax, called the basic encoding rules (BER), which codes each field of traffic with (a) syntax type, (b) length of contents field, and (c) contents field (the information).

Finally, the OMAP and ASEs are used respectively for (a) network management and (b) user-specific services. Both are beyond this general text.

SS7 PDUs

ITU-T Recommendation Q.703 of SS7 describes the procedures for transferring SS7 signaling messages across one link. It performs the operations that are typical of layer two protocols. As shown in Figure 3–14, Q.703 has many similarities to the HDLC protocol. For example, both

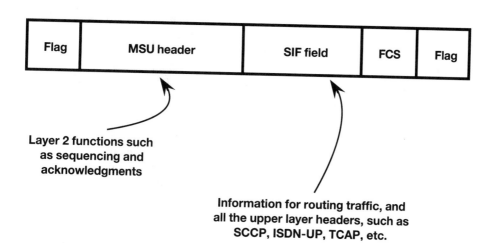

Figure 3–14
SS7 PDUs.

protocols use flags, error checks, and sending/receiving sequence numbers.

The messages are transferred in variable length signal units (SUs), and the primary task of this layer is to ensure their error-free delivery. The SUs are one of three types:

- Message signal unit (MSU)
- Link status signal unit (LSSU)
- Fill-in signal unit (FISU)

The MSU carries the actual signaling message forward to the user part (UP). Q.703 transfers the MSU across the link, and determines if the message is uncorrupted. If the message is damaged during the transfer, it is retransmitted. The LSSU and FISU do not transport UP signals; they are used to provide layer two control and status signal units between the layer two Q.703 protocols at each end of the link.

The SS7 layer 3 functions are called signaling network functions and fall into two categories:

- *Signaling message handling functions*: directs the message transfer to the proper link or user part
- *Signaling network management functions*: control, the message routing and the configuration of the SS7 network.

A message is generated at a UP function, known as the originating point and is sent to another UP function, known as the destination point. The intermediate nodes are signaling transfer points (STPs).

The STPs use information in the message to determine its routing. A routing label contains the identification of the originating and destination points. A code is also used to manage load sharing within the network. The routing label is used by the STP in combination with predetermined routing data to make the routing decisions. The route is fixed unless failures occur in the network. In this situation, the routing is modified by layer 3 functions. The load-sharing logic (and code in the label) permit the distribution of the traffic to a particular destination to be distributed to two or more output signaling links.

Conclusions on SS7

SS7 has been a huge success in the industry. It is implemented in public telephone networks by practically all carriers throughout the world. Of course, its success was almost assured because its predecessors were woefully inadequate for supporting control signaling in telephone networks. Additionally, features of SS7 have found their way into other systems such as GSM and even satellite signaling. Later in this book, examples are cited of SS7 operating with some of the emerging technologies

SUMMARY

T1 and E1 were first implemented over thirty years ago, yet remain as the prevalent option for digital carrier systems. Their use will continue, with SONET/SDH eventually replacing them in carriers' backbone networks.

The implementation of the ISDN has been slow, but its components, especially LAPD and Q.931, have been quite successful, and are used in a variety of other systems, such as frame relay and ATM. However, the long-term viability of a 144 kbit/s 2 B+D offering is not assured, due to the increasing bandwidth needs of user work stations.

X.25, while being an old technology, remains a viable option for many user applications, especially low-speed, asynchronous systems. It is embedded in many systems and products and will remain an option for many years. Nonetheless, some X.25 users will migrate to frame relay, SMDS, and ATM in the near future.

SS7, while an emerged technology, has no competition by any of the emerging communications technologies, and will remain as the prevalent out-of-band signaling protocol in the telecommunications industry. Its use will continue to increase, and it is being adapted for use in other technologies.

Frame Relay

INTRODUCTION

This chapter examines frame relay interfaces and frame relay networks. The chapter begins with a summary of why frame relay has come into existence and the types of user applications and networks it is intended to support. The frame relay standards by ITU-T and ANSI are also examined. The later part of the chapter discusses data link identifiers and PVCs, as well as how a frame relay network manages user traffic. The chapter also explains the network-to-network interface, published by the Frame Relay Forum.

THE PURPOSE OF FRAME RELAY

The purpose of a frame relay network is to provide an end-user with a high-speed virtual private network (VPN) capable of supporting applications with large bit-rate transmission requirements. It gives a user T1/E1 access rates at a lesser cost than that which can be obtained by leasing comparable T1/E1 lines. One colleague refers to this frame relay concept as a virtually meshed network.

Additionally, frame relay is designed to give a user fast services by minimizing or eliminating a variety of functions heretofore performed in most data networks, such as X.25. In effect, frame relay is fast because it performs relatively few services for an end user. This approach may not seem logical to the reader—it may seem to be a step backwards, but it has sound reasoning behind it, so read on.

The design of frame relay networks is based on the fact that data transmission systems today are experiencing far fewer errors and problems than they did in the 1970s and 1980s. During that period, protocols were developed and implemented to cope with error-prone transmission circuits. However, with the increased use of optical fibers, protocols that expend resources dealing with errors become less important—indeed, they can be thought of as an overkill solution to the care of user traffic. Frame relay takes advantage of this improved technology by eliminating many of the time-consuming and expensive error correction, editing, and retransmission features that have been part of many data networks for almost two decades.

Another factor contributing to the increased use of frame relay is the need for higher-capacity network interfaces (in bits/second). The technology of the 1980s focused on transmission rates ranging from 1.2 kbit/s to 19.2 kbit/s, which are inadequate to support applications that need large transmissions of data, such as bit-mapped graphics and large database transfers. As stated earlier, frame relay provides T1/E1 access rates to the end-user of rates from 64 kbit/s to 2.048 kbit/s.

Last, many of the networks that are in existence today were designed to support relatively "unintelligent" user work stations, such as non-programmable terminals. Today, these workstations operate with powerful microprocessors and have many capabilities. They are able to handle many tasks that were previously delegated to network components. Consequently, a frame relay network does not perform many user QOS functions, and its data management operations (such as ACKs, NAKs, etc.) are quite sparse. Many of these operations become the responsibility of the user workstation. This "bare bones" approach results in a faster network and, at the same time, places more responsibility on the end user machine for data management operations.

PERTINENT STANDARDS

Frame relay has been in the works for a number of years. Three standards led the way (see Table 4–1). ITU-T's I.122 provided the initial framework with the publication of ISDN bearer services for additional packet services. Some of the work performed on ITU-T's Q.921 (Link

Table 4–1   Frame Relay Standards

	Service Description	Core Aspects	Access Signaling
ITU-T	I.233	Q.922, Annex A	Q.933
ANSI	T1.606	T1.618	T1.617
Frame Relay Forum	Frame Relay Network-to-Network Interface, May 7, 1992		
Bellcore	TR-TSV-001369 TR-TSV-001370		

Access Procedure for the D Channel [LAPD]) demonstrated the usefulness of virtual circuit multiplexing for data link layer protocols (layer 2 of the OSI Model). Although generally not recognized, ITU-T's V.120 also provided a valuable foundation because it resulted in a specification defining multiplexing operations across the ISDN S/T interface, and multiplexing is a fundamental aspect of frame relay.

Today, the frame relay standards are published by the ITU-T and ANSI. The ANSI standards are published as T1.606, T1.617, and T1.618. The ITU-T's specifications are published as I.233, Q.222 Annex A, and Q.933. The ANSI and ITU-T documents are somewhat in technical alignment with each other. However, their update cycles vary, and some of the sections across the documents are not consistent. The Frame Relay Forum (FRF) is also an active group in the frame relay arena. Any enterprise that wishes to promote frame relay and/or have an influence on the development of frame relay standards is welcome to join the Forum. Bellcore has also published technical references on frame relay for both a UNI and an NNI. These references are cited in Table 4–1.

TYPICAL FRAME RELAY TOPOLOGY

Figure 4–1 depicts a typical frame relay topology. A user is connected to the frame relay network through (typically) a router. The router implements the frame relay user-to-network interface (UNI) protocol in order to communicate with the frame relay switch. Nothing precludes placing this protocol in the end user device, but the common prac-

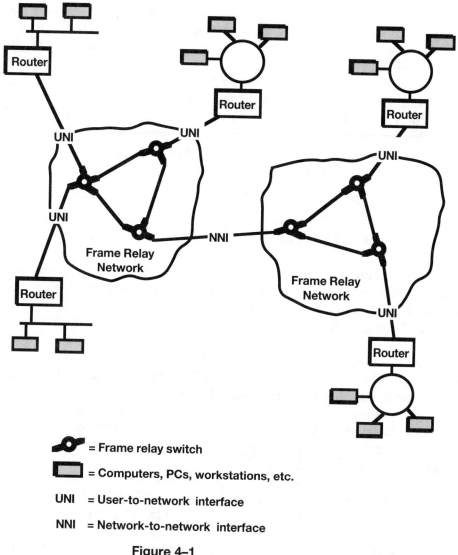

= Frame relay switch

= Computers, PCs, workstations, etc.

UNI   = User-to-network interface

NNI   = Network-to-network interface

Figure 4–1
Typical frame relay topology.

tice is to mask the frame relay operations from the user (thus providing transparent operations to the user).

If frame relay is implemented strictly in accordance with the ANSI and ITU-T standards, the physical interfaces are ISDN-based. However, most implementations use T1/E1 circuits.

The frame relay specifications also include a network-to-network interface (NNI) protocol. As of this writing, the NNI was not an approved international standard, but it has been approved by the Frame Relay Forum and is discussed later in this chapter.

The operations and topology configurations within the frame relay network are not defined in any frame relay standards nor in any working documents published by the Frame Relay Forum. Indeed, the NNI and UNI are, as their names suggest, interface specifications and the frame relay network provider is free to implement any type of protocol, configuration, etc., inside the network.

THE FRAME RELAY LAYERS

Frame relay is designed to eliminate and/or combine many operations residing in layers 3 and 2 of a conventional 7-layer model. It implements the operational aspects of statistical multiplexing found in the X.25 protocol and the efficiency of circuit switching found in TDM protocols.

The end effect of this approach is increased throughput and decreased delay and the saving of "CPU cycles" within the network because some services are eliminated. Frame relay is supposed to provide better delay performance than X.25, but it cannot match TDM performance, because TDM does little processing of the traffic.

Frame relay uses variable length PDUs (frames). This capability supports the internetworking of different types of networks (LANs and WANs), many of which employ different sizes of frames. For example, many LANs use frames of 1500 octets, and X.25 networks typically use packets of 128 to 512 octets. However, variable data units translate into variable delay, and present implementations of frame relay do not work well in systems that are delay sensitive (digitized voice, video). Consequently, the Frame Relay Forum is examining several solutions to this delay sensitivity problem with a method to accommodate voice traffic. Eventually, frame relay networks will accommodate voice traffic as well.

Figure 4–2 shows another way of viewing frame relay operations. On the left side of this figure is a depiction of a typical data communications protocol stack which encompasses the physical, data link, and network layers. These layers perform the conventional operations of the OSI bearer services discussed in Appendix A. For example, the physical layer

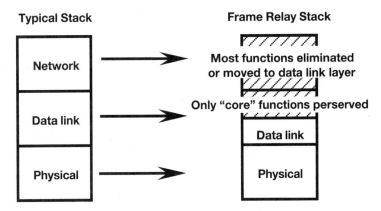

Figure 4–2
Comparison of a typical stack and the frame relay stack.

is responsible for modulation operations, encoding and decoding bits, providing timing, etc. The data link layer is responsible for the error checking and retransmission of erred traffic, and for the proper framing of the traffic at the receiver. The network layer is responsible for managing and routing the traffic within the network, establishing virtual connections, and negotiating quality of services between the network and the users.

In contrast, as shown on the right side of Figure 4–2, the frame relay stack eliminates most of the network layer and several aspects of the data link layer. The remaining network layer functions are pushed down to the data link layer. It is no wonder that frame relay is fast; it does very little.

Frame Relay and Its Relationship to ISDN Layers

In accordance with ANSI and ITU-T standards, frame relay service can be provided through the ISDN C-plane and U-plane procedures. Virtual calls may be established on an as needed basis, in which case they are negotiated during call set-up procedures through the C-plane procedures. Additionally, C-plane procedures can be established on a PVC basis, in which case, the data link connection identifier (DLCI) and associated QOS parameters must be defined by administrative specific operations.

Initial implementations of frame relay support only PVCs. But, eventually, SVCs will be supported and included in the standards. Moreover, at the present time, most implementations have not used the

ISDN technology, but have simply implemented the frame relay protocol over a T1/E1 channel. Both the ISDN and non-ISDN implementations will be explained in this chapter.

The ISDN layered architecture is shown in Figure 4–3. It is modeled as seven layers in conformance with OSI. For the UNI, three signaling

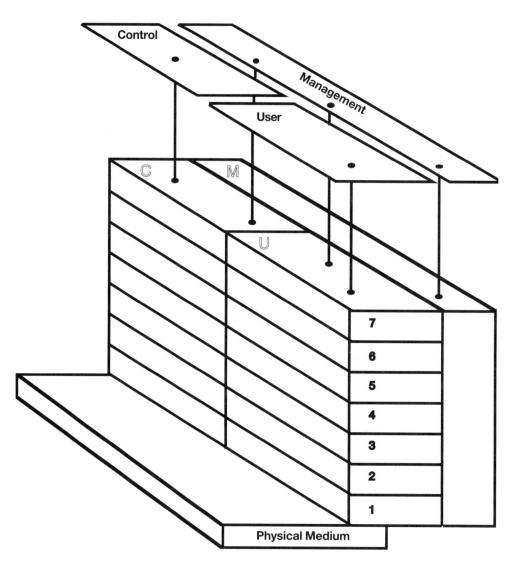

Figure 4–3.
The C, U, and M planes of the ISDN model.

planes are used. The control plane (C-plane) concerns itself with establishing calls, establishing circuits, and managing the connections. For instance, a C-plane procedure takes place with D-channel operations. The user plane (U-plane) contains the operations, service definitions, and protocols needed for exchanging user data. In an ISDN user-to-network interface, this data could pertain to applications running on ISDN D, B, or H channels. Finally, the management plane (M-plane) is used for management operations between the U- and C-planes.

The frame relay service uses the ISDN bearer service (the ISDN layer's 1–3). A bearer service through the frame relay operations provides bidirectional transfer of user traffic from the local S or T interface to the remote S or T interface. Routing through a network is provided by using an attached label, which is a PVC identifier called the DLCI.

While QOS features can be negotiated (on switched systems), the operations must occur in the C-plane only. For PVCs, the QOS is predefined and identified with a specific DLCI value. The standards require that the sequence of the frame relay service data units (SDUs) must be maintained from the sending S/T reference point to the receiving S/T reference point. In effect, the user SDUs are sent transparently and sequentially through the network.

The frame relay network detects errors but the network does not act upon them as far as sending NAKs. Errored traffic is discarded, and the network takes no more remedial action. Moreover, ACKs are not used either.

OSI and ANSI Layers

Figure 4–4 shows how the frame relay service is modeled on the OSI layered architecture and the ISDN architecture relating to the C-plane and the U-plane. The network only supports the core aspects of the Q.922 layer 2 protocol, not the full feature set of Q.922 (Q.922 is an enhancement of Q.921). Core service offerings can be made on either basic access rate or primary access rate interfaces and on ISDN channels B, D, and H, although most implementations do not use ISDN layers, except for the core functions of LAPD.

Functions that are required beyond the core functions must be implemented on an end-to-end basis and not on a UNI basis. As a consequence, protocols that reside above the layer 2 core functions are not processed by the frame relay network.

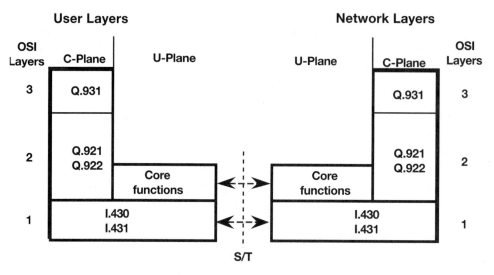

Figure 4–4
The user-network configuration.

THE FRAME RELAY PROTOCOL DATA UNIT (PDU)

Since the frame relay frame (the data link layer PDU) is derived from HDLC and the HDLC derivative, LAPD, its format is quite similar to these protocols (see Figure 4–5). Frame relay uses the beginning and ending flag, the frame check sequence field (FCS), and the I field. However, it does not have separate control and address fields. These two LAPD fields are combined into one field in the frame relay header. This part of the frame identifies a PVC, which we just learned, is called the DLCI.

The header also contains three bits that are used for congestion notification and traffic shedding. These operations are covered later in this chapter.

FRAME RELAY OPERATIONS IN MORE DETAIL

The Frame Relay Core Functions

The term core refers to the fact that only the minimum operations are implemented—those that cannot be eliminated in a data link control. Once again, this approach is in the spirit of frame relay, which is to scale

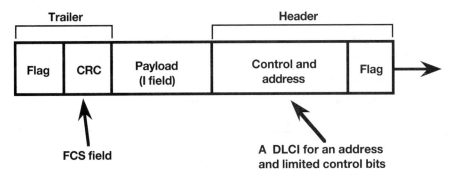

Figure 4–5
The frame relay frame.

down the number of operations a service provider performs on the user traffic. Therefore, the ITU-T Q.921 and ANSI T1.602-1988 core functions are organized around five very elementary procedures that are listed in Table 4–2.

First, a frame relay system must provide services to delimit and align frames on the channel, so flags are used at a receiver to identify the beginning and ending of a frame. The system must also provide transparency of the flags, which entails two functions: (a) bit stuffing at the sending machine, and (b) bit unstuffing at the receiving machine. Flags take the form of specific sequence of 1 and 0 bits to connote the start and end of a transmission (01111110). Bit stuffing is performed on any similar bit sequence between the flags to eliminate confusion. Bits are stuffed in the control, payload and FCS fields to prevent any of the bits in these fields from being interpreted incorrectly as a flag, which would render the traffic unintelligible. Obviously, these extra bits must be unstuffed (removed) at the receiver.

Second, the frame relay system must support virtual circuit multiplexing and demultiplexing through the use of the DLCI field in the frame. The payload fields of the frames on the channel may contain traffic from multiple users, each payload field is identified with a unique DLCI.

Third, the system must inspect the frame to make certain it aligns itself on integer number of octets prior to the zero bit insertion and following the unstuffing of the zero bit.

Fourth, the system must inspect the frame to insure that it does not exceed the maximum and minimum frame sizes (these frame sizes have not yet been defined by the standards groups).

Table 4–2. Frame Relay Core Functions

- Frame delimiting, alignment, and flag transparency
- Virtual circuit multiplexing and demultiplexing
- Octet alignment of the traffic
- Checking for maximum and minimum frame sizes
- Detection of transmission, format, and operational errors

Fifth, the system must be able to detect transmission errors (through the use of the frame check sequence [FCS] field), formatting problems, and other operational errors.

The Data Link Connection Identifier (DLCI)

While frame relay purports to eliminate the operations at the network layer, it does not eliminate all network layer operations. Figure 4–6 illustrates one network layer operation that is essential for frame relay operations: the identification of virtual connections. Frame relay uses the data link control identifier (DLCI) to identify the destination end-user. This 10-bit number is similar to a virtual circuit number in a network layer protocol. The 10-bits can be expanded with the address extension option.

The DLCIs are premapped to a destination node. This simplifies the process at the routers because they only need to consult their routing table, check the DLCI in the table, and route the traffic to the proper output port based on this address.

Inside the network, this same scheme could be used, although the frame relay switches need not maintain a strict PVC relationship in the network. Connectionless operations can be implemented to allow for dynamic and robust routing between the frame relay switches. The only requirement is to make certain the frame arrives at the destination port designated by the DLCI.

Figure 4–6 also shows an address translation operation in which a destination address created by a user application is translated into a DLCI. This service is not defined by frame relay, but is an essential part of the transmission process. In this figure, a user device is on a LAN that is connected to router 1. This LAN has a Internet Protocol (IP) network address of 128.1. Router 1 receives traffic from the user device on 128.1 which contains a destination address of 128.2, and a host address of 3.4.

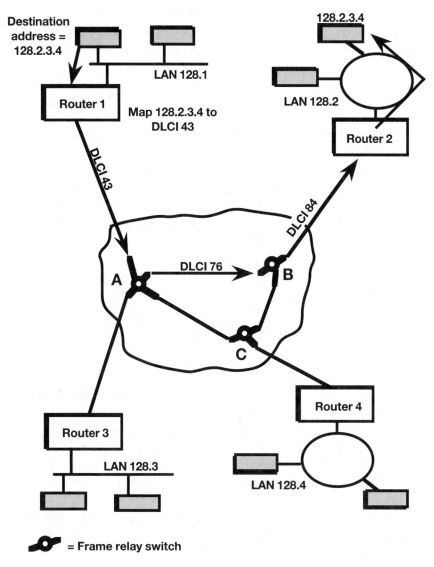

Figure 4–6
The data link connection identifier (DLCI).

Router 1 performs a table look up and determines that address 128.2.3.4 is to be mapped to DLCI 43. It then places the user traffic into the frame relay frame and codes the DLCI field as 43.

This frame is sent to frame relay switch A, which examines DLCI 43 and determines through its routing table that DLCI 43 is to be changed (mapped) to DLCI 76. Switch A also uses the routing table to determine that the next node to receive this frame is frame relay switch B. Upon receiving the frame, switch B maps DLCI 76 to DLCI 84, and relays the frame to router 2. This router need only examine the destination address of 128.2.3.4 (which was stored in the user payload and sent transparently through the network) to determine how to convey the traffic to the end user station.

Earlier discussions established that the operations within the network are not defined by the frame relay standards. Therefore, this example may or may not be used in specific implementations. Indeed, some switch vendors strip off the frame relay header and trailer at the sending UNI, and add their proprietary control fields for the transport of the payload to the receiving UNI. At the receiving interface, the frame relay header replaces the internal network headers. This approach makes sense because it allows the vendor to maintain one internal network protocol, regardless of the particular UNI protocol. For example, some vendor switches can support X.25 UNIs, frame relay UNIs, SMDS UNIs, etc. with the same internal network protocol.

Frame Relay Link Layer Error Checking

Frame relay may continue to do an error check on the frame at each frame relay switch in the network as well as at any routers that use the frame relay software. The conventional cyclic redundancy check (CRC) operation is employed with the frame check sequence (FCS) field. However, if this check reveals that the frame was distorted during transmission across the communications channel, the frame is not only discarded, but no NAK is returned to the sender.

The operation may proceed on a node-to-node basis, with each node performing the FCS check. A node also has the option of managing congestion through the use of the frame relay congestion notification bits (discussed shortly). Moreover, it may discard traffic because of congestion problems, because of a bad FCS check, or for that matter, any reason that the network so chooses. Of course, in discarding user frames, the network runs the risk of not meeting the user's QOS requirements.

Within the network, some vendors choose not to perform this node-by-node error check. In these systems, one error check is performed at the final frame relay switch. If an error is revealed, the traffic is discarded. The philosophy behind this approach is sound. Why waste resources at the intermediate nodes in the network checking for errors that rarely occur?

Potential Congestion Problems

A frame relay network must still deal with the problem of congestion, which is an operation typically handled at the network layer. Most networks today provide transmission rules for their users which include agreements on how much traffic can be sent to the network before the traffic flow is regulated (flow-controlled). Flow control is an essential ingredient to prevent congestion in a network. Congestion is a problem that is avoided by network administrators, almost at any cost, because it results in severe degradation of the network both in throughput and response time. Because of the importance of having a congestion-free network, the practice in the past has been to implement explicit flow control mechanisms (a topic covered in Chapter 2). However, frame relay takes a different approach and uses implicit flow control mechanisms (although it uses the term "explicit" in its procedures).

Queuing theory demonstrates that the offered load to the network may increase linearly with resulting throughput also increasing—but only to a point. As the traffic (offered load) in the network reaches a certain level, mild congestion begins to occur with the resulting drop in throughput. If this situation proceeded in a linear fashion, it would not be so complex a problem. However, at a point in which the utilization of the network reaches a certain level, throughput drops precipitously, due to serious congestion and the build up of queuing in the servers (buffers).

Therefore, even simple networks, such as frame relay networks, must provide a mechanism of informing users in the network when congestion is occurring, as well as a mechanism providing for flow control on user devices.

The preceding discussion on the build-up of excess queues and the resultant severe effect on network throughput also holds for its effect on response time and delay. That is to say, a continued buildup of queues will eventually result in serious delay of the delivery of user payload to the end destination.

One might think that there is a one-to-one relationship between degraded throughput and degraded response time (increased delay). While congestion degrades the QOS of both of these features, overall network throughput may actually benefit from longer queues because the network can build up these queues and use them to smooth traffic in the network over a period of time—say during the evening, when traffic may be light. However, achieving superior performance for both delay and response time requires that the network keep the queues small. Indeed, the smaller the queues, the better the response time. In the final analysis, congestion eventually degrades the QOS for both throughput and delay management.

Traffic Management

Frame relay networks employ implicit flow control techniques for the management of user traffic. The reader may recall from Chapter 2 that implicit flow control mechanisms do not require a node to stop absolutely its transmissions. Rather, the node is notified of problems, with the implicit assumption that if the node does not take some type of remedial action, such as the cessation of transmission, it risks having its traffic discarded.

Frame relay congestion and flow control options are just that: optional. Some vendors have not implemented these features. However, unless other flow control measures are implemented in the network, the use of this option is quite important.

The frame relay header shown in Figure 4–7 is used by both the user device and the network to manage and flow control traffic. The header consists of seven fields. They are listed and briefly described here and explained in more detail in subsequent discussions:

- *DLCI:* The *data link connection identifier* identifies the destination virtual circuit user (which can be any machine with a frame

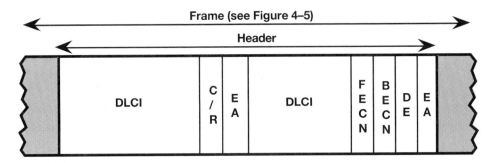

Figure 4–7
The frame relay header.

relay interface, an application, or anything else so designated by the user)

- *C/R:* The *command/response* bit is not used by the frame relay network
- *EA:* The address extension bit is used to extend the header to 3 or 4 octets
- *FECN:* The *forward explicit congestion notification* bit is used to notify the upstream user node of congestion problems
- *BECN:* The *backward explicit congestion notification* bit is used to notify the downstream user node of congestion problems
- *DE:* The *discard eligibility* bit is used to tag traffic for possible discarding

Two mechanisms are employed to notify users and frame relay switches about congestion, and take corrective action. Both capabilities are achieved by the BECN bit and the FECN bit.

Assume that the frame relay switch A in Figure 4–8 is starting to experience problems (queues are becoming full, the device is experiencing a problem with memory management, etc.). It informs both the upstream node (router 2) and the downstream node (router 1) of the problem by the use of the FECN and BECN bits, respectively. The BECN bit is set to 1 in a frame and is sent downstream to notify the source of the traffic that congestion exists in the network. This notification permits the source device to flow control its traffic until the congestion problem is solved.

In addition, the FECN bit can be set to 1 in a frame and sent to the upstream node to inform it that congestion is occurring downstream. One might question why the FECN is used to notify upstream devices that congestion is occurring downstream. After all, the downstream device is the one sending the traffic. However, it is possible that the upstream device might have a dialogue going with the downstream device that will permit this device to affect the traffic pattern.

For example, the FECN bit could be passed to an upper layer protocol (such as the transport layer). Upon receiving this notification, the transport layer could (a) slow down its acknowledgments (which in some protocols would close the transmit window at the sending device) or (b) establish a more restrictive flow control agreement with its communications source machine (which also is allowed in some protocols). For the present, since the frame relay protocol usually is terminated at the router and does not find its way to the user machine, this discussion is somewhat academic.

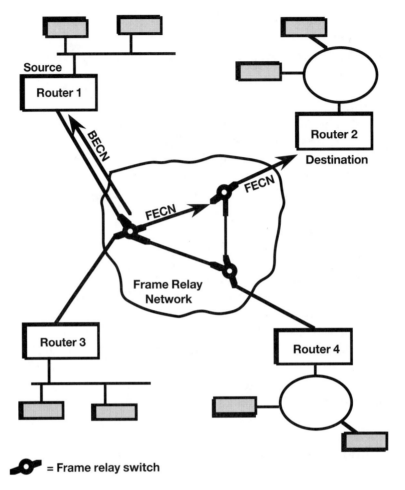

Figure 4–8
The congestion notification bits.

Vendors vary on how they operate with the BECN and FECN bits of the frame relay frame. Some vendors will read these bits but will not act upon them. Others will read the bits and accumulate statistics on them. Still other vendors will either set the bits and act upon them at the receiver or do nothing at all.

As examples, Ascom Timeplex in its Time/LAN 100 router/bridge begins to buffer frames upon receiving five consecutive frames regarding one DLCI with the BECN bit set. It will continue to reduce the output on this DLCI by one-eighth of the CIR until the output is one-half of the

CIR. Upon receiving five consecutive frames of the DLCI with the BECN bit not set, the Time/LAN router then increases its output in increments of one-sixteenth of the CIR. The idea is to reduce traffic very rapidly and then bring traffic gradually back into the network. Cisco Systems records statistics on receiving the FECN and BECN bits but takes no action on them, while a Dynatech system takes no action on FECN, BECN, nor the discard eligibility bits.

One of the reasons that the flow control bits are not implemented by some vendors is because they rely on the transport layer to flow control the end user devices. Since the transport layer typically resides in the end user machine, it is a logical tool for handling this operation. Moreover, since many applications in the transport layer (especially in TCP) are designed to operate over protocols that exercise either no flow control (such as IP) or only rudimentary flow control operations (such as ICMP), running TCP in the user device and frame relay in the router is an effective combination.

Most versions of TCP operate timers that adjust to variable delay conditions in a network. So, if a frame relay network begins to experience congestion, it is likely that the responses (ACKs from the receiver) to a TCP module's transmissions will begin to take longer to arrive at that TCP module. In turn, TCP notes the delay (after building a profile on a number of returning responses) and adjusts its timers accordingly. The end result is that the sending TCP module will automatically delay the retransmission of traffic for which it has not received an acknowledgment because of this timing operation.

The acknowledgments also contain a control field called the credit window. The value in this field sets a limit on how much more data the sender can transmit into the network to the receiver. It may give the sender the right to "open its transmit window" and send a large volume of traffic, or it may restrict the number of transmissions to a few—or even none. If a new credit is not returned back to the sending module, then this module cannot update its transmit window. Eventually, it exhausts its credits, and then must stop sending data. Its transmit window remains closed until it receives a credit window update from the remote TCP module.

Therefore, the delay in the receipt of acknowledgments from the remote machine, due to network congestion, results in: (a) the inability of the sending machine to use the credit window field to open its transmit window and send more traffic, as well as (b) the setting of the retransmission time to a longer value. The end result is a decrease in traffic flow—all without the use of the FECN and BECN bits.

Consolidated Link Layer Management (CLLM)

The frame relay standards provide for a limited number of network management and diagnostic functions, which are grouped under the consolidated link layer management (CLLM) procedures.

CLLM uses DLCI 1023 to notify the downstream node about various operations, problems, etc. The reservation of a DLCI for control purposes allows the network to send traffic to a user device or its router irrespective of receiving traffic from the upstream node. CLLM includes a diagnostic code in the frame to describe the type of problem encountered by the network. In our previous discussion, this diagnostic could identify excessive congestion, but it could also identify other problems such as failure of a processor or a link failure. The affected DLCIs are listed in a field in this frame, and they are typically the DLCIs that should flow control their traffic. The CLLM approach (at this point) is not implemented extensively in the industry.

CLLM messages.   CLLM is based on the use of HDLC XID frames and is derived from the ISO 8885 standard. The 8885 standard is one aspect of HDLC which describes the use of the XID frame information field for its contents and format. CLLM uses the XID frame to report on the problems that have been encountered in the network.

A cause code field in the XID frame, as listed below, is used for these operations. The cause code allows a node that is experiencing congestion or other problems to report the type of problem, although it can be seen from this list that the nature of the problem that can be reported is limited to reporting only a few of the events:

Network congestion, excessive traffic, short term

Network congestion, excessive traffic, long term

Facility or equipment failure, short term

Facility or equipment failure, long term

Maintenance action, short term

Maintenance action, long term

Unknown, short term

Unknown, long term

As I have noted (to my clients and in other writings), one code is particularly interesting—the unknown cause. Of course, the use of such a code is not unusual in network operations, but to state that an unknown problem is for the "short term" or "long term" seems to indicate that the reporter of the problem suffers from a lack of clairvoyance and, at the same time, possesses an ample supply of it. It seems to say, "I do not know what the problem is, but it will last a long (short) time."

Whatever the rationale for this code may be, the frame relay protocol states that the cause code is to be coded "short term" if the sender anticipates a transient problem, and it is to be coded "long term" if the sender anticipates a problem that is not just transient. In either case, the decision of how to use this field is network dependent.

The Discard Eligibility Bit

Congestion can be a problem in any demand-driven network. Frame relay handles this problem by discarding traffic to avoid potential congestion problems. In some instances, it is desirable to discern which frames of the user's traffic should be discarded.

The approach used by frame relay is the discard eligibility (DE) bit. How the DE bit is acted upon is an implementation-specific decision. However, in most instances the DE bit is turned to 1 by the user to indicate to the network that, in the event of problems, this frame (with this bit = 1) is "more eligible" for being discarded than others in which the bit is set to 0.

Of course, the DE bit need not be implemented. When congestion occurs, a network node simply may throw away traffic at random. Not only is this not fair, it may entail discarding critical data. The reader should check how this feature is implemented in their network, because vendors and networks vary on its implementation; also, some do not support the DE bit operation.

Committed Information Rate

The network may use the DE bit to aid in determining what to do with traffic. One approach is the employment of a technique used in conjunction with the DE bit called the committed information rate (CIR). An end user estimates the average amount of traffic that it will be sending to the network during a normal period of time. This measured average traffic rate is called the CIR. It is agreed upon by the user and the network and becomes part of the service contract between these two parties.

The network measures this traffic during a time interval and, if it is less than the CIR value, the network will not alter the DE bit. If the rate exceeds the CIR value during the specified period of time, the network will allow the traffic to go through unless it is congested; in which case, the congested network may tag the traffic with a DE = 1. It is likely that this traffic will be discarded. With this thought in mind, let us examine the CIR in more detail. First, the factors that make up the CIR are examined.

Committed burst rate (B_c) and excess burst rate (B_e). The committed burst rate (B_c) describes the maximum amount of data that a user is allowed to offer to the network during some time interval (T_c). The B_c is established during a call set-up or preprovisioned with a PVC. Consequently, frame relay operations at the U-plane contain no operations about the B_c value.

The excess burst rate (B_e) describes the maximum amount of data that a user may send which exceeds B_c during the time interval T_c. The value B_e also identifies the maximum number of bits that the network will attempt to deliver in excess of B_c, during an interval T_c. The B_e is also negotiated during the call set-up for SVCs and may be subject to a "lower probability" of delivery than the B_c value.

The CIR describes the information transfer rate that the network must commit to in order to support a user traffic during normal network operations. For an SVC, the CIR is negotiated during call set up under the C-plane.

The CIR must work in conjunction with the committed rate measurement interval (T_c). To iterate briefly, this measurement defines the time interval in which the user can send only a committed burst of data (B_c) and an excess amount of data (B_e). The value T_c must be computed; its value is not defined in the standards, so the network provider must establish its value. The CIR is averaged over the minimum increment of time T_c. CIR is computed as B_c/T_c.

Figure 4–9 depicts the relationships of CIR, B_c, B_e and the DE bit, and how a network can employ these tools to avoid congestion. Be aware that networks vary in how these tools are used, but this example is typical of a high quality UNI design.

If the user stays within the contracted CIR rate, its traffic is tagged as DE = 0, which means the traffic is not eligible for discarding. Also, if the user stays within the contracted CIR rate, the network will also accept bursts of B_c above the CIR (for a short duration). But, this traffic may be tagged as DE = 1, which means it is eligible to be discarded (for example, if the network is becoming congested). Most vendors do not tag

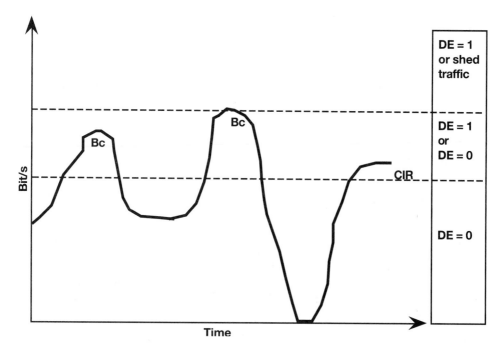

Figure 4–9
Bandwidth on demand.

B_c traffic, which makes better sense. After all, how can one reach a CIR if traffic bursts are below it?

However, if B_c continues for a prolonged period, it will be tagged as B_e. In this situation, some networks will not accept B_e traffic. Other networks accept the traffic, but will continue to tag the frames as DE = 1. Which method is the better? A prudent designer might shed this traffic to assure that the network does not become saturated. Moreover, once a network begins to accept B_e traffic, what becomes the next cutoff point? And how will the network user learn to use CIR, B_c, and B_e properly and effectively, if the network does not establish clear rules on traffic shedding? Of course, the answer is it won't. So if B_e traffic is allowed to enter the network, the user should know that this payload is leading a tenuous life through the network. For these reasons, this writer favors the shedding of B_e traffic.

It is conceivable that, as frame relay implementations mature, the network administrator will develop a cost mechanism to discourage a user from violating its CIR.

A number of vendors have adapted the Bellcore traffic policing mechanism at the UNI (see Figure 4–10). The approach entails the use of two credit pools, one for B_c traffic and one for B_e traffic. The two credit pools are maintained for every active DLCI.

When the DLCI is provisioned (as either a PVC or a SVC), the CIR parameters are established, and the following values are initialized:

- B_c_credits = B_c
- B_e_credits = B_e
- T_c = B_c/CIR
- Credit accrual timer = measurement interval (ΔT), with recommended value of 125 msec

When a frame arrives across the UNI at the network switch, it is processed in accordance with the logic chart shown in this figure. Note that arriving frames with DE = 0 and DE = 1 are handled differently. When ΔT expires, B_c_credits and B_e_credits are recomputed based on the formula shown at the bottom of Figure 4–10.

Figure 4–11 shows a more detailed view of the Bellcore traffic policing mechanism. As noted in previous discussions, values are established for CIR, B_c_credits, B_e_credits, T_c, and ΔT in relation to each DLCI. Afterwards, the DE bit in each frame is examined. Based on the value of this bit, the network node then determines if the size of the information field (I field) in octets is small enough to have the frame credited against the B_c_credit pool or the B_e_credit pool.

If the frame is externally tagged (it arrives with DE = 1), the B_c_credit pool is not decremented. The idea is that the user does not wish externally tagged traffic to reduce non-tagged traffic, because the network has more leeway in discarding this traffic.

The full use of CIR, B_c, and B_e is not supported by all networks. In contrast, a network may simply agree to provide the user with an access rate interface of DS1 or CEPT1, with the stipulation that the user can send data at that access rate, but only for a certain period of time. In other words, user traffic input is averaged over a measurement period, but the various interplays of B_c and B_e, cited above, are not implemented.

As a general rule, most public networks offer frame relay services with two prices: (a) on a fixed monthly fee based on an access rate (in bit/s), and (b) on a usage basis that is measured in volume (again, in bit/s). In the past, some networks were reluctant to discuss their general pricing structure, because of a combination of the uncertainty of this new

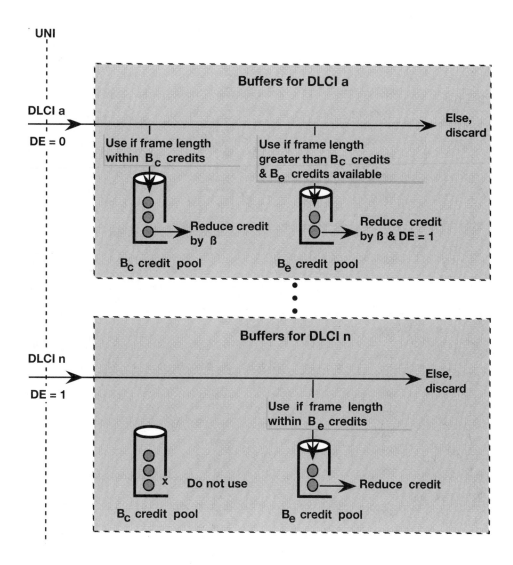

β = Information field size in octets

Upon Δ T expiring: (a) B_C_credits = Min [B_C_credits + ($B_C * (\Delta T/ T_c)$), B_C]

(b) B_e_credits = Min [B_e_credits + ($B_C * (\Delta T/ T_c)$), B_e]

Note: Each DLCI has B_C and B_e credit pools

Figure 4–10
Bellcore DLCI traffic policing.

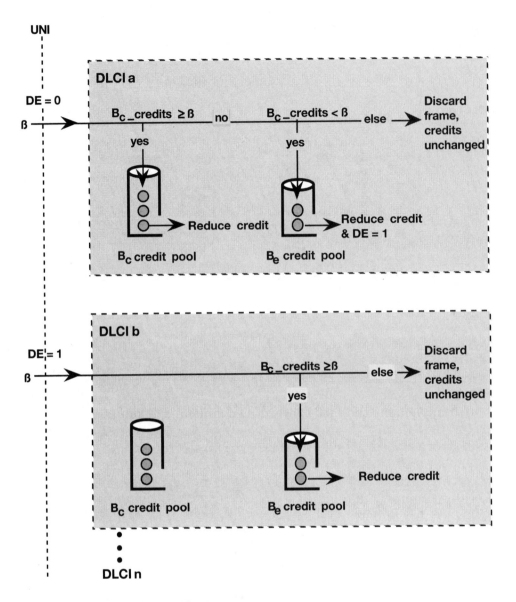

β = Information field size in octets

Note: Each DLCI has B_c and B_e credit pools

Figure 4–11
Credits with B_c and B_e pools.

technology/pricing structure, and the lack of knowledge of the competitors' pricing strategies (but this practice is changing).

Leaking CIR and Fast Forward CIR

Two methods may be employed to send data from the frame relay switch into the network. These methods are not defined in the standards; they are used at the discretion of the network provider. The first method is called by several names; I will use the term leaking CIR. With this operation, the frame relay switch matches the rate of sending frames into the network with the CIR. Since the network's communication trunks are faster than a user's CIR, the switch delays the frame transmissions in the network. That is, it "leaks" frames into the network based on the CIR speed.

I favor the fast forward CIR. With this approach, the frames are forwarded as fast as possible. The frames are received at the switch across the UNI at the CIR or B_c. Yet these frames are sent into the network as soon as possible, without regard to the CIR. They are transported through the network at the trunk speed. Fast forward CIR has two advantages over leaking CIR. First, it gives the user faster service, with less delay than leaking CIR. Second, it decreases the bookkeeping and overhead operations at the switch.

To implement the fast forward mechanism, a token pool (described earlier) is established. This pool consists of two counters, which we learned are called B_c_credits and B_e_credits. The credits reflect the available credits to a user, based on how much of the CIR has been used. Initially this pool is given a number of tokens equivalent to the subscription parameters B_c and B_e. The pool is then reduced by an amount equal to the number of bytes of data entering the network, and credited by an amount in proportion to the CIR value (i.e., B_c tokens per T_c seconds). Frames are fast forwarded as long as the pool has a positive token count.

Classes of Service Using B_c and B_e

Previous discussions have emphasized that the frame relay provider can implement traffic policing at the UNI with B_c and B_e with proprietary protocols. Some implementations have developed a class of service based on how the B_c and B_e are enforced at the UNI. This discussion illustrates a combination of several implementations.

The UNI can be provisioned for three types of service. These services are graded from a lower class of service to a higher class of service with these types:

- B_e only
- CIR and B_c
- CIR, B_c, and B_e

The B_e only class of service is considered the lowest class of service, and all frames are tagged as DE = 1. The network is permitted to discard these frames at the onset of congestion, which means that all frames thus tagged are discarded upon their arrival at a congestion point in the network.

The CIR and B_c class of service is a higher class of service. Traffic is submitted to the network from the user with DE = 0. The network is not permitted to tag frames with DE = 1 that arrive within the B_c threshold, and frames can only be discarded in the event of severe network congestion after the discarding of frames that are tagged as DE = 1 (which should be rare). The user is not permitted to send frames to the network with DE = 1, and B_e traffic is not accepted.

The CIR, B_c, and B_e class of service is yet another higher class of service. The network is obligated to receive traffic at B_c and B_e levels. The user can send traffic that is greater than CIR, and B_e levels are accepted, yet tagged for discard, if congestion occurs at any point in the network.

Requirements. Bellcore establishes several rules on the use of CIR, B_c, and B_e. They are summarized as:

$$\text{for } T = B_c/\text{CIR}: \quad \text{CIR} > 0 \quad B_c >, \quad B_e \geq 0.$$

The network node (which is termed a frame relay switching system [FR_SS]) must set T over a range of values from DT to 4 sec in increments of DT, which means that T can range from 100 ms to 4 sec in increments of 100 ms, if for example DT were set to 100 ms. Bellcore requires that the FR_SS shall support a DT of 100 ms or less. As an illustration, for a DS1 access rate, B_c can range from 0 to 768,000 octets. A B_c may span only a limited set of values within this range.

DLCIs in More Detail

Frame relay traffic is exchanged between the network users by mapping DLCIs from input lines to output lines. (Remember from previous discussions that some vendors do not perform DLCI mapping within their networks, only at the UNI.) The end user is responsible for building the frame relay frame and placing a DLCI value in the frame header. Given this information, the frame relay network must pass this traffic to a corresponding DLCI output line.

In Figure 4–12, router 1 sends traffic to three LANs connected to routers 2, 3, and 4. The traffic consists of three frames containing three different DLCIs. DLCI 100 is destined for the network at router 2; DLCI 101 is destined for the network at router 4; DLCI 102 is destined for the network at router 3.

Figure 4–12 shows the pertinent parts of the routing tables at the frame relay switches for this example. The table at switch A dictates that DLCIs 100, 101, and 102 arriving on input line number 10 are to be mapped (changed) to DLCIs 103, 104, and 105 respectively and sent to output lines 12, 13, and 14, respectively.

When the frame with DLCI = 103 arrives at switch B's input line 12, B's table dictates that DLCI 103 is to be mapped to DLCI 106 and sent to output line 15 for forwarding to router 2. When the frame with DLCI = 104 arrives at switch C's input line 13, C's table dictates that DLCI 104 is to be mapped to DLCI 107 and sent to output line 16 for forwarding to router 4. Finally, the frame with DLCI 102 is processed by switch A. It maps DLCI 102 to DLCI 105 and forwards the frame to router 3 through its output line 14.

THE FRAME RELAY NETWORK-TO-NETWORK INTERFACE (NNI)

The initial thrust of the frame relay standards development has focused on the UNI. Subsequent work resulted in the publication by the Frame Relay Forum (based on ANSI's T1.617, Annex D) of a NNI. This interface is considered instrumental to the success of frame relay since it defines the procedures for different networks to interconnect with each other to support the frame relay operations.

The relationship of the UNI to the NNI is illustrated in Figure 4–13. Obviously, the UNI defines the procedures between the user and the frame relay network, and the NNI defines the procedures between the frame relay networks.

A PVC operating across more than one frame relay network is called a multi-network PVC. Each piece of the PVC provided by each network is a PVC segment. Therefore, the multi-network PVC is the combination of the relevant PVC segments. In addition, the NNI uses the bidirectional network procedures, published in ANSI T1.617 Annex D, and further requires that all networks involved with each the PVC segment must support NNI procedures as well as UNI procedures.

Full internetworking operations between frame relay networks requires that the procedures stipulated in ANSI T1.617 Annex D be used

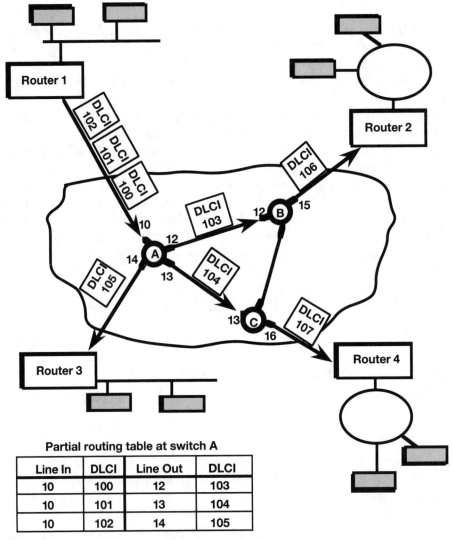

Partial routing table at switch A

Line In	DLCI	Line Out	DLCI
10	100	12	103
10	101	13	104
10	102	14	105

Partial routing table at switch B

Line In	DLCI	Line Out	DLCI
12	103	15	106

Partial routing table at switch C

Line In	DLCI	Line Out	DLCI
13	104	16	107

Figure 4–12
DLCI mappings.

at the UNI and the NNI. This concept means that a user sends a status enquiry (SE) message to the network, and the network responds with a status (S) message. The SE message, as the name suggests, is used to query the receiver about the status of PVC segments. In turn, the SE provides information about PVC segments to the user. Bidirectional procedures at the NNI require that either network be allowed to send SE or S messages.

Messages sent across the NNI are encapsulated into an HDLC unnumbered information (UI) frame. Since the UI does not have any timers or sequence numbers associated with it, these features are added into the SE and S messages. The messages are sent on DLCI 0, with the UI poll bit (P bit) set to 0. The BECN, FECN, and DE bits are not used and are set to 0.

NNI Operations

The principal operations of NNI involve these services:

- Notification of the adding of a PVC
- Detection of the deletion of a PVC
- Providing for notification of UNI or NNI failures
- Notification of a PVC segment availability or unavailability
- Verification of links between frame relay nodes
- Verification of frame relay nodes

These operations take place through the exchange of S and SE messages (defined in ANSI T1.617 Annex A) that contain information about the status of the PVCs. In essence, the NNI uses the procedures for inquiring about the status of PVCs at the UNI and simply applies them to the NNI. Many of the NNI operations resemble the LMI procedures. For example, both NNI and LMI use the ISDN S and SE messages.

The NNI messages are:

- Status enquiry (SE): A message that requests the status of a PVC or PVCs, or requests the verification of the status of a physical link
- Status (S): A message that reports on the status of a PVC, or reports on the status of a physical link. Sent in response to a status enquiry message
- Full status (FS): Message that reports on the status of all PVCs on the physical link

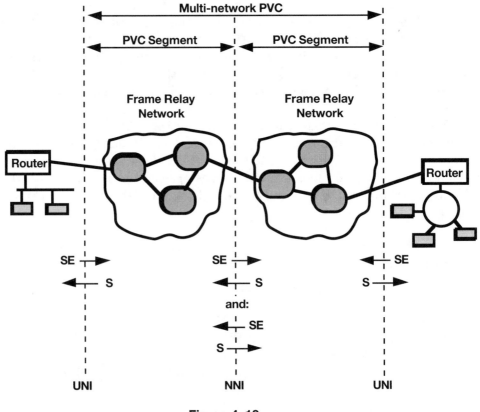

Figure 4–13
Frame relay NNI.

Bellcore Exchange Access FR (XA-FR) PVC Service

Bellcore also has published specifications for frame relay internet-working, which are cited in the section on pertinent standards. Bellcore describes this interface as the exchange access frame relay PVC service or simply XA-FR. The operations for XA-FR take place between two LECs and an interexchange carrier (IC). This interface is called the FR_ICI, which stands for frame relay inter-carrier interface. Figure 4–14 shows the topology for FR_ICI. The IC operates between the two LECs and supports the FR_ICI between each LEC. The connections between the LEC and IC occur at the point of presence (POP), which is a location within a LATA that is used for the inter-connecting of the IC facilities with those of the LEC.

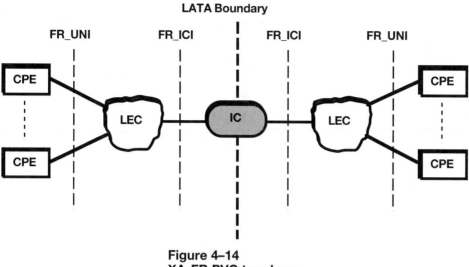

Figure 4–14
XA-FR PVC topology.

The intent of the XA-FR is to provide consistent service vis-à-vis the end-to-end PVC; that is to say, from the UNI to UNI. Given this aim, the FR_ICI supports BECN, FECN, and DE in accordance with the LEC network conventions. However, CIR is not repoliced at an ICI since it is assumed it is policed at the ingress UNI, but it is monitored by the LEC network that terminates the XA-FR service.

PVC status signaling is also supported at the ICI, in accordance with T1.606, Annex B, which is based on ITU-T Q.933 (Annex A). These procedures were explained earlier in this chapter. Last, both LECs must exchange the parameter values of CIR, B_c, B_e, and T in accordance with the Frame Relay Forum document, "Inter-Carrier Frame Relay Services: Recommendations and Guidelines."

XA-FR service performance. The XA-FR PVC service establishes several service performance parameters that must exist at the FR_ICI and some of these parameters stipulate performances from the FR_UNI to the FR_ICI. The service parameters are relevant only to the "compliant frames". This term describes those frames that are sent with in the limits of the B_c. Any frames marked with DE = 1 are considered to be non-compliant. The performance objectives described in this section include three categories: delay objectives, accuracy objectives, and availability objectives.

The XA-FR delay objectives are described in relation to (a) the FR_UNI access rate (in kbit/s); (b) the FR_ICI access rate (in kbit/s), and (c) the frame size (in octets). Delay is measured from the time the first bit of the frame (address field) is placed on the FR_UNI link to the time the last bit of the trailing flag is received at the FR_ICI. At least 95 percent of all frames should incur a delay no greater than the delay shown in Table 4–3.

The XA-FR accuracy objectives specify performance goals for (a) frames not delivered, (b) errored frames, and (c) extra frame rate. The following notations are used to describe the accuracy objectives:

N_e: Number of errored frames (frame arrives within a maximum time, and the cyclic redundancy check (CRC) computes correctly, but I field is different from original frame)

N_l: Number of lost frames (frame arrives late, or does not arrive, or CRC does not compute correctly)

N_s: Number of successful frames (frame arrives within a maximum time, the CRC computes correctly and the I field is the same as the original frame)

N_x: Number of extra frames (frames misdelivered or duplicated)

Given these notations, the accuracy objectives are computed as follows:

- Frame not delivered ratio (should be $< 1 * 10^{-5}$):
$$Nl : N_s + N_e + N_l$$
- Errored frame ratio (should be $< 5 * 10^{-13}$):
$$Ne : N_s + N_e$$
- Extra frame ratio (should be $< 5 * 10^{-8}$ frames per second):

  Total N_x observed on a connection per connection second

XA-FR PVC service specifies six availability objectives. Four parameters describe availability and are measured as long term averages. Two other parameters pertain to congestion notification, which are measured (a) beginning with the receipt of BECN or FECN, and (b) terminated at a time interval that is still "under study". The availability objectives are measured as follows:

- Scheduled hours of service: 24 hours per day, 7 days per week
- Service availability: Ratio of actual service time to schedule ser-

Table 4–3   Bellcore Frame Transfer Delay Requirements (msec.)

	FR_UNI access rate for 512 octet frames				
	56	**64**	**384**	**768**	**1536**
ICI rate					
56	161	152	99	94	91
64	152	143	90	85	82
1536	91	82	29	23	21
44210	89	80	27	22	19

	FR_UNI access rate for 1600 octet frames				
	56	**64**	**384**	**768**	**1536**
ICI rate					
56	473	444	278	261	253
64	444	415	249	232	224
1536	253	224	57	40	32
44210	245	218	52	35	27

	FR_UNI access rate for 4096 octet frames				
	56	**64**	**384**	**768**	**1536**
ICI rate					
56	1185	1112	686	643	622
64	1112	1039	613	570	549
1536	622	549	122	79	58
44210	601	528	102	59	38

vice time (criteria for outage is under study) with an objective of at least 99.95 percent

- Mean time to service restoral (MTTSR): Sum of all time between service unavailability and its restoral, with an objective of ≤ 3.5 hours
- Mean time between service outages (MTBSO): At least 3500 hours

- Fraction of time in non-congestion notification state: Time spent in non-congestion notification state, with an objective of at least 99%
- Meantime between congestion notification states: Arithmetic average of continuous time intervals in which no congestion state is entered, with an objective of at least 100 hours

Traffic measurement and billing The XA-FR defines the procedures for the generating and reporting of network statistics for OAM and billing. Guidance is provided on when to perform measurements, in what increments, and when the measurements are stored and formatted for network management use. To ensure compatibility across the differing LECs and IC, all traffic must be formatted in accordance with Bellcore's Document TR-NWT-001100, which is the well known *Bellcore Automatic Message Accounting Format* (BMAF) requirements.

Information recorded for traffic measurement and billing must include the following:

- The date that the measurement began
- The time the measurement began
- The elapsed time of the measurement interval
- Identifier of the destination interface
- Identifier of the source interface
- Destination DLCI
- Source DLCI
- Unit of measurement and appropriate counts that pertain to that measurement

OTHER NOTABLE ASPECTS OF FRAME RELAY

DLCI Values

We learned previously that the DLCI field in the frame relay frame can vary in size, and can contain two, three, or four octets. This approach allows the use of more DLCI numbers. The permissible DLCI values for the three options are listed in Table 4–4.

Added Options to Frame Relay

Since the original publication of the frame relay standards, a number of options have been added to overcome some problems that surfaced during the prototyping of the early frame relay networks. This

Table 4–4   DLCI values.

DLCI Values	Function
(a) Two-octet address format:	
0	In channel signaling
1 to 15	Reserved
16 to 991	Assigned using frame relay connection procedures (see Note)
992 to 1007	Layer 2 management of frame relay bearer service
1008 to 1022	Reserved
1023	In-channel layer management
(b) Three-octet address format with D/C = 0:	
0	In channel signaling
1 to 1023	Reserved
1024 to 63,487	Assigned using frame relay connection procedures (see Note)
63,488 to 64,511	Layer 2 management of frame relay bearer service
64,512 to 65,534	Reserved
65,535	In-channel layer management
(c) Four-octet address format with D/C = 0:	
0	In channel signaling
1 to 131,071	Reserved
131,072 to 8,126,463	Assigned using frame relay connection procedures (see Note)
8,126,464 to 8,257,535	Layer 2 management of frame relay bearer service
8,257,536 to 8,388,606	Reserved
8,388,607	In-channel layer management

Note: Some of these values may be assigned to permanent frame relay cells.

section examines two of them: (1) global and local DLCIs, and (2) multi-casting.

The DLCIs can be managed so that the numbers can be reused within a network. This approach is known as *local significance* and enables more virtual circuits to be created in a frame relay network, because DLCI values can be reused. Care must be taken that the DLCI number has only local significance and is not known to other routers.

The *global addressing* option allows a DLCI to be assigned such that a number has universal significance. This means that this number "points" to the same destination regardless of the source router.

The idea behind global addressing is to simplify addressing administration. The header size can be increased from two octets to three or four octets (see the section titled "Other Notable Aspects of Frame Relay") to allow the use of more DLCI values.

Frame relay also provides an optional feature (not yet approved) called *multicasting*. This is a "semi-broadcast" technology in which multiple routers are identified with one DLCI. A router need only send one copy of the frame with the reserved DLCI value in the header. The network is then required to duplicate the frame and deliver copies to a set of line-in DLCIs.

The Local Management Interface (LMI)

A number of vendors have implemented an additional feature for operations at the UNI called the local management interface (LMI). This operation includes some of the features of the CLLM, discussed earlier, but it has more options. Its main job is to provide status and configuration information about PVCs at a UNI. It also is used to check on the integrity of the physical link at the UNI. A network can use the LMI to inform the user CPE about the addition and deletion of PVCs, as well as if they are active (being used) or inactive (not being used).

The LMI also supports multicasting, extended addressing, and global addressing features discussed earlier in this section.

Frame Relay SVC Operations

ANSI and the ITU-T have published standards for frame relay SVC operations in ANSI T1.617 and ITU-T Q.933. The Frame Relay Forum has also developed a proposal for an SVC. It differs from the X.933/T1.617 specifications, in that it is simpler and does not use as many of the ISDN Q.931 procedures.

It is unclear, as of this writing, what the final frame relay SVC will be, but Figure 4–15 is an accurate depiction of its major operations.

The user issues a Q.933 SETUP message to the frame relay network. This message contains the DLCI to be associated with the call, and an explicit address of called party number; for example, an E.164 address (number) or its North American implementation. Private networks can use other numbering plans such as X.121, or an IP address.

The SETUP message contains the following frame relay-related information (and other administrative information):

DLCI

An explicit address

Requested end-to-end transit delay time

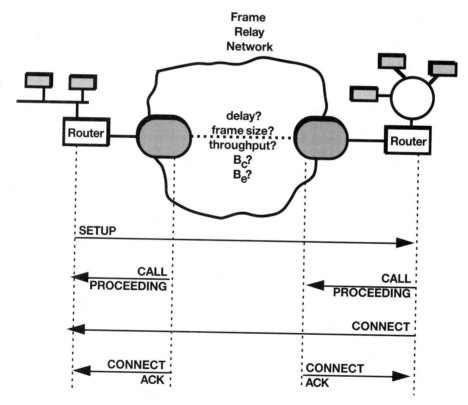

Figure 4–15
UNI SVC connection set up.

Maximum frame relay frame size that can be supported for this session

Requested throughput (incoming and outgoing)

Requested B_c (incoming and outgoing)

Requested B_e (incoming and outgoing)

The SETUP message is sent to the network, at which time the network sends a CALL PROCEEDING message to the calling user. During this period, the network examines the requested services in the SETUP message and determines if this call can be supported. If so, the SETUP message is forwarded to the remote user. This user responds with a CALL PROCEEDING message.

If the call is accepted, the user sends a CONNECT message to the network. This CONNECT message is relayed through the network to the calling party. The calling user sends a CONNECT ACKNOWLEDGE to the network. Likewise, the CONNECT issued from the called user on the remote UNI requires that the network send a CONNECT ACKNOWLEDGMENT to the remote side as well. This also places the other side of the UNI in a call active state. Once the successful completion of all these actions has taken place, traffic can then flow across the UNI.

Other Quality of Service (QOS) Options

The frame relay specifications specify a number of QOS options, in addition to the CIR. They are summarized in this section.

Residual error rate (RER) is measured through the exchange of the frame relay SDUs (FSDUs). It must be measured during an agreed upon period and is measured across an agreed upon boundary—typically, between the core functions of Q.922 and the protocol implemented above Q.922. The RER is defined as R = 1 – (total correct SDUs delivered) / (total offered SDUs).

A *delivered errored frame* is defined as a frame that is delivered when the values of one or more of the bits in the frame are discovered to be in error.

The *delivered duplicate frames* value is determined when a frame received at a destination is discovered to be the same frame as one previously delivered.

The *delivered out-of-sequence frame* describes the arrival of a frame which is not in sequence relative to previously delivered frames.

A *lost frame* is so declared when the frame is not delivered correctly within a specified time.

A *misdelivered frame* is one that is delivered to the wrong destination. In this situation, DLCI interpretation may be in error, the routing table may be out-of-date, etc.

The *switched virtual call establishment delay* and *clearing delay* refer respectively to the time taken to set up a call and clear a call across the C-plane.

The *premature disconnect* describes the loss of the virtual circuit connection, and the *switched virtual call clearing failure* describes a failure to tear down the switched virtual call.

Internetworking Frame Relay and ATM

Chapter 7 introduces the ATM adaptation layer (AAL). It is used to support multimedia traffic through an ATM network. The ATM AAL type 5 (AAL5) has been selected as the function for transporting frame relay traffic through an ATM network, and is defined in ITU-T I.363.

AAL5 supports connection-oriented protocols that use variable length formats; thus, it is a good fit for the support frame relay traffic.

Figure 4–16 shows the relationships of the frame relay and ATM protocol stacks. The user upper layer protocols (ULPs) are not involved in these operations. Their PDUs are transported transparently between the two terminals. The network layer (NW) is not defined either, and is likely IP, CLNP, or a vendor-specific network layer, such as DECnet layer 3, etc.

AAL 5 is made up of three sublayers. The frame relay service specific convergence sublayer (FR-SSCS) is the top sublayer of AAL5. Its purpose is to provide the interface between the frame relay Q.922 Core protocol and ATM. It defines how DLCIs, the DE bit and the BECN/FECN bits are translated to an ATM interface for use at the ATM UNI. The other two sublayers are more generic and operate with any AAL5 operation. I.555 defines the operations of the interworking function (IWF), which include the mapping between frame relay and ATM. I.555 also defines the operations of the frame relay BISDN terminal, and implements the Internet RFC 1294, *Multiprotocol Interconnect over Frame Relay*, the subject of the next section of this chapter.

Multiprotocol Operations Over Frame Relay

The proposed Internet standard Request for Comments (RFC) 1294 establishes the rules for how these protocols are encapsulated within the frame relay frame and transported across a network. This proposal is based on the encapsulation standards published in ANSI T1.617, Annex F.

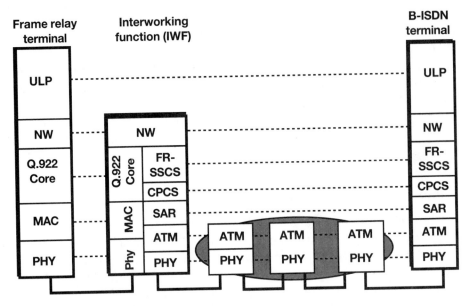

Figure 4–16
Frame relay interworking with ATM.

The Frame Relay Forum has used these documents as the basis for its Implementation Agreements (IAs) on this matter.

RFC 1294 stipulates that the protocols running over frame relay must encapsulate their PDUs within the ITU-T Q.922 Annex A frame (this frame was described in Chapter 6). The Q.922 Annex A frame must also carry a field that identifies the protocol being carried within the frame. The format for this frame is depicted in Figure 4–17a.

The notations for this figure are depicted in hexadecimal, where, for example, the notation 7E represents the flag with bits 01111110. Another convention used in RFC 1294 is shown in the control field, which is coded as 0×03, which denotes a bit structure of 00000011.

The address/control field is the conventional two-octet frame relay field containing the 10-bit DLCI and the other control bits. The control field is coded as the HDLC unnumbered information (UI), or an information field (I). Padding may be used at the convenience of the user to pad out the frame to any boundary deemed appropriate.

The network level protocol identifier (NLPID) contains values to identify common protocols that are used in the industry, such as ISO

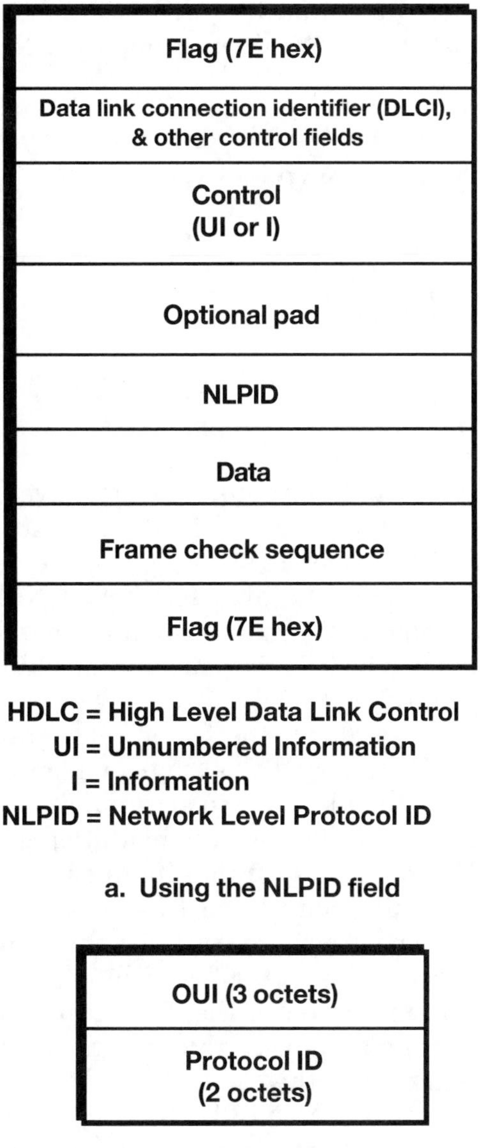

Flag (7E hex)
Data link connection identifier (DLCI), & other control fields
Control (UI or I)
Optional pad
NLPID
Data
Frame check sequence
Flag (7E hex)

HDLC = High Level Data Link Control
UI = Unnumbered Information
I = Information
NLPID = Network Level Protocol ID

a. Using the NLPID field

OUI (3 octets)
Protocol ID (2 octets)

Figure 4–17
PDUs for multiprotocol
encapsulation

b. Using the SNAP header

CLNP, the Internet's IP, etc. The purpose of this field is to inform the receiver which protocol is being carried inside the Q.922 Annex A frame. This field is ignored by the frame relay network, and is only examined by the user CPE.

The reader can obtain ISO/IEC TR9577 for the values that are currently administered by the ISO. Examples of NLPIDs values are:

0×00 Null network layer (not used by frame relay)

0×80 IEEE SNAP

0×81 ISO CLNP

0×82 ISO IS-IS

0×83 ISO IS-IS

$0 \times CC$ IP

0×8 ISDN Q.933

In the event that protocols do not have an NLPID assigned, the Subnetwork Access Protocol (SNAP) can be used to provide an NLPID value. The SNAP header is shown in Figure 4–17b. RFC 1294 requires that all stations must be able to process NLPID as well as a SNAP header. The three-octet organizational unique identifier (OUI) identifies the organization which is responsible for standardizing the content of the protocol identifier (PID).

RFC 1294 also allows the frame relay stations to negotiate certain parameters with the HDLC exchange identification (XID) operation. This negotiation must be done in accordance with Q.922 Appendix 3.

RFC 1294 also stipulates fragmentation guidelines, if this operation is needed. The fragmented frame contains an offset value which identifies the relative position of the fragmented frame within the original (whole) frame. The fragmentation for this procedure is limited to the boundaries between the two DTEs within one frame relay network.

Due to the limitation of the NLPID numbering space, some protocols do not have an assigned value. For these protocols to be routed over a frame relay network, the convention (cited in RFC 1294) is to use the SNAP format and set the NLPID to 0×80 and the OUI to 0×00-00-00. The NLPID value indicates a SNAP header follows, and the OUI value indicates an Ethertype value is to be used. After the SNAP header, the conventional Ethertype field follows. The Ethertype field has long been used to identify commonly used protocols running on top of Ethernet networks. Consequently, the same approach is simply used to indicate the protocol running on top of the frame relay.

For frame relay traffic running across LAN or MAN bridges, an approach that is similar to protocols running on top of Ethernet networks is used. The PDUs are encapsulated with the NLPID value indicating SNAP and then the SNAP header. The PDU contains a field which specifies the particular LAN/MAN header.

The Frame Relay MIB

The reader may wish to read Appendix C, if the subject of a management information base (MIB) is not familiar. The Internet authorities have published the MIB for the frame relay UNI. It is available in RFC 1315. The objects are organized into three object groups: (1) a data link connection management interface group, (2) a circuit group, and (3) an error group. These groups are stored in tables in the MIB and can be accessed by the SNMP. Figure 4–18 illustrates where SNMP operates, and lists the name of the three groups. These names are ASN.1 OBJECT IDENTIFIERS.

The frDlcmiTable contains 10 objects. Their purpose is to identify each physical port at the UNI, its IP address, the size of the DLCI header (2,3,4 octets) that is used on this interface, timers for invoking status and status enquiry messages, the maximum number of DLCIs supported at the interface, whether or not the interface uses multicasting, and some other miscellaneous operations.

The frCircuitTable contains 14 objects. Their purpose is to identify each PVC, its associated DLCI, if the DLCI is active, the number of BECNs and FECNs received since the PVC was created, statistics on the number of frames and octets sent and received since the DLCI was created, the DLCI's B_c and B_e, and some other miscellaneous operations.

The third table is the frErrTable. It contains 4 objects. Their purpose is to store information on the types of errors that have occurred at the DLCI (unknown DLCI, illegal DLCI, etc.), and the time the error was detected. One object contains the header of the frame that created the error.

FRAME RELAY WORKSHEET

The worksheet for frame relay is provided in Table 4–5. As suggested in the introduction to this book, it is recommended that the reader attempt to fill out a blank sheet (provided in Appendix D), and then read this section to check your answers.

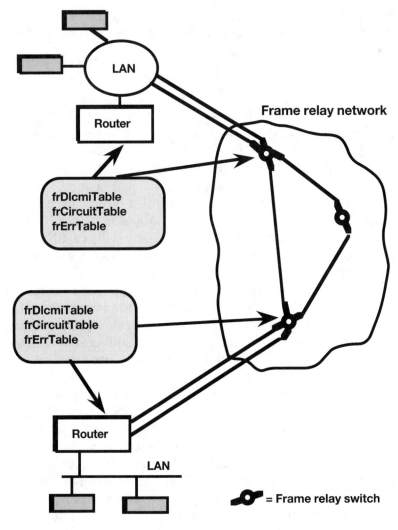

Figure 4–18
The frame relay MIB.

In answering the question as to whether or not frame relay is a new technology, the answer is no. Frame relay (from the engineering standpoint) is nothing more than a scaled down version of LAPD, with a few features added in, such as congestion notification and discard eligibility. However, from the standpoint of a service, the idea of bandwidth on demand is relatively new in the industry. So, we can conclude that (tech-

Table 4–5. Frame Relay Worksheet

Technology name:	Frame Relay
New technology?	No, uses a scaled-down version of LAPD; idea of bandwidth on demand is somewhat new
Targeted applications?	Bursty data, with high capacity requirements
Topology dependent?	Strictly speaking no, but current implementations are point-to-point
Media dependent?	No, can operate over wire, optical fiber, etc.
LAN/WAN based?	WAN-based; it is designed as high-speed WAN to interconnect LANs
Competes with:	Private leased lines, X.25-based networks, SMDS, and ATM
Complements:	LAN internetworking across wide areas
Cell/Frame based?	Frame-based
Connection management?	Yes, PVCs, with SVCs a likely addition
Flow control (explicit/implicit)?	Yes, implicit, with FECN and BECN bits
Payload integrity management?	No, ACKs, NAKs, and sequencing are a user responsibility
Traffic discard option?	Yes, with the DE bit
Bandwidth on demand?	Yes, with CIR, committed and excess burst rate operations
Addressing/identification scheme?	Yes, uses labels for identifying PVCs, which are called DLCIs

nically speaking) frame relay is not an emerging technology, but its service offering is new.

The targeted applications are data transmission (typically high capacity) applications. In the future, we shall see voice running on frame relay networks as well. We learned also that frame relay is not topology dependent. No technical reason exists why it could not operate on a multipoint link, but implementations (thus far in the industry) have focused on point-to-point connections.

The technology is not media dependent. It can operate on channels such as twisted pair, coaxial cable, optical fiber, etc.

The answer to the question "Is frame relay LAN or WAN based?" is that it is WAN based, because it is designed to interface LANs to each other through a WAN.

One of the more difficult questions dealing with these worksheets is "What do these technologies compete with?" Since some of them are so new, the answers are not clear at this writing. However, it is likely that frame relay will compete with SMDS, ATM, and certainly with the private leased line industry, as well as X.25-based networks. This writer believes that frame relay does not pose a direct threat to the X.25 users because many of these users are quite low-end as far as capacity needs—operating in the low-kilobit range. Frame relay addresses users that need higher-capacity interfaces.

Frame relay, we learned, complements LAN technology in its attempt to provide seamless interconnections between LANs.

If you miss the next question, do not pass GO. Obviously, it is frame based. We learned also that frame relay implementations are with PVCs, and SVCs are a likely addition in the near future (the draft specification for SVC is finished).

We learned also that flow control is provided. It is implicit flow control using the FECN and BECN bits. It does not use explicit flow control such as receive not ready (RNR).

Frame relay is not responsible for the accountability of user traffic. As we learned, it does not provide any ACKs or NAKs. Sequencing is not provided. Thus, all these activities are moved out of the network and into the user equipment.

Frame relay provides a traffic discard option with the DE bit, and it provides bandwidth on demand with the CIR concept.

Finally, the last question from the worksheet deals with addressing and identification schemes. We learned that frame relay has an implicit identification scheme to tag PVCs with values called data link connection identifiers (DLCIs).

SUMMARY

Frame relay systems are designed to support users increased demand for low delay and high throughput. Frame relay is so-named because most of the operations (from the user's perspective) occur at the frame data link layer. The basis for frame relay is HDLC and other HDLC-derived protocols that operate at the data link (frame) layer.

Frame relay technology assumes a network with high integrity and smart user devices. End-to-end data integrity is the user's responsibility. The frame relay network employs the CIR mechanism to govern the flow of traffic.

Fiber Distributed Data Interface (FDDI)

INTRODUCTION

This chapter examines the LAN specification called the fiber distributed data interface (FDDI). The purposes of FDDI are explained as well as the topologies used. We analyze how the FDDI protocol behaves and how the FDDI PDUs are managed at both the sending and receiving stations.

FDDI is not a new and emerging technology. However, some vendors are offering internetworked FDDIs in metropolitan areas, as a solution for implementing MANs, and this approach may have effects on frame relay and SMDS offerings. As of this writing, it is too soon to know. The last part of the chapter focuses on FDDI II, which is a new and enhanced version of FDDI, and another reason FDDI is included in this book.

THE PURPOSE OF FDDI

The FDDI was developed to support high-capacity LANs. To obtain this goal, the original FDDI specifications stipulated the use of optical fiber as the transport media, although FDDI is now available on twisted pair cable. FDDI has been deployed in many enterprises to serve as a high-speed backbone network for other LANs, such as Ethernets and token rings.

Conceptually, the standard operates with a 100 Mbit/s rate. Dual rings are provided for the LAN, so the full speed is 200 Mbit/s, although the second ring is used typically as a back-up to the primary ring. In practice, most installations have not been able to utilize the full bandwidth of FDDI. The standard defines multimode optical fiber, although single mode optical fiber can be used as well.

FDDI was designed to make transmission faster on an optical fiber transport. Due to the high-capacity 100 Mbit/s technology, FDDI has a tenfold speed increase over IEEE's 802.3, and obviously, an even greater capacity impact on 802.5's 4 Mbit/s and 16 Mbit/s LANs. FDDI was also designed to extend the distance of LAN interconnectivity. It permits the network topology to extend up to 200 Km (124 mi.).

PERTINENT STANDARDS

The FDDI standards are published by both the ISO and ANSI. They are organized around the FDDI layers, discussed in the section in this chapter titled "FDDI Layers". Listed below are other pertinent standards that should be helpful to the reader, if more detail is needed about FDDI. Other standards are being developed for operations on twisted pair, single mode fiber, and low cost fiber.

- Physical layer medium dependent interface: X3.166/ISO 9314-3.
- Physical layer protocol: X3.148/ISO 9314-1.
- Media access control: X3.139/ISO 9314-2.
- Station management: X3T9.5.

A TYPICAL FDDI TOPOLOGY

Figure 5–1 shows a typical FDDI network. Some stations are configured on FDDI with two rings and some are configured with one. The dual attachment stations (DASs) use two fibers, and the single attachment stations (SASs) use one fiber. In earlier releases of FDDI, these sta-

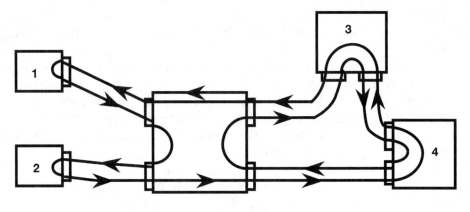

Wiring Concentrator (WC)

Figure 5–1
Typical FDDI topology.

tions were identified as class A and class B, respectively. This terminology has been eliminated from the new version. The wiring concentrator connects the stations. The class A stations can take advantage of their two fibers by using optical bypass. In the event of the failure of a line or port, the other fiber can be used to continue operations.

As a general rule, FDDI LANs are installed in organizations to serve as backbone networks to other local area networks. This means that the FDDI LAN is used to interconnect lower-speed LANs such as 802.3, Ethernets, and 802.5 networks. Due to the relatively high capacity of an FDDI network (100 Mbit/s), it serves as a cost-effective and efficient means for internetworking other types of LANs. Figure 5–2 illustrates a typical FDDI-based internet.

The FDDI network can recover from a single fault, as shown in Figure 5–3. The two stations between the break will redirect the traffic from the primary ring onto the secondary ring and vice versa.

As depicted in Figure 5–4, each dual ring node contains the same physical interfaces. They consist of transmitters and receivers, media connectors, bypass switches, and configuration switches. Of course, each station also contains the media access control (MAC) layer.

These components jointly allow the network to: (a) recover from a failed line (a media break), or (b) bypass a node. The latter capability permits a station to be bypassed while maintaining the dual rings with the other FDDI stations.

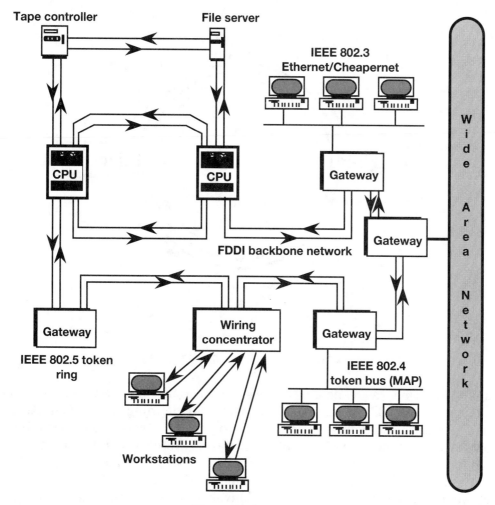

Figure 5-2
FDDI Backbones.

THE FDDI LAYERS

As depicted in Figure 5-5, FDDI is organized around the OSI Model and the IEEE LAN model. It consists of the conventional data link layer as well as the physical layer.

The FDDI physical layer is divided into two sublayers: (a) the physical layer (PHY) protocol and (b) the physical layer medium dependent (PMD) interface. The physical layer was divided into two sublayers due to the fact that issues such as coding, clocking, etc. were not controver-

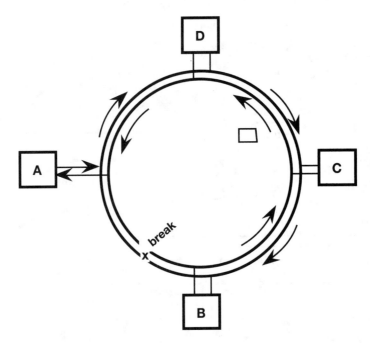

Figure 5–3
Reconfiguration after a media break.

sial, but issues such as fiber types, connections, etc. were subject to change. The PMD interface is responsible for defining the transmitting and receiving signals, providing proper power levels, and specifying cables and connections. The physical layer protocol is intended to be medium independent. It defines symbols, coding and decoding techniques, clocking requirements, the states of the lines, and data framing conventions. Therefore, the PMD sublayer was separated from the more stable PHY sublayer.

The data link layer is divided into the conventional IEEE 802.2 (a) logical link control (LLC) and (b) media access control (MAC). MAC is defined in X3.139/ISO 9314-2, which provides the procedures used for frame formatting, error checking, token handling, and managing the data link addressing.

In addition, FDDI has a station management function (SMT). The station management standard provides the procedures for managing the station attached to the FDDI. It provides for node configuration, ring initialization, error statistics, error detection and recovery, and connection management.

This section examines FDDI through an analysis of each layer. We begin with an examination of the physical layer.

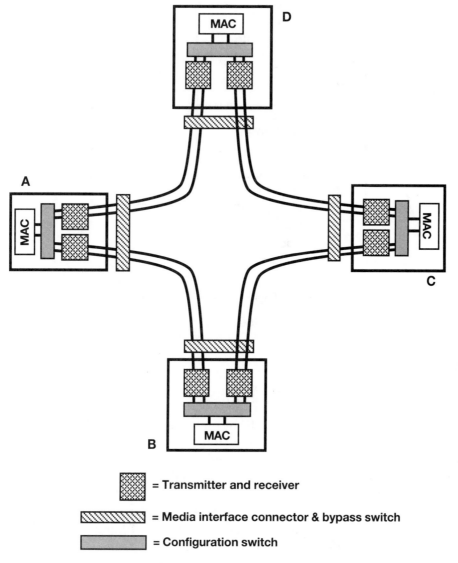

= Transmitter and receiver

= Media interface connector & bypass switch

= Configuration switch

Figure 5–4
The components of the FDDI station.

PMD Sublayer

The lowest layer of the FDDI standard is the PMD. This sublayer constitutes the lower half of the IEEE/OSI physical layer. The purpose of the PMD is to define the optical transmitters and receivers, the media interface connector, the optical fiber cable, and an optional optical bypass relay operation.

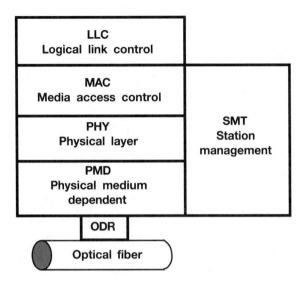

ODR = Optical drivers and receivers

Figure 5–5
FDDI layers.

The optical transmitters convert electrical signals into a modulated light signal. This layer is then responsible for sending the signal into the fiber cable. This layer defines a light emitting diode (LED) for the transmitter. LEDs are usually coupled with multimode fibers. For single mode operations, laser diodes (LDs) are used. The optical receiver is responsible for converting the incoming optical signal back to the electrical signal with a photodetecter.

The optical fiber cable, as specified by FDDI, emits light at a nominal wavelength of 1300 nanometers (NM). The fiber is 62.5/125 (core diameter/cladding diameter) micron multimode graded index cable or 85/125 micron. The draft standard (DS) also include a 50/25 micro fiber and a 100/140 micro fiber, as alternatives. In addition, the media interface connector is used to align the fiber properly with the connecting node.

The FDDI standard specifies a light signal at or near the nominal wavelength value of 1270-1380 NM. The term *nominal wavelength* is also called the center wavelength. Figure 5–6 shows the ranges used with either a LD or a LED.

The nominal wavelength is an approximate wavelength only because

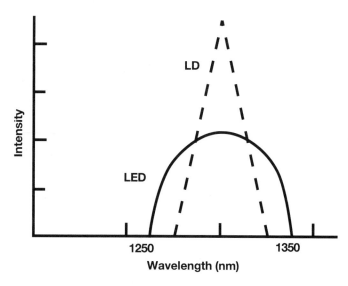

Figure 5–6
Nominal wavelength.

light rays emit signals in various wavelengths (various colors). The range for the nominal wavelength varies depending on whether a LD or LED is employed. The purpose of defining a nominal wavelength is to ensure that components from various manufacturers interwork without problems.

FDDI also specifies permissible attenuation for the light. Attenuation refers to the signal loss as the light travels through the optical fiber cable. The standard specifies cable attenuation of 1.5 dB/km with the 1300 NM wavelength.

The maximum permissible cable attenuation is simply determined by subtracting all losses of the splices and connectors from what is called a *link loss budget*. The link loss budget is the difference between the minimum transmitter power and the receiver sensitivity (receiver sensitivity is the minimum power required for the receiver to detect the signal).

Four FDDI ports are defined as shown in Figure 5–7. The ports are used to make certain that the FDDI topologies are correct. Presently, FDDI topologies permit four port types: A, B, M, and S. Port A is used to connect the incoming primary ring and the outgoing secondary ring of a dual ring FDDI topology. This port also forms part of the dual attachment station (DAS) or a dual attachment concentrator (DAC). Port B is similar to port A in that it connects the outgoing primary ring and the

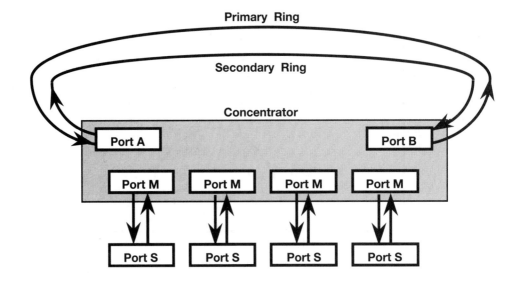

- Port A: Connects to incoming primary ring and outgoing secondary ring
- Port B: Connects to outgoing primary ring and incoming secondary ring
- Port M: Connects concentrator to an SAS or another concentrator
- Port S: Connects a SAS to a concentrator

Figure 5–7
FDDI port types.

incoming secondary ring, therefore it performs a reverse function of port A. This port is also part of the DAS or the DAC. Port M supports a signal attachment station (SAS), a DAS, or another DAC. However, port M is only implemented in the concentrator. Finally, port S connects workstations through an SAS or SAC to the concentrator (DAC).

FDDI provides for a bypass operation (which is optional and not required in the standard). The bypass option provides for connectivity on FDDI in the event of a problem with a DAS or during the absence of power in a node. The bypass relay allows the light signal to bypass the faulty node's receiver, which maintains continuity on the FDDI ring.

Be aware that relays will consume power which can limit the number of relays on the ring. They also introduce additional loss in the network because they do not perform repeater functions.

A closer look at fiber.   Typically, a light signal is transmitted down the optical fiber in the form of on/off pulses representing a digital bit

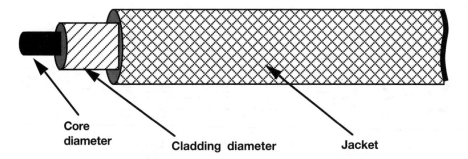

Core
diameter
Cladding diameter
Jacket

Diameters:
- **Core: 62.5 microns (micrometers)**
- **Cladding: 125 microns**
- **Or: 62.5/125**
- **Others: 50/125**
             100/140

Figure 5–8
FDDI and optical fiber.

stream. The fiber consists of three concentric cylinders of dielectric material, the jacket, the cladding, and the core (see Figure 5–8). The core and cladding are made up of transparent glass (some systems use plastic) which guides the light through the core. As the light propagates, it is reflected between the core and the cladding.

The core and the cladding have different refractive indices (refraction is the ratio of the velocity of a light wave in free space to its velocity in a medium, such as core and cladding). Since the refractive index of the two differ, light in the cladding propagates faster than it does in the core. As the light moves toward the cladding (a region of higher velocity) it is bent back toward the core and guided along the fiber (hence, another name for optical fiber is lightguide cable).

Optical fiber is usually identified by the diameters of the core and the cladding. For example, a 62.5/125 micron (micrometers) fiber is a fiber that has a core with a diameter of 62.5 microns and a cladding of 125 microns.

Since optical fiber signals consist of different frequencies of light, the fiber must be designed to handle the light entering the fiber at different angles. For example, Figure 5–9a shows that the light source is presented into the cable with the rays entering at different angles. The light rays that enter the cable directly, travel directly through the center to the end destination. However, the other light rays enter the cable at

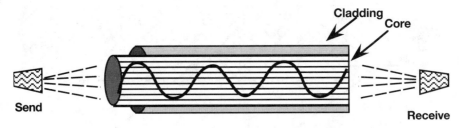

a. Light into and out of the fiber

b. Light traveling through step-index fiber

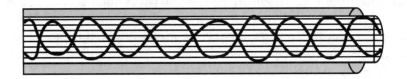

c. Light traveling through graded-index fiber

Figure 5–9
Optical fiber transmission.

slightly different angles and "ricochet" through the media. These light rays meet at the other end and combine to produce a slightly distorted pulse which is reduced in amplitude.

The effect of this operation is shown in Figure 5–9b, with two light waves traveling through the fiber at different angles. Figure 5–9b is an example of a step index fiber. This fiber is so named because the refractive index (the measure of the speed of light in the material relative to the speed of light in the vacuum) has sudden changes at the cladding/core boundary.

Figure 5–9c shows the effect of the signal through a graded index fiber. The approach with this technology is to correct some of the distortions that occur with step index fiber. The design is used to compensate for the radically different path directions of a step index technology and to reduce the difference in the delay of certain signals arriving at the receiver. Graded index fiber has less distortion and higher capacity. As this figure shows, the light ray follows a more gentle path than the zigzag path of the step index fiber.

PHY Sublayer

The physical layer (PHY) protocol defines the physical layer operations that are media independent. The physical layer defines the clocking for the network. Each FDDI station must have an autonomous clock, and the receiving station is designed to synchronize its receiver clock to the incoming data. The station uses its clock to decode the data. It then retransmits the data with the new clock that is generated locally at the station.

The PHY defines an FDDI symbol. As shown in Figure 5–10a, a symbol is received from the MAC at the data link layer. In turn, the PHY encodes the MAC symbol into a 4B/5B format. This layer also has symbols showing the status of a frame, such as errors detected, frame copied, and address recognized. The PHY is also responsible for a preamble, which contains some idle signals, and which is used for synchronizing receiver clocks. The PHY also has a repeat filter which is used to enhance the signal so that proper codes and valid line states continue to be transmitted.

The symbol from the MAC layer is taken through the 4B/5B encoder. This encoder is responsible for translating the MAC signal into a five-bit code (Figure 5–10b). The purpose of this code is to provide for enhanced clocking and control functions. The use of this code results in a data rate of 100 Mbit/s with a signaling rate of 125 megabaud, which translates into less expensive and less complex equipment.

After the 4B/5B encoding process occurs, this sending layer then translates this code into a non-return to zero inverted (NRZI) scheme which is sent into the optical fiber.

The NRZI coding scheme (Figure 5–10b) is used as follows: The transition between on and off power equals a 1, and no transition equals 0. This approach reduces the complexity of FDDI because it cuts down the number of transitions that are found in the data stream. It still provides for clocking functions, but it allows less expensive components to be used.

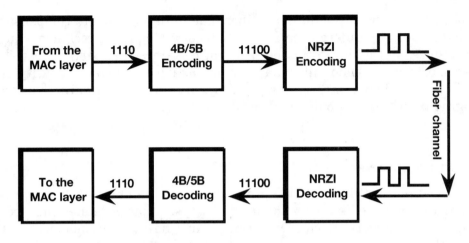

a. 4B/5B and NRZI Relationships

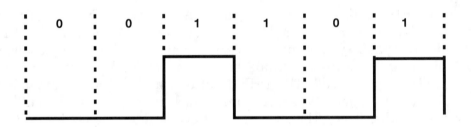

b. The NRZI Code

Figure 5–10
Physical layer encoding/decoding.

Table 5–1 shows the MAC four-bit symbols in the lefthand column and the resulting 4B/5B code in the righthand column. This code guarantees that a certain number of transitions occur on the line, which allows the receiver to synchronize itself to the traffic.

The 5-bit code allows for 32 combinations ($2^5 = 32$). Certain patterns are not needed. Therefore, the patterns selected guarantee that (a) at least two signal transitions are present in each 5-bit code, and (b) no more than three 0s in a row are allowed.

FDDI stipulates a unique approach to timing and clocking in the network. The best code to be used in a network is one which provides fre-

Table 5–1   FDDI 4B/5B Code

Data symbols	Value
0000	11110
0001	01001
0010	10100
0011	10101
0100	01010
0101	01011
0110	01110
0111	01111
1000	10010
1001	10011
1010	10110
1011	10111
1100	11010
1101	11011
1110	11100
1111	11101

Control Symbols

Quiet (Q)	00000
Idle (I)	11111
Halt (H)	00100
Starting Delimiter (J)	11000
Starting Delimiter (K)	10001
Ending Delimiter (T)	01101
Reset (logical zero)	00111
Set (logical one)	11001

quent signal state changes. The changes provide the receiver with the ability to continue to adjust to the incoming signal, thereby assuring that the transmitting device and the receiving device are synchronized with each other. The Manchester code, used in the IEEE 802.3 standard, is only 50 percent efficient because every bit requires a two-state transition on the line (i.e., two baud). Using the Manchester code, a 100-megabit transmission rate requires 200 MHz rate. In other words, Manchester code requires twice the baud for its transmission rate.

ANSI, recognizing that the 200 MHz rate would be more expensive in manufacturing the interfaces and clocking devices, devised a code called 4B/5B, in which a four-bit code is used to create a five-bit code. For every four bits transmitted from a DTE, FDDI creates five bits. Consequently, the 100 Mbit/s/s rate on FDDI requires only 125 MHz of band.

The 4B/5B is transmitted onto the fiber with differential signaling techniques: binary 1 = reverse of previous bit, and binary 0 = same as previous bit. FDDI detects the absence or presence of light pulses, based on previous pulses. This technique is more reliable than comparing the signals to an absolute threshold.

The MAC Sublayer

The FDDI data link layer consists of the MAC and LLC. MAC is responsible for the construction of the frame at the transmitter and the interpretation of the frame at the receiver. This layer is also responsible for proper addressing, delivering and receiving traffic from LLC, providing for access to the ring through a timed token protocol, ring initialization, and fault isolation.

The FDDI MAC protocol uses a timed token media access procedure. Users gain access to the network by receiving a token (a control PDU) that is circling around the ring. If the user device has traffic to send, it places this traffic in front of the token and sends it onto the ring. As the token and the data are sent around the ring, each receiving station checks a destination address in the header of the user traffic. If the destination address is for the station, it copies the payload and forwards it onto the ring. The MAC sublayer is described more fully in the section titled, "FDDI Operations in More Detail".

The LLC Sublayer

The logical link control sublayer (LLC) runs on top of all IEEE and ANSI LANs. It is not part of the FDDI standard, but can be used to provide additional services to a user workstation. It is beyond the scope of this book; an analysis of this sublayer is provided in Ulyess Black's *Data Link Protocols,* Prentice Hall, 1993.

FDDI PROTOCOL DATA UNIT

The FDDI frame (protocol data unit) is shown in Figure 5–11. The preamble (P) field precedes each frame and is used for synchronization procedures. The starting delimiter (SD) field is used to uniquely identify

P	SD	FC	DA	SA	I	FCS	ED	FS

P = Preamble
SD = Start delimiter
FC = Frame control
DA = Destination address
SA = Source address
I = Information
FCS = Frame check sequence
ED = End delimiter
FS = Frame status

Figure 5–11
The FDDI frame.

the beginning of the frame (it performs functions similar to the HDLC flag). The frame control (FC) field defines the type of frame such as a token, data frame, administrative frame, etc. The next two fields are the destination and source address (DA and SA). They may be either 16- or 48-bits and as the name suggests, they contain the sender of the frame and the intended receiver of the frame. The I field carries user information and headers from upper layers. The frame check sequence (FCS) field is a 32-bit field which calculates a cyclic redundancy check (CRC) on the frame. The frame status (FS) field indicates if the addressed station has recognized its address, if the frame was copied successfully by the destination, or if an error occurred in the detection operations at the station.

FDDI OPERATIONS IN MORE DETAIL

Figure 5–12 shows how MAC operates. A station on an FDDI network can transmit traffic after it detects the presence of a token. It must capture the token, absorb the token, and then send data. Once its data is sent, it releases the token onto the ring for the next station's use. In turn, this station can receive the data, place its own traffic in the stream, and then place the token onto the ring after it has finished sending traffic.

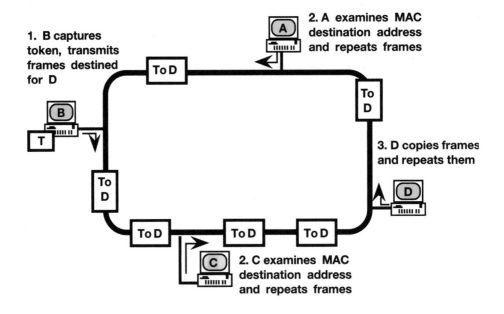

1. B captures token, transmits frames destined for D

2. A examines MAC destination address and repeats frames

3. D copies frames and repeats them

2. C examines MAC destination address and repeats frames

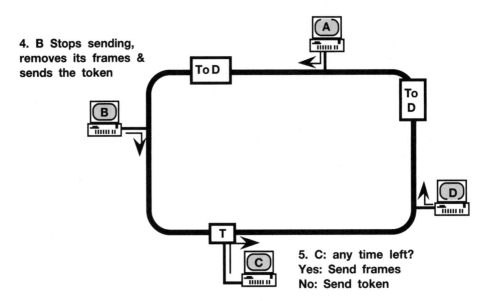

4. B Stops sending, removes its frames & sends the token

5. C: any time left?
Yes: Send frames
No: Send token

Figure 5–12
FDDI MAC operations.

Each station that originates a frame is responsible for removing this frame as it returns to the sending station. If an error is detected, as noted in certain status indicators within the frame, there is no attempt at the MAC layer to take remedial action. Error correction is the responsibility of an LLC type 2 function, or perhaps an upper layer transport protocol.

Each station on a LAN may be allocated a certain amount of capacity for synchronous traffic. If the synchronous allocation is not all used, the station then may use the excess capacity for the transmission of asynchronous traffic. Synchronous bandwidth means the station is guaranteed the right to send data, whereas asynchronous bandwidth means the network offers no such guarantee, but requires the station to contend for available resources.

Each station negotiates to determine the target token rotation time (TTRT). This value determines the amount of bandwidth per token rotation. Additionally, each station maintains a token rotation timer (TRT) which is used to measure the lapse between the receipt of successive tokens. Based on its evaluation of TTRT and when it receives a token, a station uses a token holding timer (THT) to determine how long it can transmit frames.

The THT begins after the station has sent all of its synchronous traffic. After which, if the station detects idle time is available, it is permitted to send asynchronous traffic. It is always permitted to send its synchronous frames, which are limited by the configured synchronous allocation parameter.

When a station has sent its data, it must place the token back onto the ring. The downstream station, upon receiving the token, determines if any time remains from the last time it received the token. If so, it sends frames; if not, it must pass the token to the next station on the ring.

FDDI employs a number of timers, variables, and flags to control the traffic on the ring (see Figure 5–13). The ones of interest for this overview are the token rotation time (TRT) and the token holding time (THT). TRT measures the time a station receives the token relative to the previous time that it received the token—that is to say, from the last token rotation around the ring. If the intervening upstream stations have used the token beyond the value of TRT, then the receiving station cannot use the ring and must pass the token to its downstream neighbor. One exception exists. A station can be configured so that a certain amount of bandwidth is guaranteed for that station. In this situation, a station can *always* transmit within this limit.

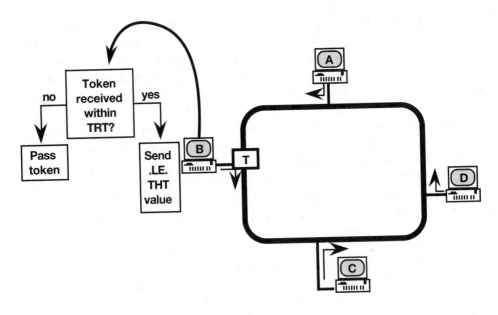

TRT = Token rotation time: time between successive arrivals of token at station

THT = Token holding time: time station can send asynchronous frames

Figure 5–13
The TRT and THT functions.

On the other hand, if the token arrives sooner than TRT, it means the intervening stations have not utilized the ring to its configured capacity; for example, the stations did not have much traffic to transmit. In this situation, the station is allowed to send frames within the bound of THT, plus any traffic that is guaranteed within a reserved bandwidth.

Figure 5–14 provides an example of the operations on an FDDI ring. For purposes of illustration, it is assumed the TRT is 15 units, and each station has been configured for synchronous THT of 3 units each. Synchronous THT is guaranteed bandwidth. Asynchronous allocations are based on demand, and the availability of unused bandwidth. It is also assumed that initial operations begin with station A, which received a token, and notes the TRT is 0 (it initializes the token operations). The reader may wish to use the table at the bottom of the figure while studying this example.

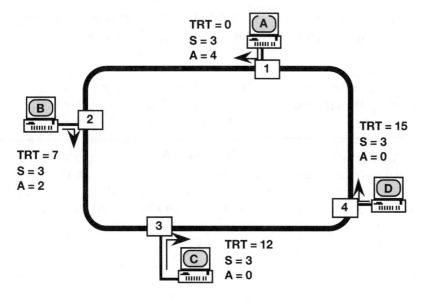

TRT = 0
S = 3
A = 4

TRT = 15
S = 3
A = 0

TRT = 7
S = 3
A = 2

TRT = 12
S = 3
A = 0

[n] = Order of the receiving of the token

TRT = 15 units

S = Synchronous frames

A = Asynchronous frames

A			**B**			**C**			**D**		
	Send			*Send*			*Send*			*Send*	
TRT	*S*	*A*	*TRT*	*S*	*A*	*TRT*	*S*	*A*	*TRT*	*S*	*A*
0	3	4	7	3	2	12	3	-	15	3	-

Figure 5–14
First Rotation of Token.

Station A sends frames to use the 3 synchronous units, and also sends frames to use 4 asynchronous units. It could be that it had more frames to send, but its THT parameter limited it to 4 asynchronous units.

Station A relinquishes the token by placing it on the ring. The token is received by B, which notes a TRT of 7. This value reflects station A's 3 + 4 units. Station B sends frames to use the 3 synchronous units and also sends frames to use 2 asynchronous units.

Station B relinquishes the token by placing it on the ring. The token is received by C, which notes a TRT of 12. This value reflects station A's 3 + 4 units, and station B's 3 + 2 units. Station C sends frames to use the 3 synchronous units. It has no other traffic, so it does not use any asynchronous units.

Station C relinquishes the token by placing it on the ring. The token is received by D, which notes a TRT of 15. This value reflects station A's 3 + 4 units, station B's 3 + 2 units, and station C's 3 units. Station D sends frames to use the 3 synchronous units. It cannot send any asynchronous traffic since the TRT limit is reached.

Figure 5–15 provides an example of the operations on the FDDI ring with the second rotation of the token. The effect of the previous rotation of the token is key to understanding how the FDDI MAC controls traffic.

FDDI-II

FDDI-II is an enhancement to the original FDDI (AKA FDDI-I). It adds the ability to transmit multimedia images on an FDDI network. The original FDDI was designed for data services. FDDI-II provides circuit-switched services for isochronous transmissions. In meeting this demand for images such as voice, video, and audio, FDDI-II supports a technique to apportion bandwidth exclusively for isochronous streams. FDDI-II also provides for an additional operation called the hybrid ring control (HRC) which allows the multiplexing of packet MAC traffic and isochronous MAC traffic. As of this writing, very little interest has been shown for FDDI-II.

The FDDI designers were cognizant of the need to stay compatible with older technologies. Consequently, traffic is formatted in the 125 microsecond cycle used with T1 technology.

Figure 5–16 shows the layers of FDDI-II. The architecture is similar to that of FDDI-I, with considerable enhancements relating to circuit-switched services. The principal difference lies in the ability for FDDI to support an isochronous MAC (IMAC). Circuit-switched multiplexed signals can be input or output, into or out of IMAC, and into a hybrid box (which is responsible for multiplexing LLC traffic as well as circuit-switched traffic).

The hybrid ring control (HRC) operation is optional and does require the use of the second generation FDDI boards (FDDI-II). The boards are now designated as MAC-2 and PHY-2.

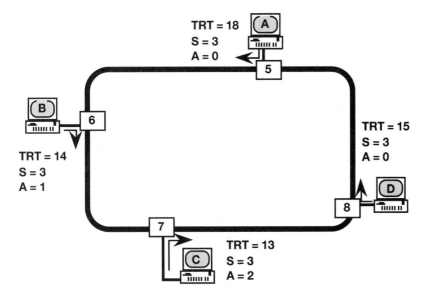

TRT = 18
S = 3
A = 0

5

TRT = 15
S = 3
A = 0

6

TRT = 14
S = 3
A = 1

8

7

TRT = 13
S = 3
A = 2

n = Order of the receiving of the token

TRT = 15 units

S = Synchronous frames

A = Asynchronous frames

A			B			C			D		
	Send			Send			Send			Send	
TRT	S	A	TRT	S	A	TRT	S	A	TRT	S	A
18	3	-	14	3	1	13	3	2	15	3	-

Figure 5–15
Second Rotation of Token.

The station management function (SMT) is still provided to all the sublayers of both the data link and physical layers.

Another enhancement to the FDDI standard was the provision for mapping to SONET. Figure 5–16 also shows the mapping capability at the physical layer where mapping occurs to and from SONET STS-3c.

The FDDI-II protocol employs a PDU called a cycle. It repeats itself continuously, thus its name. The cycle resembles (somewhat) the SDH/SONET envelope in that it carries payloads, control information,

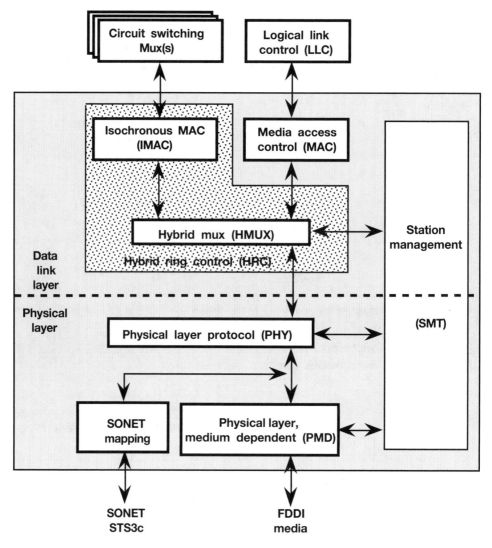

Figure 5–16
FDDI II layers.

and it is based on the 125 microsecond (µsec.) time slot. The cycle rotates
around the ring, thus it can be used by all stations.

    A master station is responsible for generating the cycle every
125 µsecs. to fulfill the need for the Nyquist criterion of 8,000 samples
each second. The cycle is absorbed (stripped) by the master station.

Since the FDDI ring operates at 100 Mbit/s, a cycle is 12,500 bits (100,000,000/8000).

The cycle contains the following control information. The *preamble* is a 20-bits (5 symbols) nondata field, and it is used for synchronization operations. The cycle *header* describes how the cycle is being used. It contains several fields to perform several functions. It indicates the beginning of a cycle, and whether or not synchronization has been established. It provides a sequence number to sequence each cycle, as well as a field to allow stations to contend for becoming the cycle master. It contains fields to identify the traffic (workload) as data or isochronous traffic. It also provides a channel for the maintenance of isochronous traffic. Finally, the *dedicated packet group* (DPG) is used for token controlled data transfer operations.

Figure 5–17 shows the structure of the FDDI-II cycle. It resembles the envelope of the SDH/SONET system in that it is organized in rows and columns and is based on a 125 μsec. cycle.

Sixteen wideband channels are available for user payload. Each channel uses 96-octets per cycle. Therefore, the channel has a capacity of 6.144 Mbit/s (96-octets * 8-bits per octet * 8,000 = 6,144,000). A channel can be used for data as well as voice. The traffic in the channel is controlled by the conventional FDDI IMAC protocol.

OTHER NOTABLE ASPECTS OF FDDI

As a general practice, FDDI LANs are installed in organizations to serve as backbone networks. The FDDI LAN is used to interconnect lower-speed LANs such as 802.3, Ethernets, and 802.5 networks. Due to the relatively high capacity of an FDDI network (100 Mbit/s/s), it serves as a cost-effective and efficient means for internetworking other types of LANs. It should be noted that, while FDDI-I has proven itself in the marketplace, FDDI-II is (at this writing) essentially dead.

Operating FDDI as a MAN

While FDDI is distance-limited to a ring of 200 km, it is certainly possible to extend FDDI operations to a larger area by connecting individual FDDI LANs to each other through bridges/routers. Indeed, Bell Atlantic has implemented this concept with its FDDI Network Service (FNS) in the Washington DC metropolitan area, where selected central

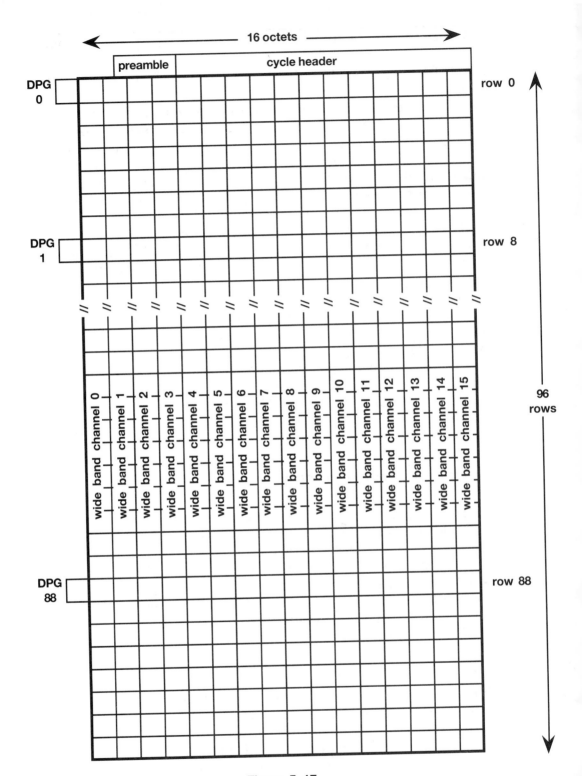

Figure 5–17
The FDDI II cycle.

offices are connected to each other and to customers' LANs. Customer interconnection at the network demarcation point is through a wire closet at the customer site. The Bell Atlantic plan allows FNS to operate (potentially) on an inventory of 12,000 miles of single mode fiber in this area.

FDDI versus ATM

A debate has begun in the industry regarding the use of FDDI or ATM for internetworking LANs and WANs. Some proponents of ATM believe FDDI will not compete well against ATM because of its limited bandwidth. Proponents of ATM also state that multimedia applications do not work well on a shared media such as a ring, and that fast cell switching with an ATM switch will support voice and video better than a shared ring technology.

It remains to be seen (at the time of this writing) how the marketplace will fall out. Certainly FDDI is well entrenched in the industry (at least FDDI-I). However, the very slow development of FDDI-II has (1) turned some users off to the technology, and (2) allowed the ATM standards to mature and catch up with FDDI.

Moreover, there is concern in the industry regarding ongoing debates about using FDDI on twisted-pair cable. Indeed, the standards groups are having trouble agreeing on the specification and some vendors are moving forward with their own approaches. For example, IBM is now supporting FDDI on both fiber and shielded twisted pair cable. The latter implementation is supposed to conform to FDDI over shielded twisted pair specifications (SDDI). However, this standard is not yet formally approved.

FDDI proponents claim that their technology is well understood and the pricing is stable. In contrast, ATM is a relatively new technology with untested performance and unstable prices.

We shall see in the next few months to the next few years how this debate will be settled. A likely scenario is for an organization to have an ATM backbone with FDDI LANs connected to this backbone.

FDDI WORKSHEET

The worksheet for FDDI is provided in Table 5–2. As suggested in the introduction to this book, it is recommended that the reader attempt to fill out a blank sheet (provided in Appendix C), and then read this sec-

Table 5–2   FDDI Worksheet

Technology name:	FDDI
New technology?	FDDI-I is well-entrenched, but FDDI-II is new, and not well-received
Targeted Applications?	Any application on a LAN. FDDI-I for data, and FDDI-II for voice and data
Topology Dependent?	Yes, dual and dual ring of trees
Media Dependent?	Yes, at the PMD sublayer, but can operate over different types of fiber as well as copper; no at PHY and MAC layers
LAN/WAN based?	LAN-based; it is designed to provide a high-speed LAN and to serve as a backbone to interconnect LANs
Competes with:	ATM, and increasingly, the lower speed LANs, such as Ethernet, token ring, IEEE 802.3
Complements:	Ironically, it also complements its potential competitors (see last question)
Cell/Frame based?	Frame-based
Connection management?	No for FDDI-I, which is connectionless. Yes for isochronous traffic on FDDI-II
Flow control (explicit/implicit)?	Yes, explicit, with token rotation timers, and SA/ASA conventions, but no congestion notification operations
Payload integrity management?	FCS checks for errors, but ACKs, NAKs, and sequencing are a user responsibility
Traffic discard option?	No
Bandwidth on Demand?	Somewhat, in that each station can be configured differently, allocation is dynamic as bandwidth is available
Addressing/identification scheme?	Yes, uses explicit addressing with the 48-bit MAC addresses

tion to check your answers. Several of the answers require no further explanation; others are further explained in this section.

FDDI is both a new technology and an old technology: FDDI-I is certainly not a new technology; it has been in operation for several years.

However, FDDI-II is a new technology. Its support of voice applications is considerably different from the FDDI-I operations. It should be emphasized that FDDI is LAN-based only. Its distance prohibits any type of wide area network application, unless routers are attached to internetwork multiple FDDI LANs.

FDDI will most likely compete with the technologies emerging that provide internetworking with LANs. Therefore, ATM will certainly emerge as a competitor. It is expected that as the component costs for FDDI equipment decrease, it could become a competitor for current LAN technologies such as IEEE 802.3 and Ethernet. Ironically, it currently complements these technologies by providing a high-speed backbone for the interconnection of these types of networks.

SUMMARY

FDDI is a dual-ring-type of token protocol with each ring operating at 100 Mbit/s. With the use of a dual-ring architecture, self-healing operations are available in the event a line or network node becomes inoperable. FDDI is designed to act as a high-speed network and/or as a high-speed backbone network for other networks. FDDI-I was implemented principally for data transmission. FDDI-II was designed to support multimedia applications. In addition, FDDI-II provides an interface into SONET/SDH. FDDI-II has received little interest in the industry.

6

Metropolitan Area Networks (MANs) and Switched Multi-Megabit Data Service (SMDS)

INTRODUCTION

This chapter examines both the metropolitan area network (MAN) and the switched multi-megabit data service (SMDS) protocols. The two technologies are included in the same chapter because SMDS uses the MAN protocol and adds service features for an end user.

THE PURPOSE OF A MAN

The MAN is designed to support LAN internetworking operations and host-to-host data transfers. The term metropolitan is used to convey the idea of interconnecting an enterprise's LANs and host machines to each other within a metropolitan area. For example, many companies have offices that are located in different parts of a metropolitan area. A MAN can interconnect these offices in a seamless manner[1].

---

[1]With the advent of suburbia and its many attractive features, many corporate offices now reside in suburban areas. The IEEE probably meant the term MAN to encompass suburbia. Notwithstanding, perhaps a better term for this technology is SAN—for

The U.S. telephone companies also view MAN as a technology to provide for the interconnecting of LANs in one metropolitan area to (a) the central office, (b) the telephone company's wide area switching facilities, as well as (c) another central office and end-user MAN in a different metropolitan area.

The MAN also forms the basis for the switched multi-megabit data service (SMDS) which is now being touted as the solution to the "WAN bottleneck." From the perspective of the U.S. telephone companies, the LAN/WAN internetworking bottleneck problem is solved with the use of MAN/SMDS technology for interconnecting LANs.

PERTINENT STANDARDS

The MAN standard evolved from the Queued Packet Synchronous Exchange (QPSX) which was developed in Australia. It is published by the IEEE as 802.6, which is also recognized by ANSI as an approved standard. The ISO publishes the MAN standard as ISO 8802/6.

A TYPICAL MAN TOPOLOGY

The MAN standard is organized around a topology and technique called the distributed queue dual bus (DQDB). This term "dual bus" means that the network uses two separate communications channels, which are called busses. The term "distributed queue" means that a user places traffic in a queue, for later transmission onto a bus, and a queue is maintained for each bus. Both busses run at the same speed, and each network node attaches to both busses.

The DQDB protocol provides for two types of access to the channel. One type of access is called pre-arbitrated (PA) access which guarantees a certain amount of bandwidth. PA access is employed for isochronous

---

the suburban area network, or SMAN—for the suburban metropolitan area network, and so on. But these terms probably won't work, because sociologists tell us that metropolitan and suburban areas are melding and becoming (in their terms) a megalopolis. The megalopolis encompasses urban and suburban landscapes that have converged on each other. They span hundreds of miles before a cow is sighted—which could reveal that a rural area has been reached (except in Houston). Perhaps a good distinction between a MAN and a WAN is that a cow can be found at their boundaries. Anyway, it is as good as any other definition that I know. I ask the reader to present a better one.

services such as voice and video. The second type of access is called queued arbitrated (QA) access which provides services based on demand. QA is designed to accommodate bursty applications such as data transmission. The provision for continuous bit rate (CBR) traffic for isochronous/asynchronous applications, and variable bit rate (VBR) traffic for bursty, asynchronous applications implies that a MAN has extensive convergence functions.

As stated earlier, the MAN is designed with two unidirectional busses. Each bus is independent of the other in the transfer of traffic. Each node has two attachments to each bus; one attachment reads the bus "slots" and the next attachment writes in the bus slots. The slots are time division multiplexed onto the bus by the head of the bus. The topology can be designed as an open bus or closed bus configuration. Figure 6–1 shows the two alternatives. The arrows depict the direction of the traffic flow and whether or not a node is upstream or downstream from other nodes on the bus. For example in Figure 6–1a, node A is upstream to node B on bus A, but downstream to node B on bus B. An upstream node has access to a slot before the downstream nodes.

Each bus is managed by a node called the headend or head of bus. The head of bus is responsible for generating fixed-length slots for use by the downstream nodes. Initially, these slots have no user payload. Through the use of several allocation schemes (discussed later), the nodes gain access to the slots and place their payload into these slots for downstream nodes to receive.

The traffic on bus A is not placed on bus B; all user payload terminates at the headend of the bus. Therefore, a source node must know the location of the destination node in order to place the traffic onto the proper bus. Later discussions in this chapter explain how nodes discover each other's position on the busses.

Topology Reconfiguration with Self-Healing Networks

In the event of a severance in the media on the DQDB closed bus topology, or a failure at the interface of the bus and the node, the faulty link between the two nodes is bypassed (see Figure 6–2a). The network is reconfigured to an open bus so all nodes still have connections to each other.

In the event a connection is lost in the DQDB open bus topology, the network reconfigures itself into two disjoint networks (see Figure 6–2b). This approach means the two networks cannot communicate with each other, but the nodes within the networks can communicate. The choice of

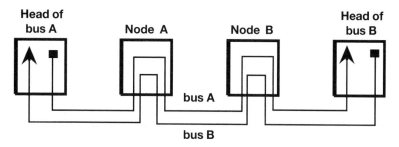

a. Open bus architecture

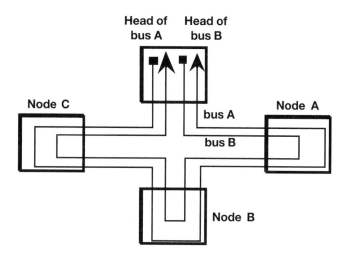

b. Closed bus architecture

Figure 6–1
The MAN bus architecture.

which topology to use is quite important to the enterprises that are connecting networks together across campuses and buildings in industrial parks. Most MANs are implemented with the closed bus topology, because of its ability to completely self-heal.

THE MAN LAYERS

As shown in Figure 6–3, the MAN layers are modeled loosely on the IEEE 802 architecture. The DQDB layer is considerably more extensive than the IEEE MAC layers, and contains several convergence services

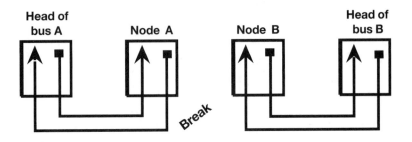

a. Open bus

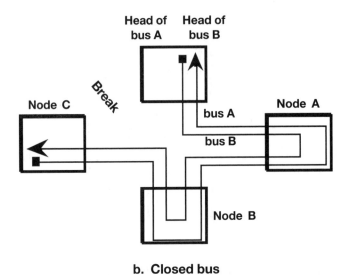

b. Closed bus

Figure 6–2
Reconfigurations.

(which closely resemble the convergence services of the asynchronous transfer mode [ATM]).

The top of Figure 6–3 shows the types of services that are supported by MAN functions. The LLC services describe any services for an LLC sublayer of a LAN node. The connection-oriented services are designed to support asynchronous virtual circuit services such as X.25 traffic. The other services shown in Figure 6–3 are under study and are not defined. The isochronous services are designed to support applications that need continuous bandwidth such as voice and video.

The convergence functions at a transmit node are responsible for

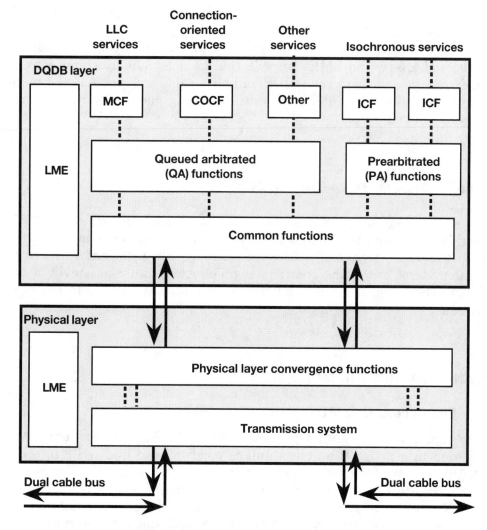

LME = Layer management entity (the MAN network management component)
MCF = MAC convergence function (services to accept LAN traffic)
COCF = Connection-oriented convergence function (virtual circuit data traffic)
ICF = Isochronous convergence function (delay sensitive traffic)

Figure 6–3
The MAN layers.

receiving user traffic and changing the traffic to DQDB formats. The reverse operations occur at the receiving node.

At the physical layer, the physical layer convergence functions sublayer assumes that each transmission system has a different convergence function. Therefore, this operation provides a consistent service to the DQDB layer. The transmission system sublayer supports the physical media such as DS3, G.703, STS-3, etc.

Depicted on the left side of Figure 6–3 are the layer management entities (LMEs). The LMEs provide the management functions for the MAN functions at both the DQDB and physical layers, such as slot management, allocation of PA/QA bandwidth, assignment of virtual circuit identifiers (VCIs), etc.

MAN PROTOCOL DATA UNITS (PDUs)

In previous chapters, a separate section described each of the PDUs used by the emerging technologies. It is necessary to make an exception with the MAN technology, because many of the principal features of these systems entail the creation, segmentation, and reassembly of the PDUs. Therefore, it is more efficient to examine the PDUs in the next section.

MAN OPERATIONS IN MORE DETAIL

The Access Unit (AU)

Figure 6–4 presents a general view of the MAN access unit (AU) residing in a node. The access unit is used to read and write into the DQDB slots, or a portion of a slot. It can support a single node or a cluster of devices that use one access unit. For example, a terminal server on a LAN can be configured with one DQDB port for access to the MAN. Its other ports are conventional EIA-232-E, RS-422, and V.35 interfaces that connect to the terminals. Another example is the configuration of a router; one of its ports contains the DQDB interface, and others connect to LANs; such as Ethernet, token ring, etc. This approach allows the network administrator to keep the end user devices and other LANs isolated from the DQDB activities and vice versa.

The DQDB signals are read and written every 125 μsec. This approach reflects ongoing technology (for example, T1/E1) in that: 1 second/8000 = .000125. The value of 8000 reflects the conventional sampling rate (in seconds) for a voice channel.

The nodes write traffic to and read traffic from the slots on the bus.

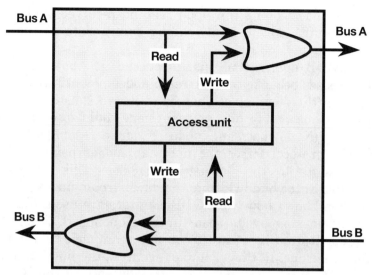

Note: Slots read and written every 125 microseconds

Figure 6–4
Read and write operations in the access unit.

They gain access to the network by writing into the slots. The read function is placed in front of the write function on each bus, which allows a node to read slots independently of writing into them. Once the slot has been written into, it is read by the downstream nodes to determine if the payload in the slot is to be copied at that node and passed to an upper layer protocol or ignored.

Overview of the DQDB Protocol

The DQDB protocol is elegantly simple. It is based on the use of counters. These counters operate on each bus to determine if a slot is available for use. One of the key points to remember during this discussion is that a DQDB node examines and reserves slots on one bus in order to use the slots on the *other* bus.

Each time a node detects a slot that has been reserved by an upstream node, it increments a counter. As slots pass by on the other bus that are not busy (the slots have no traffic in the information field), a counter is decremented. When this counter equals 0, the node is allowed to seize the next free slot and place traffic into the information field of the slot. So, a node "counts up" upon reading reserved slots on one bus (bus A) and "counts down" upon reading empty slots on the other bus (bus B).

If the DQDB operates as just discussed, it would not be a very democratic process because the upstream nodes potentially have a better chance of using the slots, since they can reserve slots before the downstream nodes. Conceivably, an upstream node could reserve all the slots and prevent the downstream nodes from accessing the network. Therefore, the protocol forces the upstream nodes periodically to pass non-reserved slots to the downstream nodes. This arrangement is designed to guarantee bandwidth to the downstream nodes.

Figure 6–5 shows how the DQDB counters are used. Node A examines each slot passing by its read function on bus B. Each slot has a reservation bit (R) that is set to 1 by an upstream node that has traffic to send on the other bus. In this example (event 1), the upstream nodes relative to node A have reserved three slots by setting the R bit to 1 in

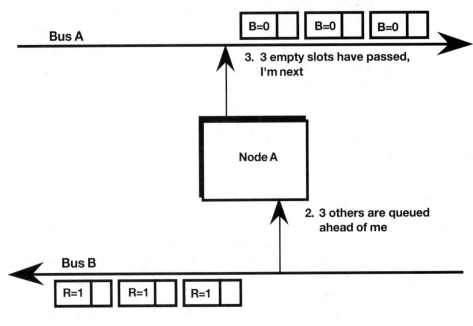

Figure 6–5
General review of the DQDB operations.

three slots. Node A remembers that three requests are queued ahead of it (event 2), by incrementing a counter by one for each reserved slot. In effect, node A knows that "Three others are ahead of me."

The R bits on bus B affect the use of slots on bus A (and vice versa). Since node A is fourth in the queue behind three requests, it examines the slots passing by on bus A (event 3). The slots contain a busy bit (B) that is set to 0 if the slot is empty (i.e., has no user payload in it). After node A has counted three empty slots passing by on bus A, it knows these three slots will be used by the nodes that queued their requests ahead of A. Therefore, it is allowed to use the next (fourth) empty slot. In effect, node A knows "My counter is zero; three empty slots that have passed are to be used by upstream nodes; I'm next."

The term distributed queue refers to the fact that each node on the dual bus maintains a queue of requests (and traffic) for each bus. One queue is for bus A and the other queue is for bus B. The two queues are identical in how they operate, but they are independent of each other. The approach allows a deterministic access approach to what is essentially bursty traffic (if the traffic is data).

The DQDB Counters

The following description assumes the use of queued access (QA), with a single priority. Multiple priorities can be obtained, but we will keep it simple at this point of the analysis.

As Figure 6–6 shows, two counters are employed by the DQDB protocol, and two bits are used in each slot on the bus. As just explained, the busy (B) bit indicates that use of the slot is being used, and the reserved (R) bit indicates that use of the slot has been requested—it has been queued for later access. The request (RQ) counter is used to indicate the number of requests that have been queued ahead of the node on bus B. The count down (CD) counter keeps count of the number of free slots on bus A that are queued ahead of a node. The effect of these operations is to create a logical, single access queue that is distributed across the DQDB network.

If a node has no data to send, it is in an idle state and is not part of the distributed queue. When a node has traffic to send, it waits until it encounters a slot on the reverse bus (bus B) that has R = 0, it sets the R bit equal to 1 in the reserve field of the slot. The effect of this operation is to signal that a QA segment has been queued on the reverse bus (bus B). Once the node joins the distributed queue, it changes from an idle state to a CD state.

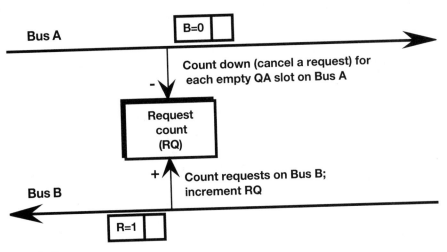

Why?
• Keeps track of requests queued

a. First operation.

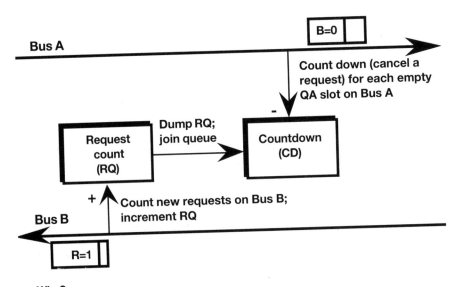

Why?
• Places QA segment in a queue
• Ensures no segment can queue out of order
• Effect is a FIFO queue on Bus A

b. Second operation.

**Figure 6–6
Queuing operations.**

Once the reserve bit field is set to 1, the node then puts the value that is in RQ into the CD counter. Next, it sets its RQ counter to 0. This action loads the CD counter with the number of slots that are queued in front of this node and places the node and its traffic into the distributed queue.

After the CD counter is loaded from RQ, CD is then decremented on the forward bus (bus A in this example) for each passing empty slot. The effect of this action is to allow the node to gain access to the bus and use a slot when its CD equals 0.

This explanation captures the essence of the DQDB protocol. Its basic concepts are surprisingly simple, yet it provides a flexible and deterministic process for accommodating a variety of user payloads.[2] Examples thus far have concentrated on how data applications (bursty applications) are supported. Later sections explain how isochronous applications use the DQDB network.

Location Discovery

Figure 6–7 shows how nodes can discover their locations relative to each other on the two busses. The MAN standard does not stipulate a protocol for location discovery, but suggests the application illustrated in this figure. The idea is quite similar to the approach used by learning tree bridges on LANs in that the node monitors traffic on both busses and examines the source address in the header of each PDU. Through the examination of this address, it infers whether a station is upstream or downstream on the two busses. In this example, station B monitors a PDU and discerns the source address is E. Since this traffic is running on bus B, node B infers correctly that node E is upstream on bus B. Therefore, in order to reach E, node B must transmit E's traffic on bus A.

The MAN standard also does not stipulate how nodes discover each other when the network starts up for the first time or when it is reinitialized. One approach is to establish a procedure, as part of the start-up process, in which each node sends its source address on both busses. The destination address in this frame is a broadcast address. Therefore, all nodes copy the frame, examine the source address field, and "glean" the relative location of the sender.

---

[2]On the other hand, this general explanation has only touched on a few of DQDB's procedures and capabilities. The interested reader should refer directly to the standard for a more detailed description of the MAN operations. Also, the journal titled *Computer Networks and ISDN Systems* devotes several of its papers to MAN discussions.

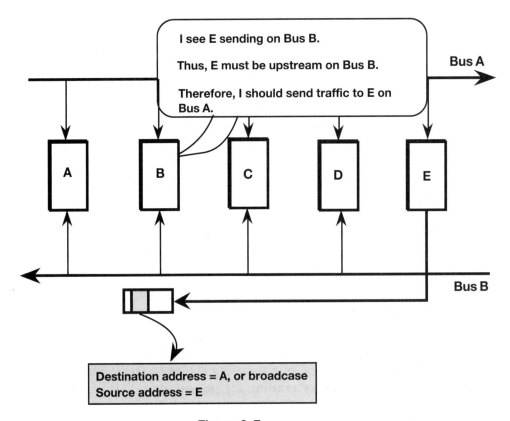

Figure 6–7
Location discovery.

Segmentation and Encapsulation Operations

A mild warning is in order regarding this section of the chapter. It is somewhat detailed, and must delve into the bits and bytes of the traffic. The reader who is interested in an overview of the MAN can skim this part of the chapter. However, the material is essential for those readers who need a more detailed understanding of how user traffic is segmented into the DQDB slots (AKA cells).

802.6 can process user traffic of up to 9188 octets. In accordance with OSI rules, this traffic is called a MAC service data unit (SDU). The MAC SDU has a considerable amount of information added to it in the form of a header and trailer (see Figure 6–8). The header is 24-octets and the trailer is 4-octets. This unit is now called an initial MAC protocol data unit (IMPDU).

After this encapsulation process, Figure 6–8 shows that the IMPDU goes through three major operations: (a) the IMPDU is segmented into

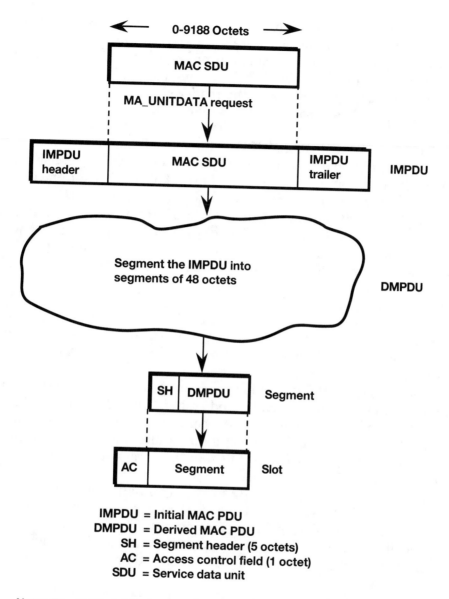

IMPDU = Initial MAC PDU
DMPDU = Derived MAC PDU
SH = Segment header (5 octets)
AC = Access control field (1 octet)
SDU = Service data unit

Note: The MAN slot is similar to the ATM cell

Figure 6–8
IMPDU creation, segmentation, and encapsulation.

48-octet units to create a derived MAC PDU (DMPDU), (b) a 5-octet header is added to create a 52-octet *segment,* and (c) another 1-octet header is added to create a 53-octet *slot.*

The purpose of these segmentation and encapsulation functions is to provide sufficient information for the data unit to traverse a MAN and to be reassembled into an accurate copy of the original MAC SDU. Additionally, some of these headers are designed for the eventual inter-working with emerging technologies such broadband ISDN (ATM).

For the ensuing discussions, the reader should refer again to Figure 6–8. It will provide a good point of reference.

The initial MAC PDU (IMPDU). Figure 6–9 shows the header and trailer that are prepared by the MAC convergence function (MCF) before segmentation operations are invoked. Several of the fields in the headers and trailers are coded in accordance with the particular technology that is using the format. For example, destination and source address values depend on what application is being applied. These addresses would likely be telephone numbers if this technology is applied to SMDS. Conversely, if the traffic resides within a LAN, these addresses might be MAC addresses.

The IMPDU header contains a common PDU header and a MCP header. The former is used in all DQDB PDUs that support bursty data services; the latter is concerned with the actual transfer of the MAC SDU. The common PDU trailer is also used in all DQDB PDUs that support bursty data services.

The purpose of the BEtag is to associate the header and trailer and all the octets between the header and trailer. The BEtag resides in both the header and trailer. This tag is necessary because a large data unit may be divided into smaller units, and this identifier provides a means to reassemble the pieces. The BEtag value is incremented by one for each sequential IMPDU sent by a node.

The buffer allocation size (BAsize) is a value to indicate the size of the original IMPDU. The BAsize is used by the receiver to determine how much buffer to allocate for the incoming traffic.

The destination address (DA) and source address (SA) fields are 64-bits in length and can contain any value deemed appropriate by the net-work administrator.

The protocol identification (PI) field is used to identify an upper layer protocol that is creating and receiving this traffic. Currently, the value is set to 1 in the MAN specification to identify an LLC entity. The PAD length (PL) field indicates the length of the PAD field.

The values of the quality of service (QOS) field depends on the application using this protocol, but the MAN standard defines two QOS operations (and subfields within the QOS field) that identify (1) delay and (2) loss of traffic. The delay subfield indicates the delay in an

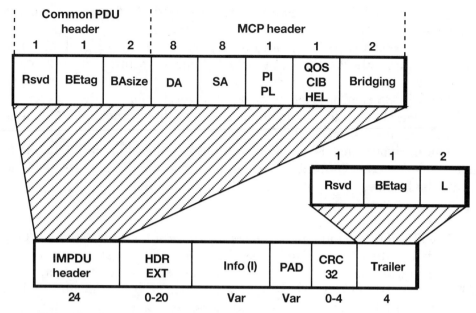

Note: Numbers represent bytes.

Rsvd = Reserved
BEtag = Beginning-end tag
BAsize = Buffer allocation size (length)
DA = Destination address
SA = Source address
PI = Protocol id
HLPI = Higher layer protocol Id
PL = PAD length
QOS = Quality of service
CIB = CRC32 indicator bit
HEL = Header extension length
PAD = Pading
CRC32 = 32 bit cyclic redundancy check
L = Length
HDR EXT = Header extension

Figure 6–9
IMPDU header and trailer.

IMPDU accessing the network. As of this writing, the loss of traffic sub-field is reserved, and not used. The CRC 32 indicator bit (CIB) indicates the presence or absence of the CRC32 field, where 0 = no CRC 32 field, and 1 = CRC 32 field is present. The HEL is used to signify the length of the header extension field (HDR EXT). The header extension field is not defined in the MAN specification. It can be used for further expansion and is used by SMDS to identify additional services.

The information (I) field contains the MAC SDU (user payload). The PAD field aligns the I field and the PAD field to an integral multiple of four octets.

The CRC32 field is used to detect the corruption of any bits in the MCP header, the header extension field, the I field, and the PAD field. CRC32 is not used when CIB = 0.

The bridging field is used (potentially) to assist in 802.1 bridging operations. Its exact contents have not been defined. The length (L) field equals the BAsize.

The DMPDU.   The IMPDU is segmented into a 48-octet PDU. This data unit is called the derived MAC PDU (DMPDU) (see Figure 6–10). The information portion (I field) of this PDU consists of 44-octets. The full PDU consists of the 44-octets plus four octets of header and trailer. Since the IMPDU is broken into small pieces, it is necessary to identify if a specific PDU is the end of the message (EOM), beginning of the message (BOM), the continuation of the message (COM), or a single segment message (SSM). The segment type (ST) value provides this information. The term "message" refers to the IMPDU.

The complete IMPDU header can reside in the BOM-indicated segment I field, because this header is 24-octets and the DMPDU I field is 44-octets. This IMPDU overhead is not so onerous if the MAC SDU consists of many octets, because the ratio of overhead to payload is small. In such a case, the first DMPDU segment contains the IMPDU header and 20-octets of user traffic. In contrast, if the user payload (MAC SDUs) consists of only a few octets, the overhead consumes substantial bandwidth, because of the high ratio of overhead to payload.

The subsequent COM-indicated segments only contain payload (MAC SDUs). The EOM contains the remaining payload and the IMPDU trailer. For the last segment of an IMPDU, any octets not used in the 44-octet I field are set to zeros.

For a single segment message (SSM), the first 24-octets of the I field contain the common PDU header and MCP header, and the remaining 20-octets can carry user traffic. The common PDU trailer is not needed in a SSM.

The sequence number (SN) is used at the destination node to reassemble all segments of the IMPDU in the correct order. The MCF (see Figure 6–3) associates a sequencing operation with *each* message

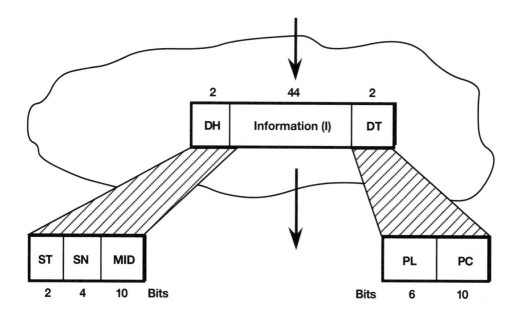

ST = Segment type (COM, EOM, BOM, or SSM)
MID = Message ID (segmentation control)
SN = Sequence number
PL = Payload length
PC = Payload CRC

Figure 6–10
DMPDU segment with header and trailer.

identifier (MID) field and virtual circuit identifier (VCI) (both are discussed shortly). Thus, a MID/VCI value uniquely identifies the payload and SN sequences this payload.

Each IMPDU is associated with a message identifier (MID). All DMPDUs of the same IMPDU must carry the same MID value. Obviously, this value allows the network to uniquely identify all traffic. Of course, don't forget that the IMPDU DA and SA are also available on the DMPDU BOM segment to cross identify the traffic.

The payload length (PL) contains a value that indicates how much of the information field actually contains data. The value can be up to the value of 44-octets. Finally, the payload CRC (PC) is an error check field that is used to check on the integrity of the data (I field) and the DMPDU header and trailer.

QA/PA segments. After the derived MAC PDU has been created and headers added, the traffic is next passed to the queued arbitrated function where an additional header is appended to the segment payload (see Figure 6–11). This header is four octets long, thus yielding a PDU of 52-octets.

The header is broken into these fields. A 20-bit virtual channel ID (VCI) is used to identify this traffic to and through the network. This field associates the segment with a virtual channel.

The control field (CF) consists of payload type and segment type. Two bits are used to identify the payload type (the type of information residing in the segment payload I field). For example, the type = 00 indicates the presence of user traffic (instead of some type of control traffic). All other values are subject to further study. The next two bits are used to indicate a segment priority which provides guidance for how the network is to handle the traffic. The exact method for handling priority segments is not defined in the standard, so MAN designers are free to implement this service in any way deemed appropriate.

The eight-bit header check sequence (HCS) field is used for error control. It checks for bit distortions in the header, and can correct a one-bit error in the PA segment header. The HCS is generated at the head of bus, and checked based on the following rules: Error detection is mandatory for each node that reads and writes the PA segment payload, but error correction is optional.

Some of the fields in the headers and trailers described thus far in this section are used at the receiver to reassemble the traffic into a copy of the initial MAC PDU (IMPDU). Since traffic is multiplexed onto the DQDB, the derived MAC PDU (DMPDU) from different IMPDUs will arrive in an interspersed manner at the receiver. Each separate IMPDU has associated with it a unique reassembly process, a process which is associated with each VCI/MID pair on the DQDB network. Therefore, reassembly operations use the VCI and MID, and VCIs arriving with two different MIDs are considered parts of two different IMPDUs. The sequence number (SN) is incremented by one for each DMPDU of a IMPDU. It is used to ensure that segments within the VCI/MID are reassembled in the proper order at the receiver.

The slot. The MAN/DQDB PDU is called a slot and (finally) it is the complete PDU that is transferred on the bus (see Figure 6–12). It consists of the segment and a one-octet control field known as the access control field (ACF). This header is added at the transmit common func-

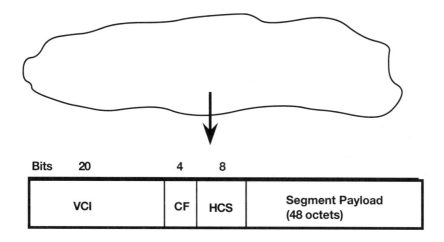

Bits	20	4	8	
	VCI	CF	HCS	Segment Payload (48 octets)

VCI = Virtual channel ID
CF = Control fields
  2 bits for payload type
  2 bits for segment priority
HCS = Header check sequence

Figure 6–11
QA/PA segment.

tions entity (see Figure 6–3) and interpreted by the receive common functions entity of the DQDB layer.

The slot header consists of five fields. The first field is a busy bit field which is set to one to indicate the slot is busy, or zero to indicate there no information residing in the slot. For QA traffic, the head of bus sets the busy bit to 0.

For PA traffic, the head of bus sets the busy bit to 1. This action prevents QA applications from using the slot. In addition, PA applications (voice, video, etc.) examine the VCI field (Figure 6–11) to determine if this slot belongs to this application. If so, the application has had this slot reserved and can use it. Thus, continuous bit rate traffic (CBR) is assured of receiving a fixed and predictable part of the bandwidth in the DQDB network. Later discussions explain how the PA slots are set-up and reserved for CBR traffic.

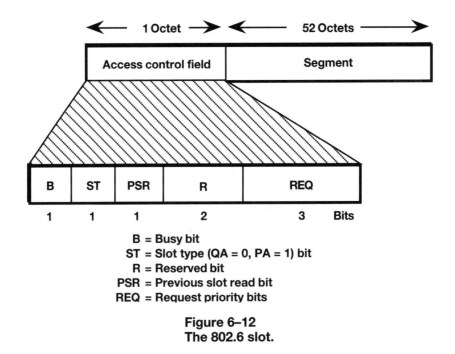

B = Busy bit
ST = Slot type (QA = 0, PA = 1) bit
R = Reserved bit
PSR = Previous slot read bit
REQ = Request priority bits

Figure 6–12
The 802.6 slot.

The slot type (ST) field identifies the slot as containing queued arbitrated (QA) or prearbitrated (PA) traffic. The permissible combinations of the busy and type bits are shown in Table 6–1.

The previous slot read (PSR) bit is used to clear a segment in the previous slot. A node turns this bit to 1 if the previous slot's payload was destined only for that node. The exact method in which the PSR slot is utilized varies, and a number of papers have been written about the subject. One proposal suggests the utilization of erasure nodes on the busses (Yokotani, Sato, & Nakatsuka, 1993). These nodes are responsible for decoding the PSR of a current received slot, and possibly releasing a stored previous slot. Therefore, the erasure node must be able to buffer a slot. If the PSR = 1, the erasure node resets the busy bit of the previous slot, and resets PSR of the current slot which results in a free slot being generated.

Interestingly, both PA and QA operations can use the reserved (R) field, but the MAN specification does not define how PA operations would use priorities vis-à-vis reserved slots. The request (REQ) field contains three bits to support three priority queues (2, 1, 0). A node can request access to a bus by setting one of these bits. Access is granted

Table 6–1   Busy and Type Bits

Busy	Type	Slot State
0	0	Empty QA slot
0	1	Reserved
1	0	Busy QA slot
1	1	PA slot

according to counters. In effect, the counters behave like the RQ and CD counters explained earlier, but allow three queues to be used.

For the PA payload, each octet represents a 64 kbit/s channel, and the slot can contain octets from more than one node, or multiple PA segments can be combined to create higher-speed channels (Yap & Hutchison, 1993). The only requirement is that, since the bandwidth is being shared by PA and QA traffic, care must be taken not to overallocate services vis-à-vis the available bandwidth.

Other Notable Aspects of the MAN

Bandwidth balancing.   At first glance, the DQDB protocol appears to favor upstream nodes over their downstream neighbors in relation to bandwidth allocation. After all, upstream neighbors are allowed to reserve slots and these slots cannot be used by the downstream neighbors on the opposite bus. To remedy this situation, DQDB employs techniques to ensure that non-reserved slots are sent downstream periodically. This approach precludes upstream nodes from monopolizing the bus.

Setting up VCs for PA traffic.   Continuous bit rate (CBR) traffic is assured of bandwidth through the use of prearbitrated (PA) slots. The headend (head of bus) is responsible for reserving these slots for nodes that are sending isochronous traffic, and the headend must guarantee these applications slots on a continuous basis.

Slots are allocated by the nodes examining virtual circuit identifiers (VCIs) inside the segment header. If the VCI is the same as that of the VCI mapped into the node, this slot (or a portion of it) is reserved for this node's use. Since multiple nodes may be allowed to use a slot, the headend provides a set of values to identify which portions of the slot can be used by a node.

Figure 6–13 shows how isochronous traffic capacity is allocated. A bandwidth manager and VCI server are responsible for coordinating isochronous traffic with the head of the bus and a node that provides Q.931 signaling termination. The process begins by an end user device issuing an ISDN SETUP request message to a node with ISDN channel management responsibility (called Q.931 signaling termination).

After analyzing the SETUP message, Q.931 signaling termination sends a request to a node (designated as the bandwidth manager and VCI server) to request an isochronous channel. Next, the bandwidth manager and VCI server then generate the VCI value and information about the required bandwidth and send this information to the head of bus B. The effect of this operation is to give bus B the needed information to reserve slots for the session. Since the head of bus is responsible for slot generation, it builds tables which will allow it to preallocate fixed offsets for the session within a PA slot or slots, and the number of bytes allocated depends on the required channel capacity—x octets for x times 64 kbit/s (Yap & Hutchison, 1993).

The head of bus B confirms the VCI required bandwidth by sending this confirmation to the bandwidth manager and VCI server. Next, the verified VCI and octet offsets (relative position of payload in the slot) are sent by the bandwidth manager to node A which confirms this information. Finally, the bandwidth manager and VCI server send a confirmed isochronous channel message to the Q.931 signaling termination, which then sends back a ISDN Q.931 SETUP ACK to the end user station.

A similar process occurs for the called device. For example, if the called device were node C. It would be informed through these entities of the required bandwidth and VCIs.

Once the allocations have been made and all nodes notified of these allocations, each node examines each slot, then examines the VCI and compares it to a VCI table (created during the allocation process). If a match occurs, it then writes into the slot based on an offset value in the table. Additionally, a connection endpoint identifier (CEPID) is also placed in the table to identify each PA user within a node. Once a VCI match occurs, and an offset value read, a CEPID is accessed to determine from which user buffer a read is to occur. The CEPID entry also contains the x octets for x times 64 kbit/s capacity requirement.

Traffic policing. Even though DQDB provides a scheme to support both PA and QA traffic, it does not define a traffic policing method beyond the discussion provided with Figures 6–5, 6–6, and 6–13. DQDB still needs a congestion control mechanism to allocate the bandwidth

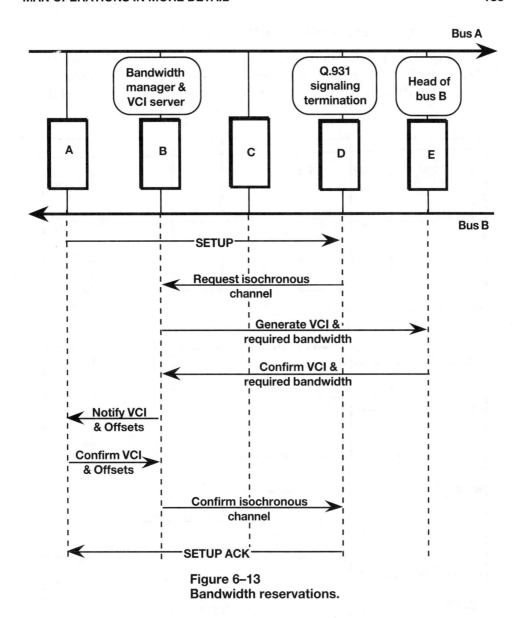

Figure 6–13
Bandwidth reservations.

that is provided with the basic QA and PA operations. Because DQDB is quite similar in this respect to the asynchronous transfer mode (ATM), we shall defer the discussion of traffic policing to the ATM chapter. The bandwidth enforcement and policing mechanisms can be applied equally well to an ATM or an SMDS/MAN architecture.

DQDB Layer Management Interface (LMI)

The 802.6 specification contains a layer management interface (LMI). Figure 6–14 provides a general overview of the relationships of the LMI operations with the layered management entity (LME) and the

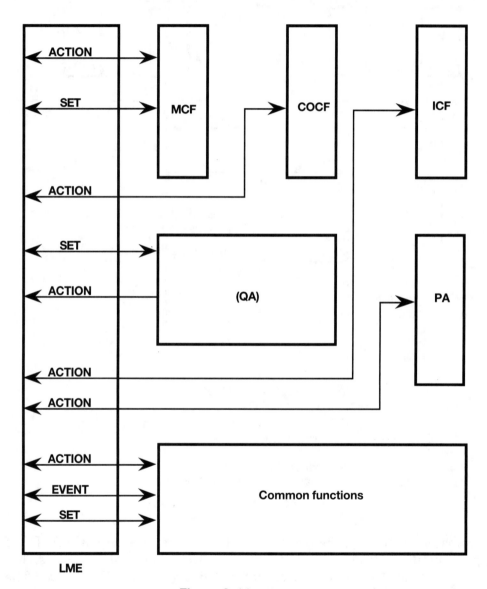

Figure 6–14
Overview of the LMI.

various DQDB functions. LMI is responsible for two operations: (a) management within the local node, and (b) the management of a remote node.

LMI uses primitives which are implemented according to the OSI Model. The LMI uses three primitive types: *Invoke* is passed from a network management protocol (NMP) to the LME to ask for a service; *reply* is returned from the LME to the NMP to provide the results of invoke; and *notify* is passed from the LME to the NMP to notify the NMP about something that happened in the DQDB layer.

The primitives are largely self-explanatory, so this paragraph provides a general summary of their functions. The *set* operation is used to set a parameter to a given value. This parameter, for example, could reside in a management information base (MIB). The *compare and set* operation uses Boolean constructs to set a parameter, given that a certain test has been made and verified. The *get* operation obtains a value of a given parameter, once again, typically from a MIB. The *action* operation is used to force the DQDB layer to form some type of action, or to move from one state to another. Finally, the *event* operation is initiated locally by a DQDB LME to report about an occurrence happening in the DQDB layer; in effect, the event is usually a status report or an alarm.

These general primitives are designed to invoke very specific functions based numerous parameters that are used with each primitive. For example, the action primitive is used to initiate VCIs for PA channels. As another example, it is used to perform slot marking functions at the head of the bus for generating PA slots. It is used to invoke a header extension, establish a list of MID values that are used at a node, and many other services which are beyond this general overview.

If the operations entail activities at other nodes, DQDB also provides a layer management protocol. It specifies management PDUs and their formats and the rules for their transmission and reception. Many services are provided by the protocol, and several of them involve head of bus operations, as well as MID allocations. The protocol also provides configuration operations and directions on how timing is to be obtained on the network.

MAN SUMMARY

The DQDB protocol provides a powerful service for both CBR and VBR applications. Through the use of pre-arbitrated (PA) and queued arbitrated (QA) services, DQDB can support multimedia traffic. The IEEE 802.6 MAN standard is based on the DQDB protocol.

INTRODUCTION TO SMDS

This section of the chapter examines the switched multi-megabit data service (SMDS) and its relationship to the MAN standard. It also examines the implementation of SMDS by the U.S. telephone companies and several European enterprises, and includes an update on the local exchange carriers (LEC) trials and future plans.

THE PURPOSE OF SMDS

SMDS is a high-speed connectionless packet switching service which extends LAN-like performance beyond a subscriber's location. Its purpose is to ease the geographic limitations that exist with low-speed wide area networks (WANs). SMDS is designed to span across LANs, MANs, and WANs.

The major goals of SMDS are to provide high-speed interfaces into customer systems and, at the same time, allow customers to take advantage of their current equipment and software. Therefore, the SMDS operations are not performed by the end user machine; they are performed by customer premises equipment (CPE), such as a router.

SMDS is positioned as a service. If SMDS is considered unto itself, it is not a new technology; it offers no new method for designing or building networks. This statement is emphasized because SMDS uses the technology of DQDB, and then offers a variety of value-added services, such as bandwidth on demand, to the SMDS customer.

SMDS is targeted for large customers and sophisticated applications that need a large communications pipeline, but do not need the pipe for a sustained period. Generally, SMDS is targeted for data applications that transfer "lots of bits" in a bursty manner.

However, applications that use SMDS can be interactive. For example, two applications can exchange information interacting through SMDS, such as an x-ray, a document, etc. The restriction (if it is fair to use this term) of SMDS is based on the fact that SMDS is not designed for real-time, full-motion video applications. Notwithstanding, it does support an interactive dialog between users, and allows them to exchange large amounts of information is a very short time. For example, it takes only about one to two seconds for a high-quality color graphic image to be sent over an SMDS network. For many applications, this speed is certainly adequate.

So, SMDS supports applications that operate with bulk data, CAD/CAM, source code transfer, and most forms of imaging. It is designed to be a niche service, albeit a large niche.

The U.S. LECs are deploying SMDS nationwide with interexchange services provided by the interexchange carrier (IXCs). All major metropolitan areas are to be interconnected through an SMDS subscriber network interface (SNI) and an interexchange carrier interface (ICI).

PERTINENT STANDARDS

SMDS is based on the MAN DQDB standard published as IEEE 802.6. The SMDS specifications are published by Bellcore as shown in Table 6–2.

A TYPICAL SMDS TOPOLOGY

SMDS uses the DQDB protocol at the SMDS SNI. With only a few variations to the IEEE 802.6 standard, SNI defines the procedures for a user's CPE to interface with an SMDS network. Figure 6–15 shows a typical SMDS topology. The SNI, as shown in the figure, can also be referred to as the SMDS interface protocol (SIP), or as an "access DQDB." The SNI operates between the CPE and the SMDS switching system (SS). The term SS replaces the MAN switching system (MSS) term. The SS is a high-speed packet switch supporting operations within a Local Access and Transport Area (LATA). The SS topology can vary and either a centralized or a distributed routing control can be implemented.

The SS must be able to support SMDS on various types of media, but the focus of SMDS is DS1 or DS3 access lines. Each SS communicates with OSs (BOC OSs) via a generic interface across a data communications network (DCN). This interface is called the Operations System/Network Element (OS/NE) interface and its purpose is to provide a standard for SMDS network management operations such as data collection and reporting, naming conventions, memory management, and billing. Multiple SSs may also be interconnected across an Inter-Switching System Interface (ISSI).

The interexchange carrier interface (ICI) interworks two LECs

Table 6–2.   SMDS Standards

TR-TSV-000773, *Local Access System Generic Requirements, Objectives, and Interfaces in Support of Switched Multi-megabit Data Service* (SMDS), Bellcore Technical Reference, Issue 1.

TA-TSV-000774, *SMDS Operations Technology Network Element Generic Requirements,* Bellcore Technical Advisory, Issue 3, February 1991, and Supplement 1, April 1991.

TR-TSV-000775, *Usage Measurement Generic Measurements in Support of Billing for Switched Multi-megabit Data Service* (SMDS), Bellcore Technical Reference, Issue 1.

TA-TSV-001059, *Inter-Switching System Interface Generic Requirements in Support of SMDS Service,* Bellcore Technical Advisory, Issue 1, December 1990.

TA-TSV-001060, *Exchange Access SMDS Service Generic Requirements,* Bellcore Technical Advisory, Issue 1, December 1990.

TA-TSV-001061, *Operations Technology Network Element Generic Requirements in Support of Inter-Switch and Exchange Access SMDS,* Bellcore Technical Advisory, Issue 1, May 1991.

TA-TSV-001062, *Generic Requirements for SMDS Customer Network Management Service,* Bellcore Technical Advisory, Issue 1, February 1991.

Internet Request for Comments, RFC 1209, *The Transmission of IP Datagrams over the SMDS Service,* March 1991.

TA-TSV-001237, *SMDS Generic Requirements for Initial Operations Management Capabilities in Support of Exchange Access and Intercompany Serving Arrangements,* Bellcore Technical Advisory, Issue 1, June 1993.

TA-TSV-001239, *Generic Requirements for Low Speed SMDS Access,* Bellcore Technical Advisory, Issue 1, June 1993.

TA-TSV-001240, *Generic Requirements for Frame Relay Access to SMDS,* Bellcore Technical Advisory, Issue 1, June 1993.

through an interexchange carrier. This interface provides inter-LATA service and is also called Exchange Access SMDS (XA SMDS). ICI is not used within a LATA.

Each SS communicates with a Bellcore Operating System (OS) through generic interface operations (GIOs). At this interface, the operations, administration, maintenance, and provisioning (OAM&P) are defined and include activities such as network management, data collection, testing, and troubleshooting.

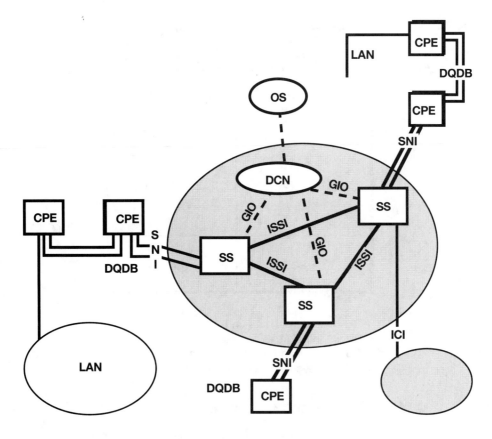

CPE = Customer premises equipment
DCN = Data communications network
OS = Operations systems
ISSI = Interswitching system interface
SNI = Subscriber network interface (AKA SIP or access DQDB)
SS = Switching system
GIO = Generic interface of operations
DQDB = Distributed queue dual bus
ICI = Interexchange carrier interface

Figure 6–15
Typical SMDS topology.

THE SMDS LAYERS

Figure 6–16 shows the SMDS layers. These layers are (with a few minor exceptions) identical to the MAN layers illustrated in Figures 6–3 and 6–7. However, different terms are used. For convenience. The MAN

| SMDS | MAN |
| SIP Layers | 802.6 Layers |

SMDS SIP Layers	MAN 802.6 Layers
L3_PDU	IMPDU
L2_PDU	DMPDU
	Segment
	Slot
L1_PDU	PLCP

PDU   = Protocol data unit
IMPDU = Initial MAC protocol data unit
DMPDU = Derived MAC protocol data unit
PLCP = Physical layer convergence procedure

Figure 6–16
SMDS and 802.6 layers.

layers are also shown in this figure. The reader will notice that the SMDS SIP layers do not divide layer 2 into three sublayers as is done in MAN. However, the same functions are still performed.

SMDS PROTOCOL DATA UNITS

As we did with the discussion of the MAN PDUs, a detailed discussion of the SMDS PDUs is deferred because many of the functions of SMDS revolve around the creation, segmentation, and reassembly of the PDUs. Therefore, these PDUs are discussed in the next section.

SMDS OPERATIONS IN MORE DETAIL

The biggest challenges at the SNI are to control traffic and resolve potential or real congestion problems, provide the user with a high quality of service, and make effective use of all network resources. In describing how SMDS (from the perspective of Bellcore) approaches these challenges, this section explains SMDS's view of congestion, how it is defined, and how it is measured. The next section describes how SMDS controls traffic and allocates bandwidth to the user to meet these challenges.

Defining and Measuring Congestion

Congestion occurs when the network degrades in its performance either in throughput or in delay. Congestion occurs not only by an unexpected increase in user traffic, but also from network failures. In such an event, the network node will either increase its queues in the buffer (which results in increased delay) or, if the buffers become full, discard traffic. Either case is undesirable.

Congestion severity is described by its *time signature,* which is a term to describe how badly performance degrades over a measured time. Time signatures are established over two periods: (a) the onset, and (b) the abatement of congestion. *Congestion onset* refers to the time in which the load at the network increases until it exceeds the capacity of the network to absorb the traffic. In contrast, *congestion abatement* is the time that is reached when the offered load is less than the capacity of the network to absorb it.

The types of onset and abatement are quite important in determining how the network is to react to traffic. As explained in previous parts of this book, each user session with the network likely displays different traffic profiles, yet the network must accommodate to all of these profiles.

Figure 6–17 depicts the ideas of onset and abatement in various combinations. Bellcore believes that five combinations of these scenarios are most likely to occur. The easiest to handle is slow onset with slow, medium, or fast abatement. The slow onset allows the network to adjust to the relatively gradual increase of traffic.

The second type of time signature is the experience of medium onset with slow, medium, and fast abatement. This also is not deemed to be a serious problem because the network will have time, like slow onset, to make adjustments.

The third type is fast onset with fast abatement. This time signature presents significant problems. Even though the abatement is fast, it is anticipated that fast onsets will occur quite sporadically—in bursts, so quickly that a network may not have time to react. An aggregation of

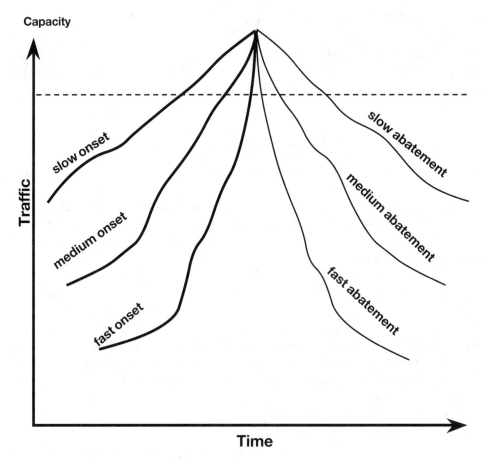

Figure 6–17
Traffic onset and abatement.

fast onsets from multiple users could severely impact the network services.

The fourth type is fast onset with medium abatement and the fifth probable scenario is fast onset with slow abatement. Both cases can present serious problems in that their presence will result in the long term degradation to the network.

Traffic management must be able to identify a time interval that effectively measures packet loss, as well as traffic onset and abatement. For the former, SMDS measures the number of L3_PDUs discarded at the SNI. For the latter, SMDS establishes a mechanism called a *thresh-*

olding strategy that is based on signaling system number 7 (SS7) operations. With this approach, multiple levels of congestion can be observed,
depending on the amount of congestion and its duration. The levels are
identified by congestion thresholds that define machine congestion (MC)
levels, as depicted in Figure 6–18. Each MC is defined by two thresholds:
(a) the *onset threshold* (OT) indicates when the congestion is building up,
and the resource is passing a measurement threshold, for example, passing from MC_1 to MC_2; (b) the *abatement threshold* (AT) indicates when
the congestion is diminishing, for example, passing from MC_2 to MC_1.
Each SMDS SS must support this multiple level thresholding strategy
(of at least K levels, with 4 proposed for K).

Given these parameters, SMDS establishes a model for an algorithm which is applicable to any of the congestion levels described previ-

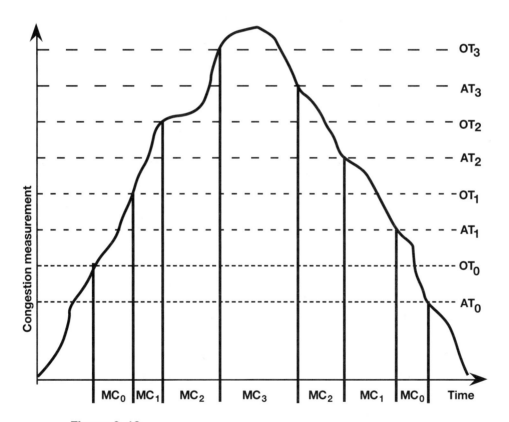

Figure 6–18
**Congestion measurement with thresholding. [Bellcore TA-
TSV-000774]**

ously. The algorithm is triggered when any OT level is passed, which means that the resource is entering into MC_n. The algorithm works as follows: Once OT_i is noted, the signaling system (SS) must send an indication to the OS that the resource has entered an MC_i. Any actions that are needed are performed at this time. Thereafter, when this resource is at MC_i the algorithm must react to three possibilities. The first possibility is that $OT_i + 1$ is crossed, which means that congestion is worsening. For this possibility, the SS must determine how long the resource has stayed at OT_i and the SS must send a signal to OS informing OS that the resource is indeed at $MC_i + 1$. At this point the resource takes any congestion measures that are needed. The second possible situation is that an AT_i is crossed, which indicates that congestion is abating. The requirement for this possibility is that the SS must determine how long the resource stayed at this MC_i and send an indication to the OS informing the OS that the resource is now at $MC_i - 1$. Thereafter, actions taken are pertinent to the situation.

The third possibility is that the resource stays at OT_i and $OT_i + 1$ longer than a specified time which means, of course, that abatement has not taken place. In this situation, the SS must inform the OS that the abatement has not occurred and remedial measures must be taken to force the congestion to abate.

It is evident that the thresholding concept is a very general model and specific procedures must be implemented vis-à-vis the particular interface (the number of users, the capacity of the network, the capacity of the UNI, the capacity of the processors servicing the UNI, etc.). Moreover, for this last situation, in which the OT_i and $OT_i + 1$ is longer than a specified time, the obvious solution is to discard traffic.

Notwithstanding the somewhat abstract approach to this problem, the model does provide a useful tool for the discussion and analysis of the significant problem of traffic management in a demand-driven network. Given these discussions, the next section examines less conceptual, more pragmatic approaches to traffic management in the SMDS network.

The Sustained Information Rate (SIR) and Access Classes

The SMDS traffic management operations are founded on the concept called the *sustained information rate* (SIR). This concept is similar to the committed information rate (CIR) found in the frame relay specification, discussed earlier in this book.

The SIR is founded on access classes which are provided for DS3 circuits, and not for DS1 circuits. The access classes are summarized in Figure 6–19. Each access class identifies the different traffic characteristics for varying applications. The access class places a limit on the amount of sustained information that the CPE can send across the SNI to the SS. It also places a limit on the burstyness of the transfer from the CPE to the SS.

The access classes do not provide a means of using a portion of a DS3 line, because when a CPE sends data across the DS3 line, it is allowed to use the maximum rate that DS3 offers (which is approximately 34 Mbit/s after overhead has been removed). Therefore, a burst of data can occur at the maximum access rate, but the CIR places a limit on the duration of the burst. The access class also places a limit on the average rate of information transferred across the interface.

When user traffic enters the SS, the SS keeps a "balance" of the traffic from each user through a *credit manager*. This credit is measured in octets. If credit is available, the network accepts the traffic. If credit is not available, the SS will not service the user payload.

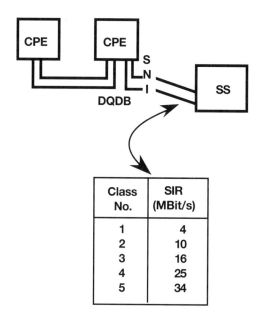

Figure 6–19
Access classes.   SIR = Sustained information rate

The credit manager parameters are set up at subscription time and are defined as follows:

- C_{mac} Maximum credit (in octets) that can accrue for a customer
- I_{inc} Interval between increments to the credit
- N_{inc} Number of octets of credit which is incremented as the credit manager counts incoming L2_PDUs (a SIP level 2 PDU as shown in Figure 6–20)

I_{inc} is measured by a credit manager by keeping count of all (empty or full) L2_PDUs. The I_{inc} and the N_{inc} work in conjunction to control the credit accumulation rate; in effect, they determine the maximum transfer rate that can be sustained for long periods [TR-TSV-000771]. In contrast, C_{max} sets a limit on the burstyness of the data transfer.

The SIR is the average rate of data transfer across the SNI and is computed as:

SIR (Mbit/s) = 8 * (12 L2_PDUs/125 µsec.) * (N_{inc} octets/ I_{inc} L2_PDUs)
where: 8 = bits per octet

Bellcore TR-TSV-000772 provides values for N_{inc}, I_{inc}, and C_{max}, as well as the SIR values associated with the SMDS access classes. This information is shown in Table 6–3. As an example of how the SIR is calculated, consider a 4 Mbit/s token ring LAN. The formula yields:

$$3,993,600 = 8 * (12/.000125) * (26/5)$$

Table 6–3   Bellcore Values for SIR and Access Classes

		Parameters		
Access Classes	**SIR**	N_{inc}	I_{inc}	C_{max}
	(Mbit/s*)	(Octets)	(L2_PDU Counts)	(Octets)
1	4	26	5	9188
2	10	13	1	9188
3	16	21	1	9188
4	25	32	1	9188
5	34	NA	NA	NA

* Approximate value

Since C_{max} establishes a limit for the accrual of a user's traffic, it allows for the user to burst a number of L3_PDUs (a SIP level 3 PDU as shown in Figure 6–20) for classes 1 through 4 before the credit limit is reached.

As depicted in Figure 6–19, classes 1, 2, and 3 will allow the customers equipment to send traffic at the sustained rate of 4, 10, and 16 Mbit/s. The approach is to support 4 Mbit/s IEEE 802.5 LANs, 10 Mbit/s 802.3 LANs, as well as 16 Mbit/s 802.5 LANs. Class 4 allows for an SIR between 16 and 34 Mbit/s which is designed to support equipment with high bit rate requirements that do not require a full DS3 channel. With class 5, there is no enforcement provided at the SNI. The SIR is the maximum bandwidth achievable across the SIP in a DS3 environment which is approximately 34 Mbit/s.

SIP Segmentation and Encapsulation Functions

The encapsulation and decapsulation functions of SIP are quite similar to MAN's 802.6 specification. Figure 6–20 shows SIP level 3 and 2 PDUs. The SMDS service data unit (SDU) is passed to layer 3. This layer is responsible for appending appropriate address information and providing for additional control information. The traffic is then passed down to layer 2 which is responsible for further segmentation and the creation of additional headers and trailers.

In conformance with the OSI concepts, the SMDS SDU is a parameter which contains the data which is passed between the layers in the same machine. In contrast, the PDU is a unit of data that is exchanged horizontally between peer entities in the same layer of different machines.

The format of the SMDS level 3 PDU is almost identical to its counterpart in the 802.6 MAN specification. Figure 6–21 shows the content of this PDU. Note that the ×+ means that this field is not processed by the SMDS network. The field resides in the PDU to insure the operability with the DQDB protocol format.

We proceed with a brief description of each field in this PDU. The *reserved* field (as its name suggests) is not used. For SMDS, these fields are set to all zeros. The *BEtag* resides both in the header and trailer. Their fields can range from 0 to 255, and are used to ensure that the first and last segments of a segmented L3_PDU can be associated with each other. The *BAsize* contains the length of octets that extends from the destination address up to and including the CRC 32 field.

The *destination address* and *source address* are 8-octet fields. They

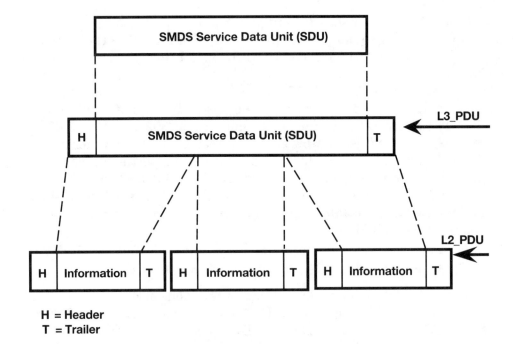

Figure 6–20
SIP L3_PDU and L2_PDU.

both contain two subfields, the *address type* and the *address*. The address type field is 4-bits in length, the address is 60-bits in length. SMDS uses a 10-digit address with a prefix of 1 before the address in accordance with the ITU-T E.164 and the North American Numbering Plan (NANP). The address type field is coded as 1100 for an individual address and 1110 for group addresses. The bits that follow the 1100 code contain a prefix of 1 and 10 digits of binary coded decimal (BCD) values. The remaining bits are coded as 1s.

The *high level protocol identifier* (HLPI) is present to ensure alignment with the DQDB protocol. The *PAD length* (PL) field indicates that the octets in the PDU are 32-bit aligned. It is used with a simple formula to obtain this service.

The *quality of service* (QOS) field is not used, and is present to ensure alignment with DQDB. The *CRC 32 indication bit* (CIB) is used to indicate the absence or presence of the CRC32 field. If the CRC32 is used, the bit is set to 1.

The *header extension length* (HEL) is a 3-bit field that indicates the number of 32-bit words that exist in the header extension field. It is set to 011 by both the SS and the CPE.

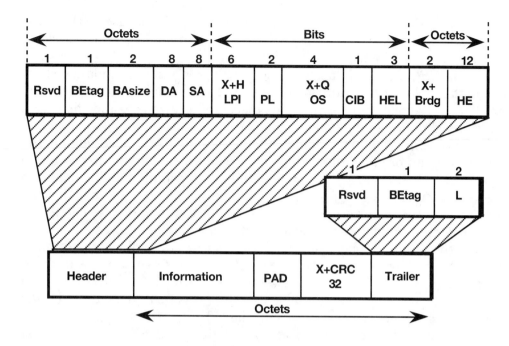

Figure 6–21
The SMDS header and trailer at level 3.

The *bridging* field is used to insure alignment with DQDB. The *header extension* (HE) field contains variable information. It is used by any protocol in any manner deemed appropriate. Discussion of the header extension field is deferred to a latter part of this section.

The information field contains end user data (the user SDU). The *PAD* field is used to make the entire PDU align on 32-bits as indicated by the *pad length* field.

The *CRC32* field is used for performing a cyclic redundancy check from the destination address through CRC32. Finally, the trailer contains the reserved and BEtag fields discussed previously, and the *length* field, which is the same value as the *BAsize* field.

To continue the discussion of the *header extension* (HE) field, it can be used by an organization in any way deemed fit for the processing of the layer 3 PDU (L3_PDU). For the SMDS protocol, the header extension is used to indicate: (a) the version of SMDS, (b) carrier selection, (c) reserved, or (d) use by other entities, such as an interexchange carrier.

The *version* part of the header extension field identifies which version of the protocol is being utilized. *Carrier selection* allows the originator to define how an interexchange carrier is selected to provide inter-LATA service. The protocol allows the choice on a preselection or an overriding basis. The overriding basis allows this field to be used and the end user can select the carrier within the PDU.

Figure 6–22 shows the SIP layer 2 PDU (L2_PDU). It is almost identical to the 802.6 MAN PDU. Therefore, only a summary description is provided of its contents. The *access control* (AC) field contains a number of bits to control the DQDB, which is not shown on the MAN PDU. A one-bit flag is used to indicate if the layer 2 PDU contains busy information or is empty (in which case the flag is set to 0). The *network control information* (NCI) is a 4-octet field which is not used in SMDS.

The *segment type* field indicates the relative position of the segmented SDU within the original data unit. It can be set to indicate the continuation of a message (00), the end of message (01), the beginning of message (10), or a single segment message (11). The *sequence number* (SN) is used to sequence the traffic.

The *message identifier* (MID) field is used to identify the segments associated with the layer 2 PDU. This value must be the same for all layer 3 PDUs that have been segmented. The *information* field contains a layer 3 PDU.

The trailer includes the *payload length* (PL) field which indicates how many of the 44-octets of the information field (the segmentation unit) contain data. Finally, the *payload CRC* (PC) field is a 10-bit field which performs a CRC check on segment type, sequence number, message ID, information field, payload length, and PC field. It does not perform the calculation on access control or the network control information fields.

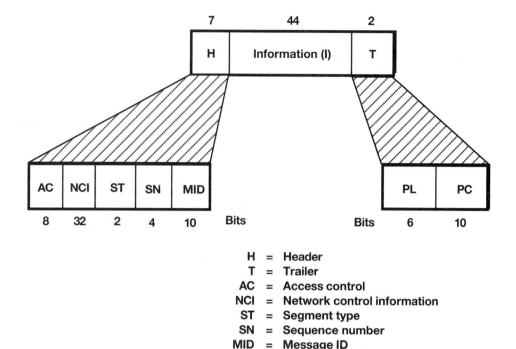

Figure 6–22
The SMDS layer 2 PDU.

At the receiving SMDS node (CPE or SS), extensive editing operations are performed on the L2_PDU and L3_PDU headers and trailers (such as CRC checks, proper payload lengths, BEtag mismatches, proper addresses, etc.). If the L2_PDU passes edit tests, the MIDs and SNs are used to reassemble the L2_PDUs into an L3_PDU. If the edit tests pass on the L3_PDU, the two BEtags are used at the L3_PDU layer to ensure further that the full SMDS SDU has arrived. The headers and trailers of the L3_PDU are then used to determine other actions (such as address screening, interexchange carrier selection, etc.).

SNI Quality of Service (QOS) Operations

SMDS defines a number of performance and QOS goals to be performed by the SMDS network on behalf of a CPE. These operations occur across the entire network within one LATA from one SNI to the other

SNI. These operations are divided into three groups: availability, accuracy, and delay.

Availability deals with the ratio of the actual service time provided to the scheduled service time. It includes an objective for a scheduled service time, which defines the length of time in hours that the network is expected to be in service. As the reader may expect, the scheduled service time is 24 hours a day, 7 days a week. It also specifies the mean time between service outages (MTBSO), which describes the average duration for any continuous interval when the service is provided. This figure is stated to be no less than 3,500 hours, which corresponds to an average of no more than 2.5 outages per year. Availability objectives also entail the mean time to restore (MTTR), which measures the time a service is lost until the time that it is fully restored. The MTTR is defined to be no more than 3.5 hours.

The *accuracy* objectives describe operations dealing with lost traffic, misdelivered traffic, duplicated traffic, missequenced traffic, etc. These statistics and services include objectives for both L3_PDUs and L2_PDUs. These ratios include the ratio of the number of incorrect data units delivered across an SNI to the number of successfully delivered PDUs. It also includes ratios of misdelivered, not delivered, duplicated, and missequenced PDUs.

The third major category is the *delay* objectives, which also cover L3_PDUs and L2_PDUs. For L3_PDUs, performance is based on individual addressing and include all PDUs that have a maximum length of 9188 octets of payload. 95 percent of all L3_PDUs delivered should traverse the SNI-to-SNI in less than 20 msec. This delay objective is for a complete DS3 connection between the SNIs. For topologies in which one CPE uses a DS3 path and the other uses a DS1 path, the delay should be less than 80 msec. Finally, for 2 CPEs both using DS1 paths, the delay should be less than 140 msec. For group addresses, the delay for these three configurations are 100 msec., 160 msec., and 220 msec., respectively.

THE INTEREXCHANGE CARRIER INTERFACE (ICI)

The purpose of the interexchange carrier interface (ICI) is to internetwork Bellcore Client Companies (BCCs) through interexchange carriers (IXCs). As depicted in Figure 6–23, the interexchange carrier switching system connects two BCCs (BCC A and BCC B) to the ICI protocol

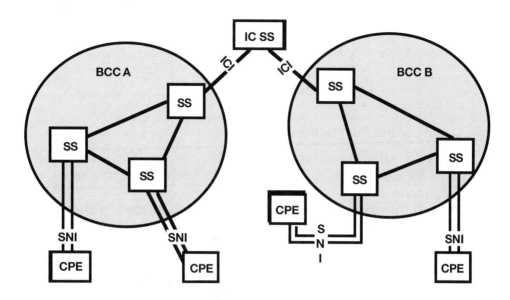

CPE = Customer premises equipment
BCC = Bellcore company
SNI = Subscriber network interface (AKA SIP or access DQDB)
SS = Switching system
ICI = Interexchange carrier interface

Figure 6–23
The Interexchange carrier interface (ICI).

(ICIP). The ICIP is responsible for providing the exchange access SMDS service (XA-SMDS).

The ICIP is based on the IEEE 802.6 standard. Layer 1, however, is based on DS3 or SONET STS-3c transmission paths. Layers 2 and 3 are almost identical to the 802.6 DQDB layers. Therefore, we shall not spend a great deal of time discussing all the headers and trailers which have been covered extensively in previous sections of this chapter.

The ICI was designed to accomplish two principal ideas: First, to process user traffic quickly and as efficiently as possible by minimizing the cell processing between the SS and the IC SS; and second, to define QOS features between the SS and the IC SS.

When a BOC SS receives SIP LU_2 PDUs, it processes these data units with minimal conversion. The first SIP LU_2 PDU is used to create a new ICIP level 3 header. This header is sent as the first ICIP L2_PDU across the ICI to the IC SS. The approach is to occupy the ICIP level 3 header in the entire 44-octets of the first ICIP L2_PDU. Thereafter, the SIP L2_PDUs become the ICIP L2_PDUs. Processing continues until the ICIP level 3 trailer is formed in the last ICIP L2_PDU. This completes the processing of the initial MAC PDU. Reviewing Figure 6–20 might aid the reader in following this discussion.

Quality of Service (QOS) Objectives

The ICIP stipulates a number of QOS and performance objectives that must be met by the IC SS and the SS. The operations dealing with these objectives are quite similar to those described earlier at the SNI. For example, performances dealing with service availability, accuracy, and delay are also stipulated for this interface. When reading this material, the reader may wish to refer back to the SNI QOS objectives described earlier in this chapter.

Briefly, the ICI requires that the mean time between service outages be no less than 2917 hours, which corresponds to no more than 3 service outages per year. The mean time to restore is to be no more than 3 hours. For accuracy objectives, the erred ICIP-SIP L3_PDU ratio should be less than $5 * 10^{-13}$. The misdelivered ICIP-SIP L3_PDU ratio should be less than $5 * 10^{-8}$. The not delivered ratio should be less than 1 to 10^{-4}, the duplicated ICIP-SIP L3_PDU ratio should be less than $5 * 10^{-8}$. And finally, the missequenced ICIP-SIP L3_PDU ratio should be less than $1 * 10^{-9}$. The delay objectives, once again, are quite similar to those existing at the SNI, and the delay ranges from 10 msec. to 160 msec. depending on if the paths used are DS3-based or DS1-based.

OTHER NOTABLE ASPECTS OF SMDS

SMDS Address Management Operations

SMDS ensures that a user cannot employ a fraudulent source address. Each SS is required to validate all SMDS addresses that are serviced at that SS. The SS maintains a table for valid addresses and uses it to verify that the source address from a sending CPE is legitimately assigned to the SNI.

In addition, SMDS provides address screening operations. These operations ensure that unwanted traffic is not sent to a CPE. The SS maintains two types of address lists for the purposes of address screening: individual address screens and group address screens. The individual address screen checks the destination addresses sent by the CPE and source addresses delivered to the CPE. A screening table consists of a set of allowed addresses for the CPE or a set of disallowed addresses.

Likewise, a table for group addresses is used for screening all destination addresses sent by this CPE to the SS. It too contains either a set of allowed addresses or a set of disallowed addresses.

Address management is divided into two categories: calling address security and called address privileges. Calling address security provides for calling identifier validation, maximum calling attempts allowed, a time-out value to terminate a session if no transactions have been sent during a measured time interval, and a maximum time allowed between successive set-up attempts. Called address privileges provide for privilege codes which identify the functions and authorizations permitted for this address (the exact functions and authorizations are implementation-specific).

The ISSI

One of the objectives of this book is to describe the SNI and network-to-network (NNI) for the emerging communications technologies. The SMDS specification contains yet another interface, known as the *interswitching system interface* (ISSI). To gain a system perspective of ISSI, refer back to Figure 6–15, which shows this interface between the SSs. Figure 6–24 shows a more detailed view of the ISSI and depicts its major functions. Stated another way, the purpose of the ISSI is to define the operations that reside between SSs within a LATA.

The ISSI operates with layers 2 and 3 of the IEEE 802.6 standard. The physical layer consists of a DS3 or SONET STS-3c. The ISSI cell consists of the 48-octet PDU and a 5-octet ISSI cell header. The ISSI cell header is quite similar to the ATM cell header, discussed later in this book. It contains fields to identify a virtual circuit ID (VCI), payload type, segment priority, and a header check. All also are supported in the ATM technology. In addition, the ISSI defines load-splitting operations that may be necessary if multiple ISSI links are deployed between SSs to meet the user's performance objectives. To complete this brief overview, route discovery between the SSs is performed with the open shortest path first (OSPF) protocol, a specification published by the Internet authorities. For more detail, the reader is referred to Bellcore TA-TSV-001059.

The generic interface of operations (GIO) is used between SSs and the DCN. The GIO is used between SSs and the DCN. The GIO is

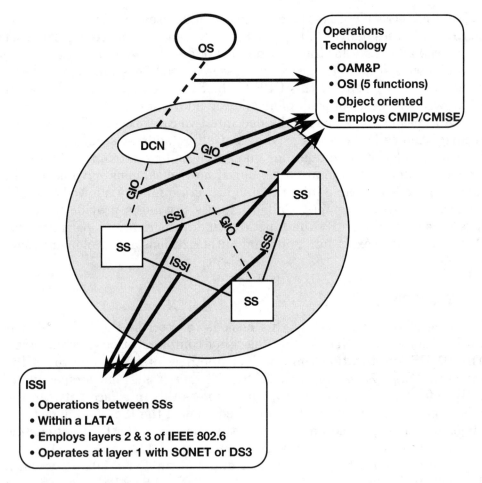

Figure 6–24
The ISSI and operations technology interfaces.

responsible for activities including network management, data collection, testing, troubleshooting, etc., which are explained in more detail in
the section on operations technology which follows.

The Operations System/Network Element (OS/NE) Interface (Operations Technology)

The SMDS specification contains still another interface, known as
the Operations System/Network Element (OS/NE) Interface. To gain a
system perspective of this interface, refer back to Figure 6–15, which

shows the operations between the OS, the DCN and the SS. Figure 6–24 also shows a more detailed view of this interface and its major functions. The purpose of the OS/NE is provide a standard for SMDS network management operations, such as data collection and reporting, naming conventions, memory management, and billing, through the use of the OSI's CMIP/CMISE (common management information protocol/common management information service element) network management protocol. All these are called operations technology.

It is beyond the scope of this general text to describe the ISSI in detail. This section provides an overview of its major attributes. For more detail, the reader is referred to Bellcore TA-TSV-000774.

Operations technology encompasses the following SMDS "domains":

- Memory administration: Managing NE databases
- Maintenance: Monitoring, diagnosing, and analyzing the equipment, path and services
- Traffic management: Monitoring network traffic for control measures
- Data collection: Monitoring network traffic for capacity planning
- Customer network management (CNM)   Providing the customer with relevant statistics

The operations technology domains are correlated to the OSI five network management functional elements. These elements are described here. First, *fault management* is used to detect, isolate, and repair problems. It encompasses activities such as the ability to trace faults through the system, carry out diagnostics, and act upon the detection of errors in order to correct the faults. It is also concerned with the use and management of error logs. Fault management also defines how to trace errors through the log and time stamping of the fault management messages.

Accounting management is needed in any type of shared resource environment. It defines how network usage, charges, and costs are to be identified. It allows users and managers to place limits on usage and to negotiate additional resources.

Configuration management is used to identify and control SMDS elements. It defines the procedures for initializing, operating, and closing down these elements, and the procedures for reconfiguring them. It is also used to associate names with managed elements and to set up parameters for the elements. It collects data about the operations in the open system in order to recognize a change in the state of the system.

Security management is concerned with protecting SMDS elements. It provides the rules for authentication procedures, the maintenance of access control routines, and the maintenance of security logs.

Performance management supports the gathering of statistical data and applies the data to various analysis routines to measure the performance of the system.

The relationship of the SMDS operations technology domains and the OSI functional areas is provided in Table 6–4, which is derived from Bellcore TA-TSV-00074.

SMDS managed object classes. Since SMDS uses the OSI Network Management Model, its managed resources are called managed objects. These objects belong to managed object classes (which are also known as object classes). The managed object is a network resource that is to be monitored and managed, such as an SNI, the state of the SNI, L3_PDUs processed at the SNI (sent, received, and discarded), etc.

A group of similar objects belong to an object class. Object classes contain attributes that are similar in their association with the objects in the class. Object classes are a representation of data in a database. The database is defined by a management information base (MIB) specification, which is covered shortly.

Bellcore's SMDS has defined a lengthy and extensive set of objects, object classes, and their associated attributes. Their examination is well beyond this general text, so Table 6–5 summarizes the major features of the SMDS object groups.

Table 6–4   OSI Network Management and Operations Technology

OSI Function	Maintenance	Traffic Management	Data Collection	Memory Administration	CNM
Performance	X	X			X
Fault	X	X			X
Configuration	X	X		X	X
Security				X	X
Accounting					X

Table 6–5   SMDS Managed Object Classes

SMDS-Specific

Subscriber access: IDs, addresses, access class, state information for each instance on the SNI

Address screening: Parameters to support calling and called address management

Maximum concurrent data units: Maximum PDUs that can be sent between the CPE and SS

Group address: Parameters to support group addressing

Access class table: Parameters for access table (see Figure 6–19)

SS trunks: Parameters associated with trunks, such as ID, CO-responsible, trunk speed, etc.

Routing selection table: Parameters for a table to route L3_PDUs through the network

Generic Memory Administration

Database change report: Parameters for report, such as source, date, time, objects affected, etc.

Database capture report: Parameters to support the storage of reports

Memory: Parameters that define the binary information stored in the system

Backup schedule: Parameters that define when a backup is taken of the software and data

Mediation service: Parameters internal to the SS supervisory activities

Security Administration

User-related security: Security parameters for all users, such as login ids, privilege codes, etc.

Calling address security: Parameters to support calling address screening

Address privileges: Parameters to support called address screening

Service-related security: Security parameters associated with CMIP/CMISE operations (see below)

Object class access security: An additional level of access control beyond CMIP/CMISE

Attribute security: Privilege codes associated with an object class instance and attributes

CMIP/CMISE operations. Earlier references were made to the employment of CMIP/CMISE for SMDS network management. The conventional CMIP/CMISE operations are employed and are used as follows:

- M-CREATE creates an instance of a managed object within the system (obviously CMIP/CMISE is object-oriented with the use of creation capabilities).
- M-DELETE deletes one or more instances of a managed object in a system.
- M-GET retrieves attributes of an existing instance of a managed object.
- M-SET replaces, adds, and removes attribute values of an existing instance in an object class. It may also set attributes to default values.
- M-ACTION requests a specialized function to be performed on an existing instance of an object class.
- M-EVENTS-REPORT reports on an occurrence of an event pertaining to an existing instance of an object class.

The reader may recognize many of these operations, as they are quite similar to the widely used simple network management protocol (SNMP). The principal difference is that SNMP provides no features for create and delete actions, nor does it have inheritance characteristics (which afford the designer to use object-oriented programming). Notwithstanding, the operations of CMIP, CMISE, and SNMP are similar.

Relationship of User Protocols and SMDS

Figure 6–25 shows the conceptual model for running TCP/IP over SMDS. TCP/IP is chosen for this example because of its wide use in the data communications environment. User traffic is sent from the end system across the LAN to (typically) a router. The router terminates the LAN physical and MAC operations by removing their headers. It may also terminate the LLC operations, but some implementations of this topology (albeit few) run LLC end-to-end.

Whatever the case with LLC, the SMDS interface protocol (SIP) resides at the router. It does not reside in the end user device. Therefore, this implementation is quite simple from the standpoint of the user, because the task of integrating SMDS rests with the router. The router

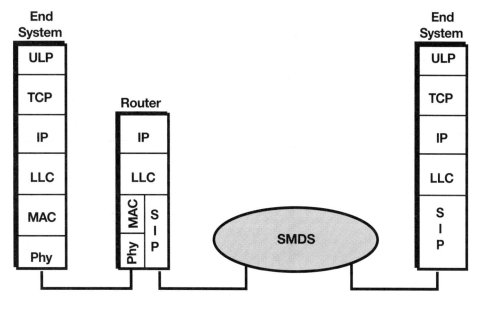

Figure 6–25
TCP/IP and SMDS.

is charged with the job of encapsulating the user traffic into a DQDB slot and transporting it to the SMDS network node. It must interpret an IP destination address in the IP datagram into an address acceptable to SIP, such as E.164. Once the user traffic is passed off to the network, and until it is reassembled at the receiving SIP, the IP, TCP and upper layer protocol headers and user payload are not examined and are passed transparently through SMDS.

The Customer Network Management (CNM) Service

SMDS provides a network management service, called the customer network management (CNM) service. Presently, CNM operates with the SNMP and a MIB published through the Internet authorities. The authentication service is provided to ensure SNMP messages are processed properly. This service and SNMP operate over the user datagram protocol (UDP) (see Figure 6–26).

Through the use of the SMDS MIB and SNMP, the SMDS user is provided with a wide array of network management messages. As examples, information is provided on the number of SIP L3 _PDUs that have been processed successfully or discarded. For those PDUs discarded,

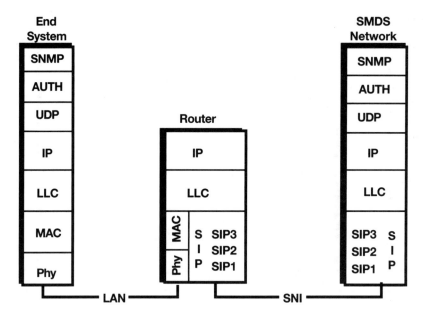

Figure 6–26
SMDS customer network management (CNM) service.

diagnostics are available to determine the reason. Statistics are also provided on SIP L3_PDUs that have not passed address screening, that have exceeded the threshold allowed for the amount of traffic transmitted, and many other network management messages.

SNMP is also used by BCC agents to signify when (a) several subscription parameters have been changed (such as access class), (b) individual addresses have been changed at an SNI, or (c) excessive authentication errors have occurred. The next section expands these examples with a discussion of the SMDS MIB.

THE SMDS MIB

Figure 6–27 shows the location of the MIB in relation to the SMDS network. The SIP layers are also shown in this figure to aid in reading the following material. The MIB is organized around managed objects

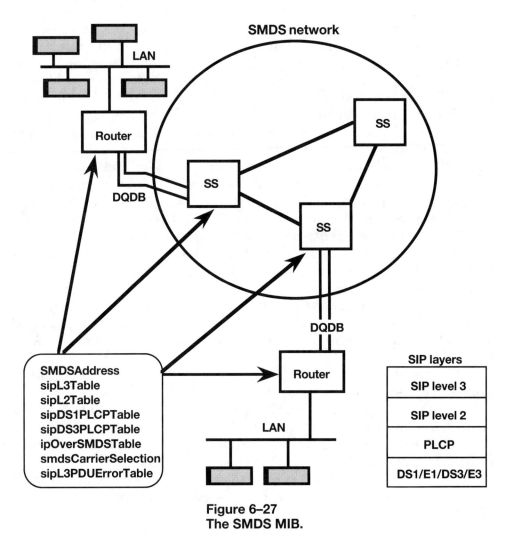

SMDS network

LAN

SS

Router

SS

DQDB

SS

DQDB

SMDSAddress
sipL3Table
sipL2Table
sipDS1PLCPTable
sipDS3PLCPTable
ipOverSMDSTable
smdsCarrierSelection
sipL3PDUErrorTable

Router

LAN

SIP layers
SIP level 3
SIP level 2
PLCP
DS1/E1/DS3/E3

Figure 6–27
The SMDS MIB.

which are contained in major groups. The groups, in turn, are defined in tables. As shown in Figure 6–27, at the bottom of the lefthand side, the major entries in the MIB are the SMDS address, which is the conventional 60-bit SMDS address preceded by the 4-bits to signify individual group addresses. Thereafter, the groups are listed with their object identifier name. These names will be used in this discussion.

The *sipL3Table* contains the layer 3 (L3_PDU) parameters used to manage each SIP port. It contains entries such as port numbers, statistics on received traffic, information on errors such as unrecognized addresses, as well as various errors that have occurred at this interface.

The *sipL2Table*, as its name implies, contains information about layer 2 (L2_PDU) parameters and the state variables associated with each SIP port. It contains information such as the interface port number, information on the amount of the number of level 2 PDUs processed, error information such as violation of length of PDU, sequence number errors, MID errors, etc.

The *sipDS1PLCPTable* contains information on DS1 parameters and state variables for each port. The entries in the table contain error information such as DS1 severely erred framing seconds (SEFS), alarms, unavailable seconds encountered by the PLCP, etc.

The *sipDS3PLCPTable* contains information about DS3 interfaces and state variables for each SIP port. Like its counterpart in DS1, this table contains information on severely erred framing seconds, alarms, unavailable seconds, etc.

The *ipOverSMDSTable* contains information relating to operations of IP running on top of SMDS. It contains information such as the IP address, the SMDS address relevant to the IP station, addresses for ARP resolution, etc.

The *smdsCarrierSelection* group contains information on the inter-exchange carrier selection for the transport of traffic between LATAs.

Finally, the *sipL3PDUErrorTable* contains information about errors encountered in processing the layer 3 PDU. Entries such as destination error, source error, invalid BAsize, invalid header extension, invalid PAD error, BEtag mismatch, etc. form the basis for this group.

As stated earlier, SNMP is used to monitor the MIBs and report on alarm conditions that have occurred based on the definitions in the MIBs.

MAN/SMDS WORKSHEET

The worksheet for MAN/SMDS is provided in Table 6–6. As suggested in the introduction to this book, it is recommended the reader attempt to fill out a blank sheet (provided in Appendix D), and then read this section to check the answers.

MAN/SMDS are relatively new technologies since they employ the DQDB technology. In addition, the cell-based PDUs have not seen much use in the industry until recently. The SMDS is not topology dependent,

Table 6–6 MAN/SMDS Worksheet

Technology name:	MAN/SMDS
New technology?	Yes, DQDB is new
Targeted applications?	MAN targets multimedia; SMDS targets high-speed data transfers and internetworking LANs
Topology dependent?	If physical layer of IEEE 802.6 used, yes; otherwise, no
Media dependent?	No
LAN/WAN based?	Neither, supports interconnections of LANs across WANS and MANs
Competes with:	Leased lines, frame relay at lower transfer rates, private networks
Complements:	LAN internetworking
Cell/Frame based?	Cell, which is called a slot
Connection management?	None, connectionless
Flow control (explicit/implicit)?	Implicit, with the SIR
Payload integrity management?	No
Traffic discard option?	Yes
Bandwidth on demand?	Yes
Addressing/identification scheme?	Yes, ISDN addresses

unless the implementor chooses to use the DQDB physical layer. Nothing prevents layers 2 and 3 of SMDS from residing on other types of topologies, such as a point-to-point SONET carrier. The technology is neither a LAN- nor WAN-based system. Indeed, it can be thought of as a hybrid, and implementations of SMDS may be used to interconnect LANs, MANs, and WANs. It is too soon to know (at the writing of this book) what technologies will be competing with SMDS. But it is quite likely that SMDS will compete with private leased lines, private networks, frame relay at DS1 rates, and perhaps X.25 interfaces that are using high transfer rates. The remainder of the answers in Table 6–6 are self-explanatory, so I shall not embellish this explanation further.

SMDS SUMMARY

SMDS is a connectionless high-speed transport service for data applications. It provides high-speed pipelines for LAN interconnections, as well as DS1 and DS3 interfaces. It supports bandwidth on demand, as well as numerous QOS features, such as the ability to negotiate delay and a certain sustained information rate. SMDS is targeted by the Bell Operating Companies to provide enterprises with a flexible VPN for data transfer.

Asynchronous Transfer Mode (ATM)

INTRODUCTION

This chapter examines the asynchronous transfer mode (ATM) protocol. The ATM cells, the UNI, and the SNI are explained, as well as the ATM multiplexing and routing operations. The rationale for the 53-octet cell size is analyzed, and the issues of delay, error correction/detection, and synchronization are examined. Emphasis is placed on the ATM Forum activities, since this body has assumed the lead in defining the ATM standards.

THE PURPOSE OF ATM

The purpose of ATM is to provide a high-speed, low-delay, multiplexing and switching network to support any type of user traffic, such as voice, data, or video applications. ATM is one of four fast relay services that are covered in this book, the other three are frame relay, MAN, and SMDS.

ATM segments and multiplexes user traffic into small, fixed-length units called cells. The cell is 53-octets, with 5-octets reserved for the cell header. Each cell is identified with virtual circuit identifiers that are contained in the cell header. An ATM network uses these identifiers to relay the traffic through high-speed switches from the sending customer premises equipment (CPE) to the receiving CPE.

ATM provides limited error detection operations. It provides no retransmisson services, and few operations are performed on the small header. The intention of this approach—small cells and with minimal services performed—is to implement a network that is fast enough to support multi-megabit transfer rates.

The ITU-T, ANSI, and the ATM Forum have selected ATM to be part of a broadband ISDN (BISDN) specification to provide for the convergence, multiplexing, and switching operations. ATM resides on top of the physical layer of a conventional layered model, but it does not require the use of a specific physical layer protocol. The physical layer could be implemented with SONET/SDH, DS3, FDDI, CEPT4, and others. However, for large public networks, SONET/SDH is the preferred physical layer.

PERTINENT STANDARDS

A number of documents are pertinent to ATM. Most of these documents are published by the ITU-T, ANSI, and the ATM Forum. They are listed in Table 7–1. The ATM standards have been in development since 1984, and the ITU-T and the ATM Forum completed many of their standards in 1993. The ATM Forum has also published a BISDN Intercarrier Interface (B-ICI) for inter-LATA operations in the United States.

AN ATM TOPOLOGY

Before an ATM topology is examined, several definitions are in order. As just stated, ATM is part of a BISDN which is designed to support public or private networks. Consequently, ATM comes in two forms for the *user-to-network interface* (UNI):

- A public UNI defines the interface between a public service ATM network and a private ATM switch.

Table 7–1   ATM Standards

T1S.1/92-185 "Broadband ISDN User-Network Interfaces: Rates and Formats Specification," ANSI Draft Standard, March 1992.

T1S1.5/92-002R3 "Broadband ISDN ATM Aspects—ATM Layer Functionality and Specification," ANSI Draft Standard, May 1992.

ITU-TS Recommendation I.413 B-ISDN "User-Network Interface," Matsuyama, December 1990.

ITU-TS COM XVIII-R91-E "Annex 3 of Report of Study Group XVIII/8," Melbourne, February 1992, pp 103–105.

ITU-TS Recommendation I.610 "OAM Principles of B-ISDN Access," Matsuyama, December 1990.

ITU-TS COM XV111-R91-E "Annex 3 of Report of Study Group XVIII/8," Melbourne, February 1992, pp 85–102.

ITU-TS COM XV111-R56-E "Report of Working Party XVIII/8—General B-ISDN Aspects," January 1991.

ITU-TS COM XV111-R70-E "Report of Working Party XVIII/8—General B-ISDN Aspects," January 1991.

ITU-TS COM XV111-R91-E "Annex 3 of Report of Study Group XVIII/8," Melbourne, February 1992, pp 85–102.

ITU-TS Q.93B "B-ISDN Call Control"

ITU-TS Q.SAAL "Signaling AAL"

ITU-TS Q.SSCOP "Service Specific Connection-Oriented Protocol"

The ATM Forum, ATM User-Network Interface Specification, Version 2.0, June 1, 1992.

The ATM Forum, BISDN Inter Carrier Interface (B-ICI), Version 1.0, June 1, 1993.

- A private UNI defines an ATM interface with an end user and a private ATM switch.

This distinction may seem somewhat artificial, but it is important because each interface will likely use different physical media and span different geographical distances.

Figure 7–1 shows an ATM topology. The reader may have noticed that the title to this section is slightly different from other chapters. I

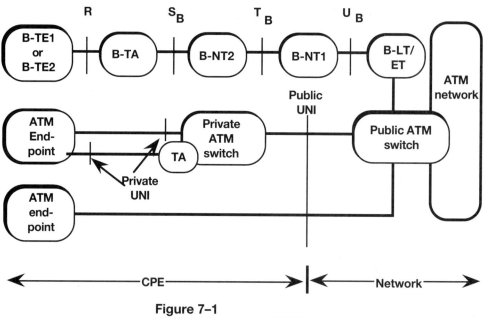

Figure 7-1
An ATM topology. [ATM92a]

have excluded the word "typical" because the topology depicted in Figure 7-1 is a conceptual model, as viewed by the standards groups and the ATM Forum. Furthermore, at this embryonic stage in the evolution of ATM, there is no such thing as a typical typology (but Figure 7-2, explained shortly, shows a likely topology).

It is obvious from a brief glance of Figure 7-1 that the ATM interfaces and topology are organized around the ISDN model. The UNI can span across public or private S_B, T_B, and U_B interfaces (where $_B$ means broadband). Internal adapters may or may not be involved. If they are involved, a user device (the B-TE1 or B-TE2) is connected through the R reference point to the B-TA. B-NT2s and B-NT1s are also permitted at the interface with B-NT2 considered to be part of the CPE. For purposes of simplicity, the picture shows only one side of an ATM network. The other side could be a mirror image of the side shown in Figure 7-1, or it could have variations of the interfaces and components shown in the figure.

Figure 7-2 shows a likely ATM-based topology. The topology is a star, point-to-point topology, but nothing precludes the use of other topologies such as multipoint configurations. As stated briefly in the

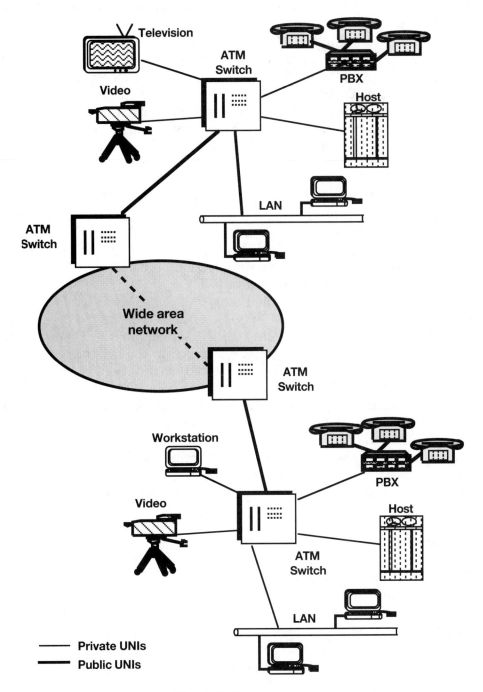

Figure 7–2
A likely ATM topology.

223

introduction to this chapter, ATM is designed to support multimedia service. Its functions permit the switching of voice, video, and data traffic through the same switching fabric. The interconnection of LANs can be supported by the ATM technology because it has convergence and segmentation reassembly operations for connectionless data. Convergence services are also provided for fixed bit-rate video and variable bit-rate voice operations. This figure also shows that ATM can act as a local area network hub or backbone, or a wide area network switch.

The VPI and VCI labels.   The user traffic at the UNI is identified by two values in the cell header: (a) the virtual channel identifier (VCI), and (b) the virtual path identifier (VPI). These fields constitute a virtual circuit identifier. Users are assigned these values when (a) the user enters into a session with a network as a connection on demand or (b) when a user is provisioned to the network as a PVC.

ATM services can be obtained as a PVC or as an SVC. Perhaps a better term to describe the ITU-T view of a PVC is that it is a semi-permanent virtual circuit technology (SPVC, see Chapter 2). That is, users of two connecting endpoints are preprovisioned in the network, and then given a session (connection) when requested by the user—*if* the session can be supported by the network. Moreover, the ATM Forum and the ITU-T have published specifications for ATM connections on demand, or SVCs (ATM Forum, 1993), which are explained later in this chapter.

The VPIs and VCIs are also used in the ATM network. They are examined by switches in order to determine how to route the cell through the ATM network. The VPI/VCI labels are similar to the data link connection identifiers (DLCIs) used in frame relay networks.

ATM LAYERS

As illustrated in Figure 7–3, the ATM layers are similar to the layers of some other emerging communications technologies (the Metropolitan Area Network [MAN] and the switched multi-megabit data service [SMDS]). ATM provides convergence functions at the ATM adaptation layer (AAL) for connection-oriented and connectionless variable bit rate (VBR) applications. It supports isochronous applications (voice, video) with constant bit rate (CBR) services.

A convenient way to think of the AAL is that it is actually divided into two sublayers, as shown in Figure 7–3. The segmentation and reassembly (SAR) sublayer, as the name implies, is responsible for sup-

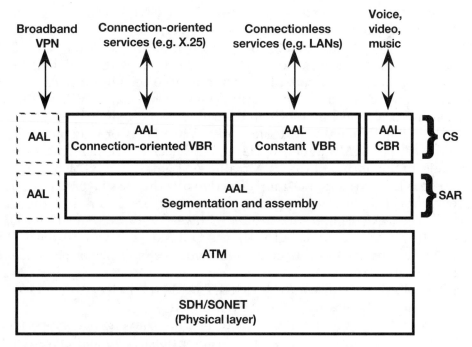

Note: Dashed box means AAL services may not be needed

AAL = ATM adaptation layer
CBR = Constant bit rate
VBR = Variable bit rate
VPN = Virtual private network

Figure 7–3
The ATM layers.

porting user PDUs that are different in size and format (headers and trailers) at the sending site and reassembling the 48-bit data units into the user-formatted PDUs at the receiving site. The other sublayer is called the convergence sublayer (CS), and its functions depend upon the type of traffic being processed by the AAL, such as voice, video, data, etc. Specific examples of SAR and CS operations are provided later in this chapter.

The SAR and CS entities provide OSI interfaces to the ATM layer with service definitions. The ATM layer is then responsible for relaying and routing the traffic through the ATM switch. At the originating machine, it is responsible for receiving the 48-byte unit from the AAL, and placing the 5-byte cell header onto the AAL unit. The cell header is used to define the operations: (a) at the local UNI, (b) in the network, and (c) at the remote UNI. The terminating ATM layer processes the cell header, removes it, and sends the 48-byte unit to the AAL. The ATM layer is connection-oriented and cells are associated with established virtual connections. Traffic must be segmented into cells by the AAL before the ATM layer can process the traffic. The switch uses the VPI/VCI label to identify the connection to which the cell is associated.

Broadband virtual private networks (VPNs) may or may not use the services of the ATM adaptation layer. The decision to use this service depends on the nature of the VPN traffic as it enters the ATM device.

The ATM layers do not map directly with the OSI layers. The ATM layer performs operations typically found in layers 2 and 3 of the OSI Model. The AAL combines features of layers 2, 4, and 5 of the OSI Model. It is not a clean fit, but then, the OSI Model is over 10 years old. It should be changed to reflect the emerging technologies, a point made in other parts of this book.

The physical layer can be a SONET or SDH carrier. It may also be other carrier technologies, such as DS3, E3, etc.

A user CPE may use the AAL to provide convergence support for different kinds of traffic across virtual connections. If virtual private networks provide services that are compatible with ATM, then the AAL services are not needed.

Whatever the implementation of AAL at the user device, the ATM network is not concerned with AAL operations. Indeed, the ATM bearer service is "masked" from these CS and SAR functions. The ATM bearer service includes the ATM and physical layers, shown in Figure 7–3. The bearer services are application independent, and AAL is tasked with accommodating the requirements of different applications. However, the network administrator can use the ATM cell header to determine how to handle the payload inside the cell. For example, the VPI/VCI might identify teleconferencing traffic, datagrams, voice packets, etc., and the network might choose to establish different queues and traffic shedding procedures for these applications. Perhaps a better way to compare the AAL and ATM layer vis-à-vis the supported applications is that AAL interacts directly with the application, and the ATM layer interacts directly with AAL. At any rate, the ATM layer cannot be completely application-independent, else the UNI operations could not manage different types of traffic.

These ideas are amplified in Figure 7–4. For user traffic, upper layer protocols (ULP) and AAL operations are not imbedded into the ATM network functions. The dotted arrows indicate that logical opera-

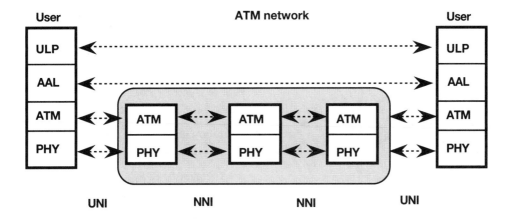

ULP = Upper layer protocols
PHY = Physical layer

Figure 7–4
Relationship of user and network layers (user traffic).

tions occur between peer layers at the user nodes and the ATM nodes. Therefore, the ULP headers, user payload, and the AAL headers are passed transparently through the ATM network.

ATM and the BISDN Model

The BISDN reference model is based on the OSI Reference Model and the ISDN standards. Figure 7–5 shows the layers of BISDN, which are quite similar to the ITU-T model for frame relay (see Figure 4–3, Chapter 4). The physical layer could consist of different media, even though the ITU-T has encouraged the use of ATM with the SONET/ SDH technology.

The BISDN model contains several planes. The user plane is responsible for providing user information transfer, flow control, and recovery. The BISDN control plane is responsible for setting up the call and managing the connections. It is also responsible for connection release.

The management plane has two functions: plane management and layer management. Plane management has no layered structure. It is responsible for coordination of all the planes. Layer management is responsible for managing the entities in the layers and performing operation, administration, and maintenance services (OAM). The shadowed

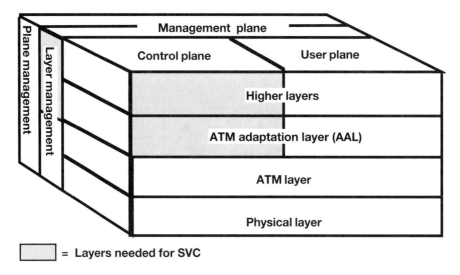

= Layers needed for SVC

Figure 7–5
ATM and the BISDN reference model.

parts of Figure 7–5 include the functions needed for the implementation of SVC services.

The rather abstract view of BISDN and ATM in Figure 7–5 can be viewed in a more pragmatic way. The three planes depicted in Figure 7–5 (control, user, and management) are shown in Figure 7–6 with the placements of likely protocols residing in the layers. The plane management functions are internal to the BISDN model and are not relevant to this discussion.

On the left side of Figure 7–6 is the control plane, which contains the Q.93B signaling protocol used to set up connections in the ATM network. The layer below Q.93B is the signaling ATM adaptation layer (SAAL). SAAL supports the transport of the Q.93B messages between any two machines that are running ATM SVCs. SAAL contains three sublayers (the full definitions of these sublayers have not been defined completely by the ITU-T). Briefly, they provide the following functions. The AAL common part (AALCP) detects corrupted traffic transported across any interface using the control plane procedures. The service specific connection-oriented part (SSCOP) supports the transfer of variable length traffic across the interface, and recovers from errored or lost service data units. The service specific coordination function (SSCF) provides the interface to the next upper layer, in this case, Q.93B. Due to the standards for SAAL not being complete (and due to the fact that this

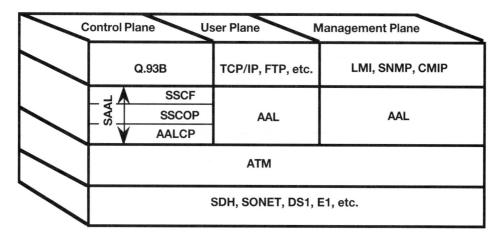

Figure 7–6
Protocol placement in the BISDN layers.

book is an overview), this chapter will concentrate on the AALs that are specific to user payload, covered later in the chapter in the section titled, "AAL Types".

In the middle of the figure is the user plane, which contains user and applications specific protocols, such as TCP/IP, FTP, etc. (examples only). The invocation of the user plane protocols take place only if (a) the control plane has set up a connection successfully, or (b) the connection was preprovisioned.

The management plane provides management services, and is implemented with the ATM Local Management Interface (LMI, discussed later in this chapter), the Internet Simple Network Management Protocol (SNMP), or the OSI Common Management Information Protocol (CMIP) (although SNMP/CMIP could be at the user plane).

ATM PROTOCOL DATA UNITS (CELLS)

The ATM protocol data unit (PDU) is called a cell. It is 53-octets in length, with 5-octets devoted to the ATM header, and 48-octets used by AAL and the user payload. As shown in Figure 7–7, the ATM cell is configured slightly differently for the UNI and NNI. Since flow control operates at the UNI interface, a flow control field is defined for the traffic traversing this interface, but not at the NNI. The flow control field is called the generic flow control (GFC) field. If GFC is not used, this four-

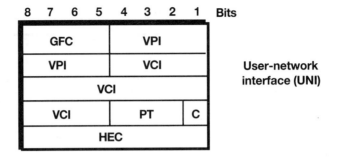

User-network
interface (UNI)

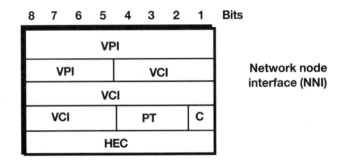

Network node
interface (NNI)

VPI = Virtual path identifier
GFC = Generic flow control
VCI = Virtual channel identifier
PT = Payload type
C = Cell loss priority
HEC = Header error control

Figure 7–7
The ATM PDUs (cells).

bit field is set to zeros. Its specific use is being defined in the standards, and the ATM Forum has defined its contents (along with some reserved bits in other parts of the header) to define a variety of control and operations, maintenance, and administration (OAM) operations. These reserved fields are also used to manage the establishment and releasing of virtual channel connections.

Most of the values in the five-octet cell header consist of the virtual circuit labels of VPI and VCI. A total of 24-bits are available with 8-bits assigned to VPI and 16-bits assigned to VCI. For the NNI, the VPI field contains 12-bits (the NNI is covered later in this chapter). A combination of the VPI and VCI field is called the VPCI field.

A payload type identifier (PT) field identifies the type of traffic residing in the cell. The cell may contain user traffic or management/ control traffic. The ATM Forum has expanded the use of this field to identify other payload types (OAM, control, etc.). One of particular interest is a payload type that indicates that (a) the cell contains user data and the receiver is notified of congestion problems, or (b) the cell contains user data and the user is notified that congestion has not been experienced. In other words, this field is now used for congestion notification operations, which is similar to the frame relay congestion notification bits (FECN and BECN bits).

The cell loss priority (C) field is a one-bit value. If C is set to 1, the cell is subject to being discarded by the network. Whether or not the cell is discarded depends on network conditions and the policy of the network administrator. Whatever the policy of the administrator may be, C set to 0 indicates a higher priority of the cell to the network. It must be treated with more care than a cell in which the C bit is set to 1. The C bit is quite similar to the frame relay discard eligibility (DE) bit.

The header error control (HEC) field is an error check field. The HEC can also correct a one-bit error. It is calculated on the ATM header, and not on the user payload. ATM employs an adaptive error detection/ correction mechanism with the HEC (discussed in more detail later).

Meta-Signaling Cells and Other Cells

The VCI, VPI, and other parts of the first four octets of the cell can be coded in a variety of formats to identify non-user payload cells. One such convention is called meta-signaling. It is used to establish a session with the network and negotiate session services. Another convention is called broadcasting. With this feature, the VPI and VCI are coded to indicate that the cell is to be broadcast to all stations on the UNI. Other conventions provide for management services, the identification of idle cells (cells that are empty), and unassigned cells. Unassigned cells are sent when the ATM module has no user payload to send. Thus, the continuous sending of cells at the sender and the continuous reception of cells at the receiver allows the network to operate synchronously, yet support bursty, asynchronous services.

The ATM Forum has defined an additional coding for octets 1–4 to identify a network management cell. This coding convention is part of the Forum's interim local management interface (ILMI) specification (available in the ATM UNI Specification, v. 2.0) (ATM User-Interface Specification, Version 2.0, 1992). The ILMI is covered later in this chapter.

ATM OPERATIONS IN MORE DETAIL

Physical Layer Interfaces

The standards groups have published specifications on how to operate ATM over (a) SONET/SDH, (b) a 44.736 Mbit/s carrier, and (c) a 100 Mbit/s FDDI system. The relationship of the ATM and SONET/SDH is covered in Chapter 8, because the reader will need more information on SONET/SDH systems before the full ATM/SONET/SDH operation can be explained. This section provides an overview of the other two physical layer specifications.

The 44.736 Mbit/s interface (DS3).   ATM is specified by the ATM Forum to run on top of a DS3 carrier system. In order for this feature to be implemented, a physical layer convergence protocol (PLCP) has been defined which is quite similar to the PLCP discussed earlier in this book relating to the IEEE 802.6 and the SMDS specification.

ATM cells are carried in the DS3 by mapping 53-byte ATM cells into a DS3 PLCP payload. The PLCP is then further mapped into the DS3 information payload. This concept is illustrated in Figure 7–8, which depicts a DS3 PLCP transmission stream. As the reader might expect, it consists of a 125 μsec. frame running within a DS3 payload. The frame consists of 12 rows of ATM cells with each cell preceded by four octets of overhead. The overhead is used for framing, error checking, bit stuffing (if necessary), alarm conditions, path overhead identifiers, and reserved octets. Cell delineation is accomplished in a simple manner by placing the cells in predetermined locations within the PLCP.

The interface for 100 Mbit/s multimode fiber systems.   The ATM Forum UNI also specifies the interface for an ATM/FDDI configuration. As we have learned earlier in this book, FDDI runs at 100 Mbit/s and uses the 4B/5B coding scheme operating with a 125 Mbaud line signaling rate. The FDDI control codes are used in a limited manner (the reader may wish to refer to Table 5–1 in Chapter 5). The mnemonic JK is used to signal an idle line, and the mnemonic TT is used to signify the begin-

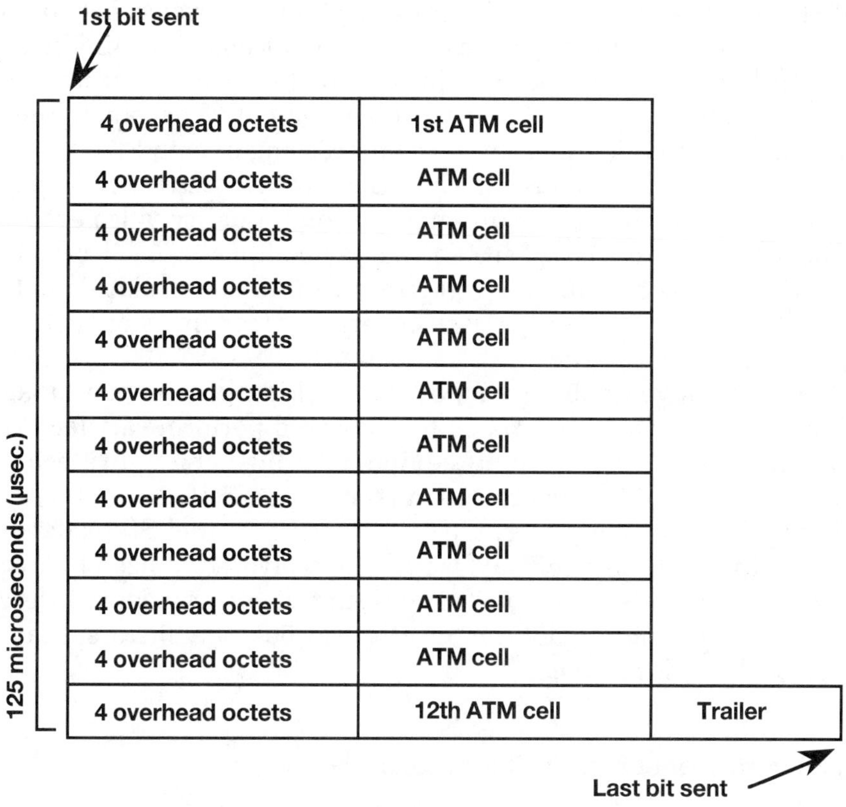

1st bit sent

4 overhead octets	1st ATM cell	
4 overhead octets	ATM cell	
4 overhead octets	ATM cell	
4 overhead octets	ATM cell	
4 overhead octets	ATM cell	
4 overhead octets	ATM cell	
4 overhead octets	ATM cell	
4 overhead octets	ATM cell	
4 overhead octets	ATM cell	
4 overhead octets	ATM cell	
4 overhead octets	ATM cell	
4 overhead octets	12th ATM cell	Trailer

125 microseconds (µsec.)

Last bit sent

Figure 7–8
Running ATM over DS-3.

ning of a cell. That is to say, TT is used for cell delineation. The cell bytes then follow contiguously on the channel, and the cell and its start of cell code (TT) must also be contiguous to each other on the channel. Other mnemonics shown in Table 5–1 in Chapter 5 are either reserved or not recommended for usage. The only other FDDI mnemonic is QQ, which is used to code a loss of signal indication. Be aware that these operations do not use the FDDI media access control (MAC) layer. The ATM layer operates directly on top of the FDDI physical layer.

Rationale for the Cell Size

The reader might wonder why a cell of 53-octets was chosen. Why not 32? Why not 64? The 53-octet size was a result of a compromise between various groups in the standards committee. The compromise resulted in a cell length that is: (a) acceptable for voice networks; (b) avails itself to forward error correction operations; (c) minimizes the number of bits that must be transmitted in the event of errors; and (d) works with ongoing carrier transport equipment. The small cell also avoids the delay that is inherent in the processing of long PDUs.

We shall see in Chapter 8 that the SONET/SDH STS-3 payload accommodates the bit rate requirement for high-quality video even when the video images are carried inside the payload of the ATM cell. In the next section of this chapter, the cell size is examined in relation to transmission errors, equipment processing errors, transmission delay, and equipment processing delay.

After extensive deliberations in the working groups, it was agreed that a cell size between 32- and 64-octets would perform satisfactorily in that it (a) worked with ongoing equipment (did not require echo cancellers), (b) it provided acceptable transmission efficiency, and (c) it was not overly complex to implement. Japan and the United States favored a cell size with 64-octets user payload; Europe favored a size of 32-octets. A compromise was reached and the 48-octet size was adapted (less the 5-octet header). Subsequent material describes the factors that contributed to these decisions.

Network Transparency Operations

This section delves into more detail about the ATM cell and its relationship to errors, delay, and the size of the cell. The service data unit (SDU) is a useful OSI term for this discussion. The reader can refer to Appendix B if the concept of the SDU is not familiar. Three concepts are examined in this discussion:

Semantic transparency:	Transporting the user's SDUs without error from source to destination
SDU size transparency:	Accommodating the user's variable size SDUs
Time transparency:	Transporting the user's SDUs with a fixed delay between source and destination

Errors and error rates. Errors are impossible to eliminate completely in a communications network. As reliable as optical fiber is, noise, signal dispersion, and other impairments will persist. Errors also will occur in equipment due to the malfunctioning of hardware, imperfections in component design, etc.

Communications networks error rates are measured with a simple formula called the bit error rate (BER). It is calculated as follows:

$$BER = \frac{\text{errored bits received}}{\text{total bits sent}}$$

It is important that BER be measured over a period of time which smoothes out the randomness of errors on communications channels. Generally, measurement periods should be several orders of magnitude more than the actual BER value.

Another useful statistic is called the block error rate (BLER) which defines the number of blocks sent in relation to the number of blocks received in error. The term block is generic; for this discussion, a block refers to a cell. BLER is calculated as:

$$BLER = \frac{\text{errored blocks received}}{\text{total blocks sent}}$$

Other error statistics are important. One that is pertinent to ATM is called the loss rate, which is the ratio of lost or inserted blocks in relation to the total number of blocks sent. Lost blocks (cells) are any blocks that are damaged and cannot be corrected, blocks that arrive too late to be useful (as in a CBR application), and blocks that are discarded by the network. Inserted blocks are blocks that are received by a user when they were supposed to go to another user.

These error rate statistics play a key role in the design of the network. As errors increase, user payload (especially data) must be retransmitted from the user CPE. The end effect is the creation of more traffic in the network. Ideally, one wishes to design a network that is cost-effective from the standpoint of its incidence of errors as well as efficient in its treatment of errors. This brings us to the subject of error rates on optical fiber and the size of an ATM cell.

AT&T (Observations of Error Characteristics of Fiber Optic Transmission Systems, 1989) conducted a study on the error rates of optical fiber. This study is summarized in Figure 7–9. Under normal operating conditions, most errors that occur on optical fiber are single-bit errors. As this figure illustrates, 99.64 percent of the errors are single bit. Although not shown in this figure, AT&T also conducted a study on the error rate during maintenance conditions. For example, during

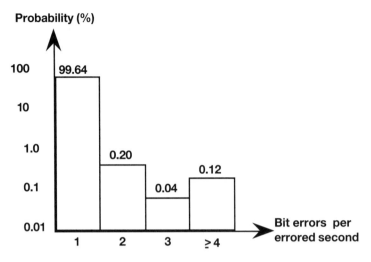

Figure 7–9
Bit error rates (BER) on optical fiber. [ATT89a]

switch-over operations on a device, the single-bit error probability dropped to 65 percent.

These studies and others have paved the way for a more detailed analysis of the effect of using forward error correction (FEC) to correct a one-bit error. Subsequent research has confirmed that small cell headers experiencing a one-bit error can be corrected, which obviates (a) retransmitting data cells and (b) discarding voice/video cells. Since errors are rare on optical fiber, ATM does not execute cumbersome retransmission schemes, it will *correct* headers with one-bit errors and *discard* headers with multiple bit errors.

Error correction and detection. The ATM HEC operations protect the cell header. The HEC field of 8-bits was selected because it allows the correction of a single bit error and the detection of multiple bit errors. Figure 7–10 depicts the general logic of the correction and detection functions. Initially, the HEC operation is in the correction mode. If it detects either a single bit or multiple bit errors in the header, it moves to the detection mode. A single bit error is corrected, and a multiple bit error results in the discarding of the cell. Once in the correction mode, all cells that have errors are discarded. When a cell is detected that has no error, the operation returns to the correction mode.

ATM uses an adaptation of the Hamming code technique known as the Bose-Chadhuri-Hocquenghem (BHC) codes. These codes provide

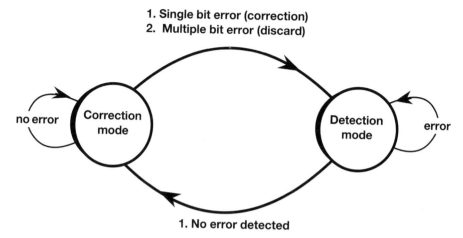

1. Single bit error (correction)
2. Multiple bit error (discard)

no error   Correction mode   Detection mode   error

1. No error detected

Figure 7–10
Header protection in the HEC field.

error correction/detection schemes based on the ratio of protection bits in the HEC field, for example to protected bits (in the cell header, for example). Without delving into error coding theory, which is best left to theoretical texts, it is instructive to note that a 40-bit field (the cell header of 5-octets * 8 bit per octet = 40 bits) needs at least 6 protection bits to correct a single bit. The 6-bits will also detect 36 percent of multiple bit errors. Improvements to an 89 percent detection rate of multiple bit errors can be attained if 8 protection bits are used (dePrycker, 1992a). Of course, this is a good arrangement if optical fiber media are employed, due to optical fiber error characteristics. The HEC field of 8-bits was selected due to these characteristics.

It can be concluded that the ATM approach of using small cells and small cell headers lends itself to efficient error correction/detection schemes. The next questions are: What size should the full cell be? And, if it is to be small, can it be variable in length or must it be fixed? These two questions are answered in the next two sections.

Variable and fixed size payload.   Ideally, one would like to have a choice of the size of a SDU, based on the quality of the circuits. For error-prone links, it is desirable to have small SDUs because the number of bits retransmitted are small. For high-quality links, it is desirable to have larger SDUs in order to increase the ratio of user payload to the cell header. Moreover, it is well-known that variable length SDUs provide greater transmission efficiency than fixed length SDUs (a demonstration of this fact will follow shortly). However, transmission efficiency is not the only criteria that should be used in making the decision on fixed or variable length SDUs. Equally important is the effect of a fixed/variable length SDU on switching speed and network delay. An analysis of these factors vis-à-vis transmission efficiency led the ITU-T to the decision to use fixed cells.

The transmission efficiency of a protocol can be calculated as: TE = $L_i/(L_i + L_0)$, where TE = transmission efficiency; L_i = length of information field (user traffic); L_0 = length of control header. Various studies reveal that the TE is better for variable length SDUs than for fixed length SDUs. It is common sense that the more bits in the information field, the better the TE for a given L_0. However, a large I field entails overhead at a switch because it takes longer for the cell to be processed and to leave the switch.

In essence, the ITU-T had to balance conflicting factors in choosing (a) a fixed or variable length PDU, and (b) the size of the SDU. After a lengthy analysis, with Europe opting for a cell size of 32-octets, and the U.S. and Japan opting for a cell size of 64-octets, the difference was split, and the size of 48-octets was selected at the 1989 Geneva meeting.

Transmission delay. Previous discussions in this book have explained how certain applications require a predictable and fixed delay of traffic between the source and destination, such as video and voice applications. The task is to design a network that provides time transparency for user payload. This concept means that (for certain applications) the user's payload has a fixed and nonvariable delay between the sender and receiver. In effect, an ATM network must emulate circuit networks, which are designed to provide a fixed delay of voice and video traffic from source to destination.

The ATM layer is not tasked with providing time transparency services to the end user application. This service is delegated to the ATM adaptation layer (AAL). It entails conditioning the received user payload (typically holding the user payload in buffer) to achieve a fixed delay from the source to endpoints. For isochronous traffic, the ITU-T requires that transmissions from source to destination shall not incur an overall delay greater than 199 msec.

As illustrated in Figure 7–11, a number of factors contribute to the delay of a cell. At the sender, delay is incurred when voice and video traf-

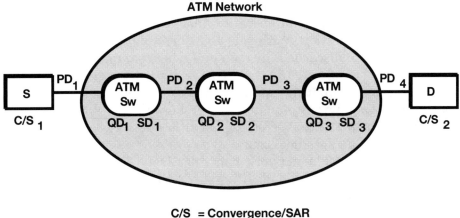

ATM Network

C/S = Convergence/SAR
PD = Propagation delay
QD = Queueing delay
SD = Switching delay
S = Source
D = Destination
Sw = Switch

Figure 7–11
Factors in delay.

fic is translated and segmented into cells by convergence services and segmentation and reassembly services (C/S). At the receiver, delay is incurred for the opposite operations.

Each transmission channel in the end-to-end path incurs a delay in the transport of the cell between the user device or the switch(es). This is shown in Figure 7–11 as propagation delay (PD). While ranges vary on the exact delay in the transmission, it is predictable because it depends on the distance between the transmitting station and the receiving station. ITU-T provides guidance with ranges running between 4 to 5 μsec. per km. The IEEE uses a 4.2 μsec. delay on CSMA/CD networks.

This compromise will work well enough when one examines what happens in the event of a rare cell loss for voice traffic. The loss of traffic of around 32–64 ms in duration is quite disruptive, because it means the loss of speech phonemes. On the other hand, cell loss of a duration of some 4–16 ms is not very noticeable nor disturbing to the listener. Therefore, a payload size of around 32- to 64-bytes would be acceptable to an audio listener, because the loss of one cell would result in a loss of only 4 ms for a 32 Kbit/s coded voice signal, and 8 ms for 64 Kbit/s coded voice signal.

Additionally, techniques are available to permit the selective shedding of parts of a sample, an operation that can be invoked during periods of network congestion. With selective shedding, the least significant bits of samples are discarded, instead of a whole sample. For example, a 32 Kbit/s 4-bit sample of 15_{10} could have its least significant bit discarded. At the terminating application, the sample of 14_{10} would be interpreted in the digital-to-analog conversion process. While this introduces distortion in the converted signal, it is not noticeable, as long as excessive discarding does not occur.

Of course, an ATM network can selectively shed many blocks of samples to reduce its workload. If its sheds the least-significant bits of these samples, a large part of a voice signal can be discarded without the listener noticing a degradation of the speech signal. Depending on the digital coding scheme used, approximately 1–10 percent of the cells of a speech signal can be discarded, if discarding is done selectively.

At the switch, two forms of delay are incurred: queuing delay (QD) and switching delay (SD). The switches are performing STDM operations, and since traffic arrives asynchronously at the switches, it is necessary to build queues to accommodate peaks in traffic. Obviously, delay varies in relation to the amount of traffic in the network.

The second factor at the switch is the speed at which traffic is relayed through the switch. SD varies depending on the type of switching operation employed, but it is a fixed delay vis-à-vis the size of the cell. Generally, switches today can process a cell within a broad range of 2 and 100 μsec., and the ITU-T's Q.507 Recommendation stipulates an average SD of smaller than 450 μsec. Additional delay may be incurred to remove jitter by queuing the traffic slightly longer at each receiver. Notwithstanding, jitter can usually be accommodated during the queuing process.

Thus, delay through an ATM network entails the summation of the delays incurred at the user devices, the switches, and on the communications links.

It should be emphasized that switching can be performed faster than the line speed. Consequently, switching speed is usually limited by the communications lines (155 Mbit/s, etc.) that are input into the switch (unless excessive queuing occurs).

Given the assumptions and estimates cited above, end-to-end delay of a cell traveling from the sending machine to the receiving machine (in Figure 7–11) is 14.297 ms, well within the ITU-T requirements of 199 msec. for (say) voice traffic. This value was derived by the following assumptions:

C/S   12000 µsec. (6000 µsec. each at source and destination)
PD   2000 µsec. (assuming a distance of 500 km between source
            and destination)
SD   72 µsec. (24 µsec. at each switch, with 3 switches)
QD   225 µsec. (75 µsec. at each switch, with 3 switches)

Several conclusions are made from this discussion. First, it is obvious that propagation delay is a significant factor when compared to some of the other ingredients of delay. In older systems with slow processors, this factor was not considered significant because of the large QD, C/S, and SD values. In newer systems, the PD remains the same, of course, but the ratios change. The second conclusion is the C/S functions may consume considerable overhead if voice and video must be converted to digital signals and then segmented at the transmitter (and have the complementary yet opposite functions performed at the receiver). The third conclusion is that in a pure ATM network, the C/S functions are performed only twice: at the source and at the destination points. If other networks reside between these two machines, like an intervening synchronous network (such as SONET/SDH), additional mapping and convergence functions must be performed to enter and exit the synchronous network. Consequently, the C/S functions play an important role at this additional interface. Finally, be aware that C/S, SD and QD will vary between machines, and QD is a function of the amount of traffic entering the switch and the size of the cell queues.

The surface has only been touched regarding the design considerations in providing the ATM user semantic, SDU size, and time transparency services. Notwithstanding, we can now move forward to a more detailed examination of the ATM virtual circuits, and the ATM multiplexing and switching operations. Later, the transparency services are revisited in their relation to traffic management in an ATM network.

ATM Labels

Earlier discussions explained that an ATM connection is identified through two labels called the virtual channel identifier (VCI) and virtual path identifier (VPI). In each direction, at a given interface, different virtual paths are multiplexed by ATM onto a physical circuit. The VCIs and VPIs identify these multiplexed connections.

As shown in Figure 7–12, virtual channel connections can have end-to-end significance between two end users. However, the values of these connection identifiers can change as the traffic is relayed through the

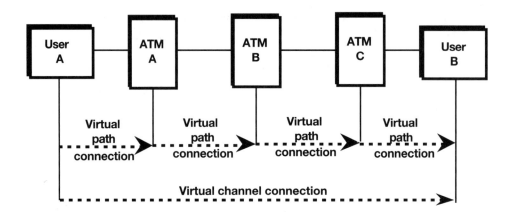

Figure 7–12
Types of ATM connections.

ATM network. For example, in a switched virtual connection, the specific VCI value has no end-to-end significance. It is the responsibility of the ATM network to "keep track" of the different VCI values as they relate to each other on an end-to-end basis.

Routing in the ATM network is performed by the ATM switch examining both the VCI and VPI fields in the cell, or only the VPI field. This choice depends on how the switch is designed, and if VCIs are terminated within the network—a topic covered in the switching section of this chapter.

The VCI/VPI fields can be used with switched or nonswitched (broadcast) ATM operations. They can be used with point-to-point or point to multi-point operations. They can be pre-established (PVCs) or set up on demand (SVCs), based on signaling procedures, such as the ISDN network layer protocol (Q.931).

Additionally, the value assigned to the VCI at the user-to-network interface (UNI) can be assigned by (a) the network, (b) the user, or (c) through a negotiation process between the network and the user.

Multiplexing VCIs and VPIs

The ATM layer has two multiplexing hierarchies: the virtual channel and the virtual path. The virtual path identifier (VPI) is a bundle of virtual channels. Each bundle must have the same endpoints. The purpose of the VPI is to identify a group of virtual channel (VC) connections. This approach allows VCIs to be "nailed-up" end-to-end to provide semi-permanent connections for the support of a large number of user sessions. VPIs and VCIs can also be established on demand.

The VC is used to identify a unidirectional facility for the transfer of the ATM traffic. The VCI is assigned at the time a VC session is activated in the ATM network. Routing might occur in an ATM network at the VC level, but VCs are usually mapped through the network without further translation. If VCIs are used in the network, the ATM switch must translate the incoming VCI values into outgoing VCI values on the outgoing VC links. The VC links must be concatenated to form a full virtual channel connection (VCC). The VCCs are used for user-to-user, user-to-network, or network transfer of traffic.

The VPI identifies a group of VC links that share the same virtual path connection (VPC). The VPI value is assigned each time the VP is switched in the ATM network. Like the VC, the VP is unidirectional for the transfer of traffic between two contiguous ATM entities.

As shown in Figure 7–13, two different VCs that belong to different VPs at a particular interface are allowed to have the same VCI value (VCI 1, VCI 2). Consequently, the concatenation of VCI and VPI is necessary to uniquely identify a virtual connection.

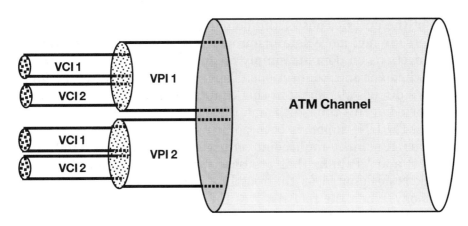

VPI   =   Virtual path identifiers
VP link =   Terminated by points where VPI is assigned, translated or removed
VCI   =   Virtual channel identifiers
VC link =   Terminated by points where VCI is assigned, translated or removed

Figure 7–13
The ATM connection identifiers.

ATM Connections on Demand

Early ATM efforts focused on PVC implementations, with user sessions granted if network capacity were available. The ATM Forum has reached a consensus that connections on demand are quite important and has issued specifications for this operation (ATM User Network Interface Specification, 1993). This section describes this specification, which is known as "Phase 1". Be aware that Phase 1 (as of this writing) is still a draft, and therefore subject to final balloting by the ATM Forum members.

Like all connections on demand (switched virtual calls [SVCs]), a user-to-user session through an ATM network requires a connection set-up procedure, and a disconnection procedure. With connection set-up, the user must furnish the network with the calling and called party addresses and the QOS that is needed for the session. In the event of a network failure, the demand connections may not be automatically reestablished, although they may remain active for an arbitrary amount of time.

The ATM connection control procedures are organized around the ISDN Q.931 layer 3 operations (explained in Chapter 3), and a subset of Q.931 that is in development, which is called Q.93B (ITU-T Draft Recommendation Q.93B, 1993). Phase 1 signaling must support 14 capabilities at the UNI; they are listed in Table 7–2, and summarized herein. Several of the descriptions of the demand connections are self-explanatory, others are not; more detail is provided for the ambiguous titles.

Connections on demand simply mean that the ATM UNI must support switched channel connections. They are established with signaling procedures discussed shortly in this chapter. Permanent connections are not supported at this time, although they are under study.

The ATM UNI supports both point-to-point and point-to-multipoint connections. A point-to-multipoint connection is defined as a collection of related VC and VP links that are associated with endpoint nodes. One ATM link is designated as the root link, which serves as the root in a tree topology. When the root receives information, it sends copies of this information to all leaf nodes on the tree. Communications must occur between the leaf nodes through the root node. Connections are established initially through the root node and then one leaf node, then other nodes can be added with "add party" operations, one leaf entry at a time.

The UNI supports either symmetric or asymmetric connections for bandwidth allocation. Bandwidth is specified independently in both directions across the virtual connection. Forward direction is from the

Table 7–2   Capabilities of ATM Demand Connections

* Connections on demand (switched)
* Point-to-point and point-to-multipoint connections
* Symmetric or asymmetric bandwidth requirements
* Single connection calls
* Specific procedures for call set-up, request, answer, clear and out-of-band signaling
* Support of class A, C and X transport services
* Non-negotiation of QOS between users
* Specification of VPI/VCI ranges
* Designation of an out-of-band signaling channel
* Mechanisms for error recovery
* Guidelines for addressing formats
* Client registration procedures
* Methods of identifying end-to-end capability parameters
* Nonsupport of multicasting operations

calling party to the called party. Backward direction is from the called party to the calling party.

The UNI also defines specific procedures for the requesting and setting up of connections and the clearing of the connections (and the reason for clearing) as well as an out-of-band channel used for control purposes.

Phase 1 supports ATM class A and class C transport services (which are explained later in this chapter), as well as class X procedures which are published by the ATM Forum. Class X is a quality of service (QOS) operation that is user defined. The reader may wish to refer to the section titled "Classes of Traffic" if more about this subject is desired at this time.

For phase 1, QOS cannot be negotiated between users. A user may request a certain level of service and can send the QOS parameters in the connection set-up request. The receiver merely indicates if these values can be accommodated. It is not allowed to return any traffic suggesting a negotiation procedure.

UNI also specifies a range for the use of VCI and VPI. For phase 1, there is a one-to-one mapping between a VPCI and a VPI; therefore values beyond 8-bits are not permitted. As part of the designation for the VPI/VCI ranges, an out-of-band signaling channel has been designated with the values of VCI = 5 and VPCI = 0 (VPCI is a VCI/VPI).

The phase 1 specification contains several mechanisms for handling error recovery procedures. Among these are provisions for the signaling of non-fatal errors (errors which can be recovered), procedures for recovering from resets, procedures for forcing VCCs into an idle state, and other diagnostic information pertaining to error recovery and call clearing.

One important capability at the UNI is the provision for addressing format guidelines. These guidelines are organized around OSI network service access points (NSAPs) which are specified in ISO 8348 and ITU-T X.213 Annex A. The phase 1 specification requires the use of these formats in accordance with registration procedures contained in ISO 10589. This feature of the standard will greatly facilitate interworking different vendors' systems.

The ATM demand connections capabilities also includes a client registration procedure which allows users to exchange address information across the UNI, as well as other administrative information. This procedure allows an ATM network administrator to load network addresses into the port on a dynamic basis.

Also supported is the ability to identify end-to-end capability parameters. Capability means that provisions are available to identify what protocol is running inside the ATM PDU; that is to say, what protocols are operating above the ATM services. This capability allows the two end users to run various types of protocol families across an ATM-based network and use this identifier to separate and demultiplex the traffic to the respective protocol families at the receiving machine.

Finally, the fourteenth capability is really not a capability, in that the phase 1 specification does not support multicast operations at this time.

Table 7–3 lists the ATM messages employed for demand connections at the UNI and their functions. Since these messages are derived from Q.931, which was discussed in Chapter 3, this section will concentrate on the use of these messages vis-à-vis the ATM UNI operation. The reader can refer to Chapter 3 for a discussion of the functions of these messages as well as the format of the Q.931 message.

The call establishment messages, as the name implies, are used to set up a connection between the users and the network. This message contains the typical Q.931 fields such as protocol discriminator, call reference, message type, and message length. The information content of

Table 7–3   ATM Connection Control Messages

Message	Function
Call establishment	
SETUP	Initiate the call establishment
CALL PROCEEDING	Call establishment has begun
CONNECT	Call has been accepted
CONNECT ACKNOWLEDGE	Call acceptance has been acknowledged
Call clearing	
RELEASE	Initiate call clearing
RELEASE COMPLETE	Call has been cleared
Miscellaneous	
STATUS ENQUIRY (SE)	Sent to solicit a status message
STATUS (S)	Sent in response to SE or to report error
Global call reference	
RESTART	Restart all VCs
RESTART ACKNOWLEDGE	ACKS the RESTART
Point-to-multipoint operations	
ADD PARTY	Add party to an existing connection
ADD PARTY ACKNOWLEDGE	ACKs the ADD PARTY
ADD PARTY REJECT	Rejects the ADD PARTY
DROP PARTY	Drops party from an existing connection
DROP PARTY ACKNOWLEDGE	ACKS the DROP PARTY

the field, of course, is tailored for the specific ATM UNI interface. This section concentrates on the SETUP message, since it is used to initiate the demand connection.

The SETUP message contains the parameters (information elements) necessary to establish the connection. The key information elements are examined in the following material.

ATM AAL parameters.   These parameters indicate the AAL values for the connection. Listed below are some of the major parameters (not all parameters are included):

- Type of traffic (1–5, or user defined)
- Subtype of traffic (circuit emulation, high-quality audio, video, etc.)
- Maximum SDU size for AAL types 3/4
- CBR rate (64 kbit/s, DS1, E1, etc.)
- Clock recovery type for AAL type 1
- Mode for AAL type 5 (message mode or streaming mode)
- Error correction method for AAL type 1

ATM cell rate.   These parameters indicate: (a) forward and backward peak cell rates, (b) forward and backward sustainable cell rates, and (c) forward and backward maximum burst sizes.

Broadband bearer and higher-layer capabilities.   Several information elements indicate a wide variety of services that are requested for the connection at the BISDN bearer service (the lower three layers of OSI), and the upper layers (the top four layers of OSI). Examples of the parameters are: (a) indication that timing is or is not required; (b) indication of CBR/VBR traffic; (c) identification of the user layers (protocol entities) that are to operate over ATM; and (d) for X.25/X.75 traffic, the default packet size of the user payload, etc.

Addresses and numbers.   Four information elements contain parameters for the called, calling party numbers and subaddresses. The party numbers are coded in accordance with the E.164 addressing standard. The subadresses are coded as OSI network service points (NSAPs).

Connection identifier.   This information element contains the values of the VPI and VCI that are to be associated with the session.

Quality of service parameter.   This parameter contains values to request (and maybe receive) certain QOS operations. QOS is requested for both directions of the connection.

Cause parameters.   This information element contains diagnostics information and the reasons that certain messages are generated. Diagnostics such as user busy, call rejected, network out of order, QOS unavailable, user cell rate not available, etc. are supported the cause parameters.

Transit network selection. This parameter is important in the United States for traffic that is sent across LATAs. It allows the user to choose the interexchange carrier (IXC) for the session.

ATM Switching

One of the most important components of ATM is the switch and how fast it relays cells (without cell loss). ATM must accept asynchronous and synchronous traffic, and queuing delay and switching delay must be minimized if the ATM network is to perform satisfactorily in its support of the input and output lines operating (perhaps) at 150 Mbit/s. Consequently, the ATM switch requires a processor that is capable of supporting high-speed ports, and switches are available today with a total switching capacity of up to 1.2 (Gbit/s).

One purpose of the ATM switch is to adapt gracefully to changing network traffic profiles with the ability to adjust to increased or decreased input traffic (within limits however). Since the switch of an ATM network must receive traffic that is somewhat unpredictable, the ATM switch must be able to adjust to changing network conditions. All these problems and their solutions are not defined in the ATM standards. Vendors are free to implement the techniques they deem appropriate for their product.

Several approaches to cell switching are examined in this section. The topic of ATM switching is worthy of an entire book, so the goal of this section is more modest, and is designed to give the reader a general overview of cell switching architecture, and the major components of an ATM switch. Be aware that there is no "best way" to build an ATM switch, and the ATM standards wisely do not address this aspect of the technology.

As described earlier, the virtual channels are aggregated through multiplexers and switched (routed) through the network. Two types of switches are used in an ATM network for the multiplexing/demultiplexing and routing of traffic (see Figure 7–14). Conceptually, a virtual path switch is only required to examine the VPI part of the header for multiplexing/demultiplexing and routing. Virtual channel switches, on the other hand, must examine the whole routing field; that is to say, the VCI and the VPI.

The intent of this approach is to speed-up switching and routing at intermediate points. Instead of requiring the switch (in some instances) to examine the entire routing field (which could entail overhead and pro-

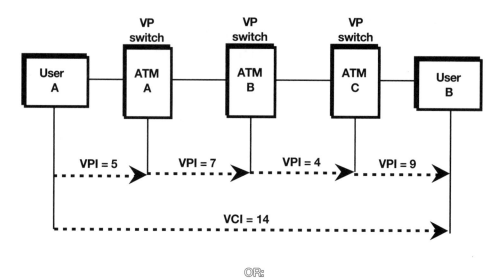

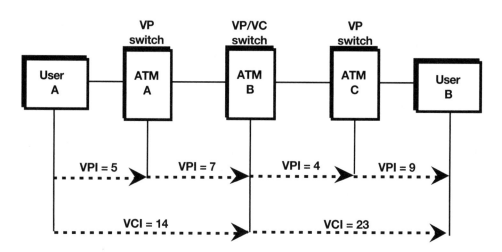

Figure 7–14
ATM switches: Virtual path and virtual channel switches.

cessing delays), the VCI value can be transported transparently through a virtual path connection. Please note that while the VPI/VCI can be used as the routing label, nothing precludes a switch designer from placing a special routing label onto the cell for its processing at the switch.

In the bottom part of Figure 7–14, two virtual path connections exist from user A to ATM B and from ATM B to user B. Virtual path connections can be thought of using leased lines for connecting nodes. The node can use the ANI identifier on any virtual path connection (with the use of the VCI). Also, as suggested in the lower part of Figure 7–14, virtual path connections exist between the user and the network and within the network itself. This approach could be applied to the North American carrier system in which the VCI 14 could exist in one operator and VCI 23 in another operator. This approach is a "clean" interface and also lends itself to relatively small look-up tables.

The VP and VC switches actually treat virtual path and virtual channel connections the same in the sense that they can be set up on a demand basis or they can be set up on a permanent basis. The network also can associate QOS features, such as traffic usage, with either a virtual channel or a virtual path.

Figure 7–15 provides an example of how VPIs and VCIs are translated and mapped at the switch. The top of the figure is an example of VPI switching. The VCIs are passed unaltered through the switch, and VCIs 14 and 15 remain bundled in a VPI. The incoming VPI 7 is translated to outgoing VPI 4, but the VCI values are not altered.

In contrast, the bottom part of the figure shows both VPIs and VCIs being translated and mapped to different values. VCI 14 (in the incoming VPI 7 bundle) is translated and mapped to outgoing VPI 7 with the VCI value changed to 23. The incoming VCI 15 (also bundled into VPI 7) is translated and mapped into VCI 88 and bundled into outgoing VPI 10.

The switching fabric. Before delving into the details of ATM switching, a couple of definitions are in order. First, a *switching fabric* describes the components of the switch, which include its hardware and software architecture. Second, a *network element* is a part of the overall switching fabric, such as a software or hardware component.

Figure 7–16 shows a general depiction of how an ATM switch operates. Later discussions examine specific implementations. The ATM machine receives a cell on an incoming port (inlet) and reads the VCI/VPI value. This value has been reserved to identify a specific end user for a virtual circuit. It also identifies outgoing port (outlet) for the next node that is to receive the traffic. The ATM switch then examines a routing table for the match of the incoming VPI number and the incoming port to that of an outgoing VPI number and corresponding outgoing port. Many implementations do not make use of a routing table, but rely on the switching fabric to be self-routing; these switches are explained later.

The header in the outgoing cell is changed with the new VPI value

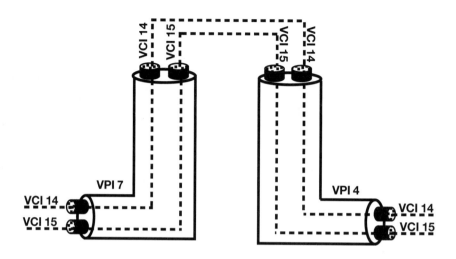

a. VPI translation

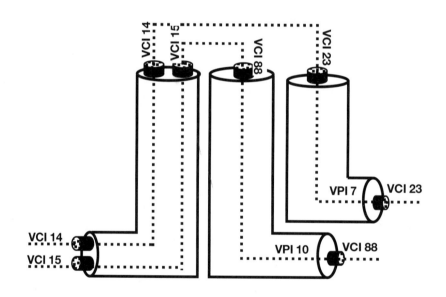

b. VPI/VCI translation

Figure 7–15
VPI/VCI switching.

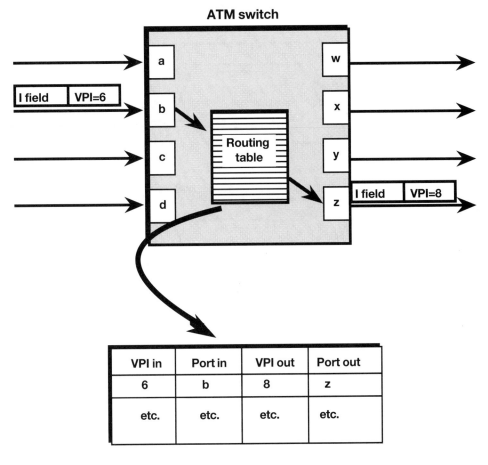

VPI = Virtual path identifier
Note: VPI and VCI may be examined at the VCI/VPI switch

Figure 7–16
ATM routing operations.

placed in the field of the cell header. The new VPI value is used by the next ATM switch to perform subsequent routing operations.

This approach requires the routing tables to be established in advance; that is, this approach uses connection-oriented operations. Nothing precludes a connectionless implementation to support dynamic switching and adaptive routing in the event of problems.

Banyan and delta switching networks. A technique that is often employed in high-speed switches is the *banyan network* (Goke &

Input   Output

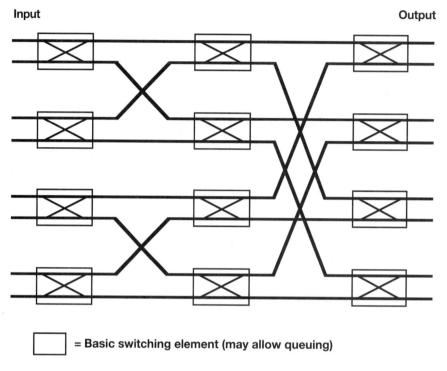

☐ = Basic switching element (may allow queuing)

Figure 7–17
The banyan-delta switching fabric.

Lipovski, 1993). As illustrated in Figure 7–17, this switch is characterized by one path only existing between an input to the final output port.
With this approach, routing operations are quite simple and straightforward, but cells may be blocked (collide) if more than one arrives at a
switching element at the same time.

Banyan networks are classified further into subgroups. An (L)-level
banyan network is one in which only the adjacent elements are connected, so each path passes through the same number of L states. A special
banyan implementation is a *delta network*. It is characterized by having
the same number of inputs to outputs, with each output identified with a
unique value (say a destination address). Each digit in an address identifies a next stage of a switching element in the switching fabric. Figure
7–18 shows an example of the delta network.

A number of switching technologies have been proposed for ATM.
Some of these proposals and implementations are based on the concept
of single state networks wherein a single stage of switching elements are
connected to the inputs and outputs of the switch. Multistage networks

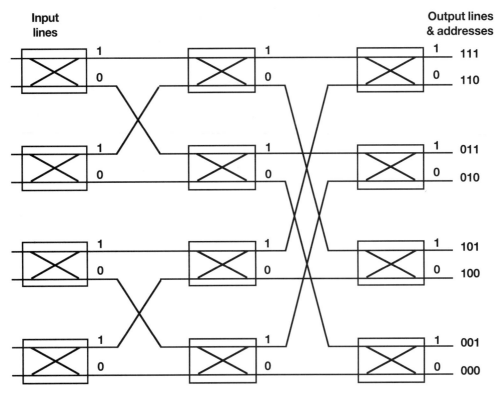

Input lines

Output lines & addresses

111
110
011
010
101
100
001
000

Figure 7–18
A three-stage folded delta network.

(in which several stages of switching elements are connected by various links to form the switching fabric) have also seen use. As of this writing, it is certain that ATM switching will be a melding of several switching techniques, and these techniques will form a switching element that is reproduced and interconnected to form the switching fabric.

Figure 7–18 shows a *multistage delta network* which is called a 3-stage folded network. With this approach, traffic that arrives through the input lines (inlets) are examined by (in this example) three decision stages (switching elements). The traffic is passed to subsequent stages on output lines (outlets) based on an analysis of the address (which could be a tag that is used at the switch). Each stage is responsible for examining one bit in the address, typically from the most significant bit (MSB) to the least significant bit (LSB). In this example, a three-bit address is used for simplicity.

The resulting address resolution is made one stage at a time in each bit of the address. As Figure 7–18 shows, the resultant output is based on the three bit address analysis of the three stages.

Figure 7–19 provides an example of an address resolution performed by the banyan multistage switch. The bold lines indicate the decision processes performed at the various stages. The destination address (010) is examined by each of the three stages on a bit-by-bit basis from the MSB to the LSB. The traffic, with its associated header, and address of 010 is routed to the output line identified as 010.

Collisions may occur in a delta network if cells arrive at a switching element at the same time. These cells may collide at an intermediate element, even though they are destined for the same output line, or they may collide at the final element. The probability of collisions occurring can be reduced by placing buffers in each switching element and allow-

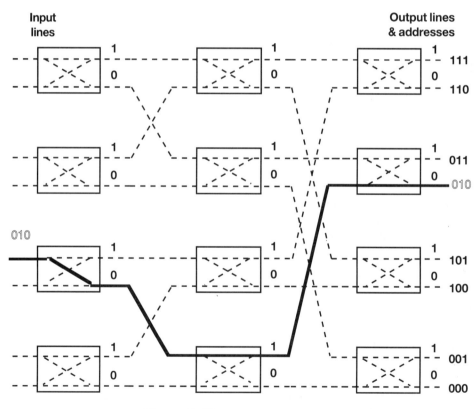

Figure 7–19
Example of address resolution.

ing the buffers to queue awaiting cells, which is an approach that is implemented in a number of high-speed switches. Alternately, the switch can operate at a higher speed than the inlets, thus giving an arrival time at the switching elements with no collision probability. Other mechanisms may be employed, and systems today can operate at a cell loss rate of 10^{-10}.

Figure 7–20 shows three alternatives for the use of buffers and queues within the switching fabric. The first approach places the buffers on the inlet side. Each inlet has a dedicated buffer which is serviced on a first-in, first-out (FIFO) operation. Logic in the buffer determines when the buffer is to be served and when the cell is to be out-queued to the bus. This approach encounters collisions if two cells at the head of the queue in different buffers are destined for the same switching element. Also, if a queued cell is awaiting the availability of a line to a switching element, subsequent cells in the queue are also blocked even though an inlet may be available.

Output buffers are yet another alternative. This approach requires that a single cell only can be served by an outlet which also could create output collisions and contention. Collisions can occur only if the switching fabric operates at the same speed as the incoming lines to the switching elements. In other words, several cells are contending for the same outlet. As discussed earlier, using different speeds within the switching fabric vis-à-vis the communications lines can reduce this problem. So, if collision is not to occur, cell transfer must be performed at n times the speed of the inlets, where n = the number of possible lines.

The last illustration shows central queuing which entails non-dedicated buffers between the inlets and the outlets; that is to say, a central memory element is used by all inlets to the outlets. The process does not work on the FIFO discipline, because the inlet traffic is placed into a single queue. To service the queues efficiently and fairly, there must be some mechanism to access the entries in the queues and to index the traffic in either sequential or random access operation.

Network switches can be developed that use parallel switching techniques. Figure 7–21 shows a parallel banyan switch, also called *vertical stacking*. While switching techniques vary with parallel switching; as a general rule, the ATM cells that belong to the same connection are passed on the same plane of the multi-dimensioned switch. How this is determined varies among switch designers. During call establishment, decisions are made as to which plane will be used for the connection. The switch contains logic which examines each incoming call, places it on the appropriate plane, and then switches it to the appropriate output line.

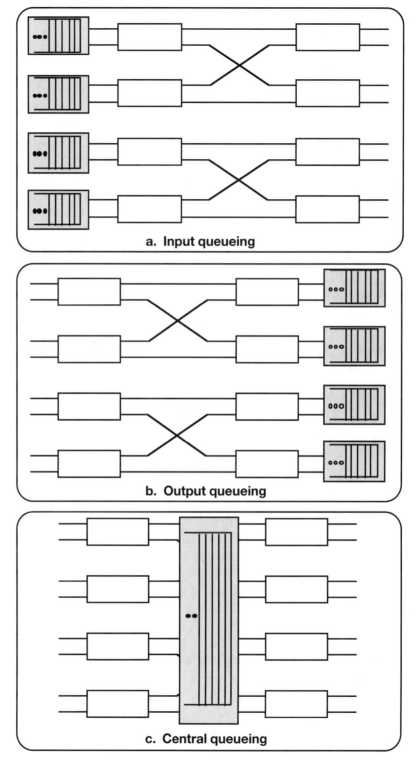

Figure 7–20
Switch buffers and queues.

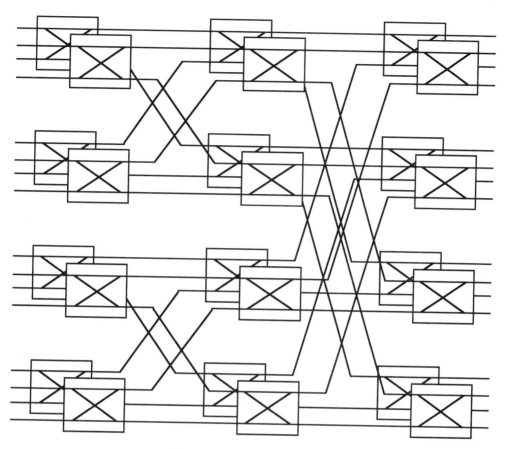

**Figure 7–21
A parallel banyan switch.**

This approach has shown to be highly efficient, and it provides very high throughput.

This general review has dealt with some of the major issues of ATM switching, but has purposely remained at a general level. The interested reader should obtain the IEEE trade journals and the ATM Forum documents for more information on an ATM switching.

Classes of Traffic

The AAL is designed to support different types of applications, and different types of traffic. It is organized around a concept called service classes, which are summarized in Table 7–4. The classes are defined with regards to the following operations (Chapter 2 provides a tutorial on these concepts):

- Timing between sender and receiver (present or not present)
- Bit rate (variable or constant)
- Connectionless or connection-oriented sessions between sender and receiver
- Sequencing of user payload
- Flow control operations
- Accounting for user traffic
- Segmentation and reassembly of user PDUs

As of this writing, the ITU-T had approved four classes, and labels them A through D. Another class of traffic for frame relay is under study. We will now summarize these classes and their major features.

Classes A and B require timing relationships between the source and destination. Therefore, clocking mechanisms are utilized for this traffic. ATM does not specify the type of synchronization—it could be a timestamp or a synchronous clock. Classes C and D do not require timing relationships. A constant bit rate (CBR) is required for class A; and a variable bit rate (VBR) is permitted for classes B, C, and D. Classes A, B, and C are connection-oriented; class D is connectionless.

If the reader studied Chapter 2, it is obvious that these classes are intended to support different types of user applications. For example, class A is designed to support a CBR requirement for high-quality video applications. On the other hand, class B, while connection-oriented, supports variable bit rate applications and is applicable for voice and variable bit rate video applications. For example, the class B service could be used by information retrieval services in which large amounts of video traffic are sent to the user and then delays occur as the user examines the information. As a result of this type of exchange, a variable bit rate service is needed.

Class C services are the connection-oriented data transfer services such as X.25-type connections. Conventional connectionless services such as datagram networks are supported with class D services. Both of these classes also support (final decisions by ITU-T still pending) the multiplexing of multiple end users' traffic over one connection. As of this writing, another class was under study and undergoing revisions through ITU-T working groups. This work has been published in ITU-T Recommendation I.363 Annex 5.

Last, the ATM Forum has specified class X, which is a connection-oriented service in which the QOS and timing requirements are defined by the user.

Table 7–4 AAL Classes of Traffic

Purpose: Convert and aggregate different traffic into standard formats to support different user applications.

Class A	Constant bit rate
	Connection-oriented, e.g., constant bit-rate video
	Timing relationship between source and destination: Required
Class B	Variable bit rate
	Connection-oriented, e.g., variable bit rate video, voice
	Timing relationship between source and destination: Required
Class C	Variable bit-rate
	Connection-oriented, e.g., bursty data services
	Timing relationship between source and destination: Not required
Class D	Variable bit rate
	Connectionless, e.g., bursty datagram services
	Timing relationship between source and destination: Not required
Class X	Traffic type and timing requirements defined by the user (unrestricted)

AAL Types

In order to support different types of user traffic and provide the classes of service just described, AAL employs several protocol types. Each type is implemented to support one or a number of user applications, such as voice, data, etc. Each type consists of a specific SAR and CS. As a general statement, the type 1 protocol supports class A traffic, type 2 supports class B traffic and so on, but other combinations are permitted, if appropriate.

The point has been made several times in this book that the very nature of emerging technologies makes it difficult to write about the subject matter as if the protocols were cast in stone. The AAL is no exception. Initially, the ITU-T standards published four AAL types to support four classes of traffic. However, as the standards groups became more attuned to the tasks at hand, it was recognized that this approach needed to be modified. Also, little interest was shown in defining type 2 traffic for class B applications, and it was also acknowledged that a user-specified class of traffic should be included, as well as a provision for interworking frame relay into ATM. So, modifications were made to the specifications to reflect these changes. This section includes the latest changes made to AAL, and at the time this book went to press, it is up-to-date.

One of the major changes is that VBR applications are serviced at the AAL by (a) a common part (CP), and (b) a service specific part (SSP). As the names of these parts imply, the CP pertains to all VBR applications, and the SSP pertains to VBR applications that require additional and specific services. With these thoughts in mind, the next part of this chapter provides a review of the AAL protocol types.

Type 1.   The AAL uses type 1 PDUs to support applications requiring a CBR transfer to and from the layer above AAL. It is also responsible for the following tasks: (a) acting as SAR of user information, (b) handling the variable cell delay, (c) detecting lost and missequenced cells, (d) providing source clock frequency recovery at the receiver. The SAR has the job of receiving a 47-octet PDU from CS and adding a 1-byte header to it at the transmitting side and performing the reverse operation at the receiving side. CS is responsible for the other operations.

As depicted in Figure 7–22, the AAL type 1 PDU consists of 48-octets with 46- or 47-octets available for the user's payload. The AAL type 1 header consists of 4-bits in the two fields. The first field is a sequence number (SN) and is used for detection of mistakenly inserted cells or lost cells. One bit in the SN field is used to identify two modes of operation: (a) the unstructured data transfer (UDT), or the structured

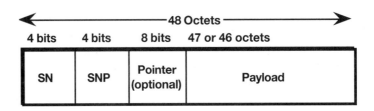

Where:

    SN = Sequence number (1 bit CSI, 3 bits sequence count)
  SNP = Sequence number protection
  CSI = Convergence sublayer indication
         (0 = no pointer; 1 = pointer)

**Figure 7–22
AAL type 1 PDU.**

data transfer (SDT). The former is a bitstream operation that has an associated bit clock, and the latter is a byte stream (8-bit octets) with a fixed block length that has an associate clock (the fixed length block is usually a n * 64 kbit/s channel). The other header field is the sequence number protection (SNP) and it is used to provide for error detection and correction operations on the SN field.

As mentioned earlier, the type 1 CS is responsible for clock recovery for both audio and video services. This operation can be accomplished through the use of real time synchronization or through the use of time-stamping.

The sequence number is not used to request the sender to retransmit the PDU, because of the unacceptable delays involved in visual systems. Rather, the detection of traffic loss with the sequence number can be used as feed-back information to the sender to modify its operations with the knowledge of the nature of the loss [WADA89].

One bit in the SN field is called the convergence sublayer indication [MINO93], and is used to indicate that an 8-bit pointer exists. This capability allows a cell to be partially filled if the user application so requires.

Type 2. AAL type 2 is employed for VBR services where a timing relationship is required between the source and destination sites (see Figure 7–23). For example, class B traffic such as variable bit-rate audio or video, falls into this category. The standards bodies have not defined AAL 2 fully, so this explanation reflects the state of its standardization.

This category of service requires that timing information be maintained between the transmitting and receiving site. It is responsible for handling variable cell delay, as well as the detection and handling of lost or missequenced cells. Since a cell may be only partially filled, due to the

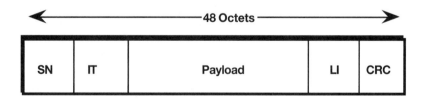

SN = Sequence number
IT = Information type
LI = Length indicator
CRC = Cyclic redundancy check

Figure 7–23
AAL type 2 PDU.

VBR operations, the SAR, in addition to its segmentation and reassembly operations, also maintains a record of how octets in the user payload area actually contain traffic.

The PDU for AAL type 2 traffic consists of both a header and a trailer. The header consists of a sequence number (SN) as well as an information type (IT) field. The length of these fields and their exact functions have not been determined as of this writing. Obviously, the SN will be used for detection of lost and mistakenly inserted cells. The IT field can contain the indication of beginning of message (BOM), continuation of message (COM), or end of message (EOM). It may also contain timing information for audio or video signals.

The AAL type 2 trailer consists of a length indicator (LI) which will be used to determine the number of octets residing in the payload field. And finally, the cyclic redundancy check (CRC) will be used for error detection (and perhaps error correction) of the user payload.

Earlier discussions in this chapter explained how an ATM network can selectively discard voice traffic. AAL2 does not define shedding operations, and the application running on top of AAL2 assumes these responsibilities. For more detailed discussions on cell shedding, refer to *ATM: Foundation for Broadband Networks*, the second book in this series.

Type 3 and type 4. The original ATM standards established AAL3 for VBR connection-oriented operations, and AAL4 for VBR connectionless operations. These two types have been combined and are treated as one type. This section of the book describes the initial types 3 and 4, and then explains the revised approach.

AAL type 3 is used to support connection-oriented variable bit-rate services. These bursty-type data services do not require timing relationships to be maintained between the source and destination. Obviously, type 3 can support class C traffic.

The AAL type 4 PDU is used to support either message-mode service or stream-mode service for data systems (not voice or video). It is designed also to support connectionless services, although the ITU-T also provides for this type to give assured operations in which lost traffic can be retransmitted. With assured operations, flow control is a mandatory feature. In addition, type 4 operations may also provide non-assured operations in which lost or discarded traffic is not recovered, nor is flow control provided.

AAL 3/4. As the AAL standard matured, it became evident that the original types were inappropriate. Therefore, AAL3 and AAL4 were combined due to their similarities, and to support SMDS.

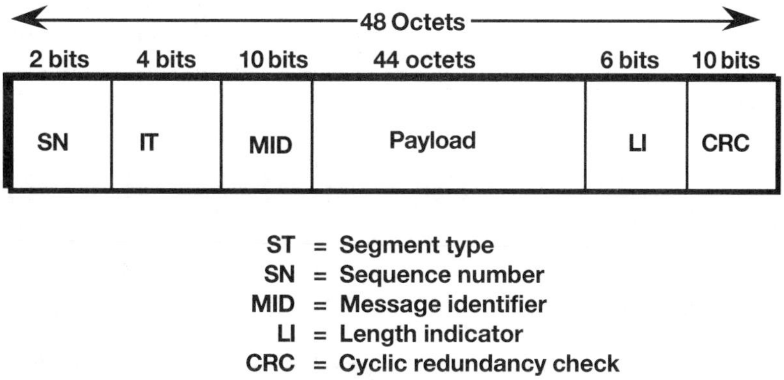

2 bits	4 bits	10 bits	44 octets	6 bits	10 bits
SN	IT	MID	Payload	LI	CRC

ST = Segment type
SN = Sequence number
MID = Message identifier
LI = Length indicator
CRC = Cyclic redundancy check

Figure 7–24
AAL type 3/4 PDU.

As shown in Figure 7–24, the AAL 3/4 PDU carries 44-octets in the payload and 5 fields in the header and trailer. The 2-bit segment type (ST) is used to indicate the beginning of message (BOM), continuation of message (COM), end of message (EOM), or single segment message (SSM). The sequence number is used for sequencing the traffic. It is incremented by one for each PDU sent, and a state variable at the receiver indicates the next expected PDU. If the received SN is different from the state variable, the PDU is discarded. The message identification (MID) subfield is used to reassemble traffic on a given connection. The length indicator (LI) defines the size of the payload. Finally, the cyclic redundancy check (CRC) field is a 10-bit field used to determine if an error has occurred in any part of the cell.

The AAL 3/4 layers and their functions closely resemble their counterparts in MAN and SMDS. As depicted in Figure 7–25, AAL 3/4 consists of a CS/SAR. CS is responsible primarily for error checking, and SAR is responsible for segmentation and reassembly operations. AAL 3/4 can accept and process a user_PDU of up to 65,535-octets. Like MAN and SMDS, the traffic is segmented into 53 octet cells for transmission on the media.

AAL 3/4 operations support two types of data transfer requirements: (a) a message-mode service and (b) a stream-mode service. The message-mode service allows a single SDU to be segmented into smaller pieces for transmission. In the stream-mode service, one or more fixed size SDUs are transported as one AAL convergence function PDU. The stream-mode will allow a SDU to be as small as one octet.

The CS_PDU headers and trailers shown in Figure 7–25 are derived from the MAN PDU header and trailer (see Figure 6–9), and the SMDS

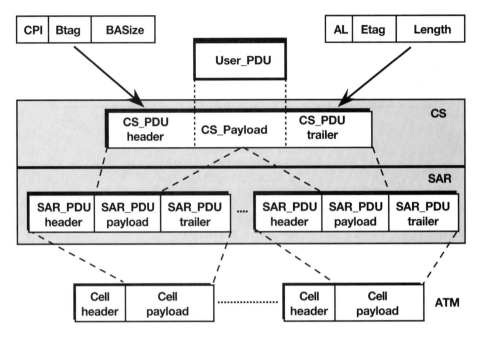

Figure 7–25
AAL 3/4 layers and PDUs.

header and trailer at level 3 (see Figure 6–19). The header contains three fields. The common part identifier (CPI) field (one octet) identifies the type of traffic and certain values that are to be implemented in the other fields of the headers and trailers. The Btag field (one octet) is used to identify all CS_PDUs that are associated with a session. The buffer allocation size field (BAsize, two octets) defines the size of the buffer to receive the CS_PDU at the receiver. The trailer also contains three fields. The alignment field (AL) is a filler to 32-bit align the trailer. The Etag (one octet) is used with the Btag in the header to correlate all traffic associated with the CS_payload. Finally, the Length field specifies the length of the CS_payload (in octets).

Type 5.   The purpose of AAL5 is to provide guidance for transporting upper layer protocols over ATM. It is a convenient service for frame relay because it supports connection-oriented services. In essence, the frame relay user traffic is given to an ATM backbone network for transport to another frame relay user. These operations are described in more detail in Chapter 4. For this chapter, it is sufficient to note that the following requirements must be met:

Frame relay QOS will remain intact from end-to-end

The frame relay user shall not be aware of the ATM operations

The frame relay DLCI must be mapped to the ATM VPI/VCI and vice versa

The frame relay DE bit must be mapped to the ATM CLP bit

QOS remains intact end-to-end

Order of frames will remain intact end-to-end

Congestion and flow control operations must be consistent end-to-end

Traffic Management in an ATM Network

The digitizing of analog images is an easy task with today's technology. These techniques have been around for over 30 years, and are now found in common devices such as personal computers, and even children's toys. A more formidable task is the management of the digital images of voice, video, and data applications in a coherent and integrated fashion.

A properly constructed ATM network must manage traffic fairly and provide effective allocation of the network capacity for different sorts of applications, such as voice, video, and data. The ATM network must also provide cost effective operations relative to the service level (QOS) stipulated by the network user and it must be able to support the different delay requirements of the applications, an important function known as cell delay variation (CDV) operations.

The ATM network must be able to adapt to unforeseen situations. For example, if unusual bursts of traffic occur at the various end user devices, the ATM network must adjust. The network must be able to shed traffic in certain conditions to prevent or react to congestion. In so doing, it must be able to enforce the "maximum peak cell rates" of each VPC/VPI connection. This means that user traffic beyond the maximum peak rate may be discarded.

In addition to these responsibilities, the network must be able to conduct surveillance operations on users' traffic. It must be able to monitor (police) all VPI/VCI connections, and verify their correctness (that they are properly mapped and operating effectively). It must be able to detect and emit alarms when problems are encountered on a VPI/VCI.

An effective ATM network must be designed with an understanding that both the user and the network must assume responsibility for certain QOS operations between them. The user is responsible for agreeing to a service contract with the ATM network that stipulates rules on the use of the network, such as the amount of traffic that can be submitted in a measured time period. In turn, the network assumes the responsibility of supporting the user QOS requirements.

To give the reader an idea of the diversity of traffic that an ATM network must support, Table 7–5 shows some characteristics of different kinds of variable bit-rate traffic. One column lists the mean burst traffic pattern in cells for these applications. The other column lists the average arrival time of the cells (stated in either kilobits or megabits per second). It is obvious that traffic patterns vary widely not just between VBR and CBR traffic but also within the different types of VBR traffic, and the ATM network must support them all (that is, when the final ATM "network in the sky" becomes a reality).

The natural bit rate.   All applications exhibit a natural bit rate (dePrycker, 1992a). It is the rate at which the application generates and/or receives a certain number of bits per second (bit/s) based on its "natural requirements." For example, compressed voice using conventional ITU-T standards exhibits a natural bit rate of 32 kbit/s. Large data transfer systems have a fluctuating natural bit rate depending on the nature of the traffic, ranging anywhere from a few kbits/s to several hundred kbit/s. High-definition television (HDTV) has a natural bit rate of approximately 100-150 kbit/s.

The challenge of the ATM network is to support the natural bit rates of all applications being serviced. Due to widely variable traffic profiles, it may be imperative to discard traffic from certain users if the network experiences congestion problems. This situation is illustrated in Figure 7–26a. For a brief period of time, the user exceeds the transfer

Table 7–5   Traffic Characteristics of VBR Sources (Stated in cells)

Source	Mean Burst	Average Arrival Rate
Connectionless data	200	700 kbit/s
VBR video	2	25 Mbit/s
Connection-oriented data	20	25 Mbit/s
Background data/video	3	1 Mbit/s
VBR video/data	30	21 Mbit/s
Slow video	3	6 Mbit/s

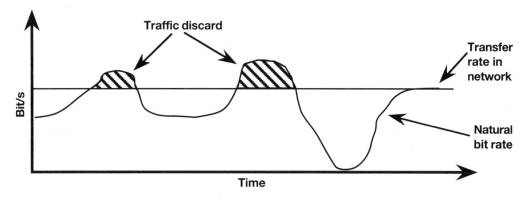

a. User exceeds transfer rate

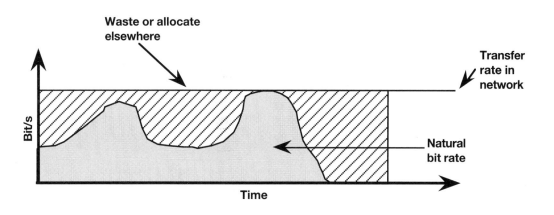

b. Transfer rate greater than user bit rate

Figure 7–26
Bandwidth allocations. [DePr91]

rate permitted in the network with a burst of data. The shadowed portion of the graph is the burstyness of the traffic which exceeds, momen-
tarily, the permitted rate of the network. With ATM networks, this traffic will be tagged for possible discard, as in frame relay networks.

An application's natural bit rate (because of its burstyness) may not
use the full rate allocated by the network to this application. The possibility of wasted bandwidth occurs, as is shown in the striped area of
Figure 7–26b. Therefore, AAL can stuff bits into the application's data
stream such that it appears to be the same as the transfer rate in the

network. Or, the network can attempt to allocate this bandwidth to other applications.

With these discussions in mind, the following material examines some approaches to the management of this complex situation.

ITU-T Recommendation I.35B defines a number of performance parameters for ATM-based networks. Figure 7–27 summarizes these categories, which are: (1) successfully delivered cells, (2) lost cells, (3) inserted cells, and (4) severely damaged cells.

As shown in Figure 7–27a, any error-free cell that arrives (Δt) before its maximum allowed time T is considered to be a successfully delivered cell ($\Delta t < T$) . In Figure 7–27b, cell loss occurs when a cell arrives later than time T ($\Delta t > T$). Even though the cell arrived safely, it is still considered lost, because in some applications, such as voice and video, a late-arriving cell cannot be used in the digital-to-analog conversion process.

In Figure 7–27c, the cell is defined as lost when it is either lost or discarded in the network. In Figure 7–27d, the cell is defined as lost when the cell has an error of more than one bit in the header. The header error correction function cannot correct an error that encompasses more than one bit.

In Figure 7–27e, an inserted cell error occurs when a cell arrives at the destination from a source other than the virtual connection source. In Figure 7–27f, a severely damaged cell is one in which the user I field contains bit errors.

A wide number of proposals exist for the management of traffic in an ATM network. Figure 7–28 shows one proposal that has met with general approval from the ITU-T. This analysis was provided by Bell Laboratories and its Teletraffic Theory Group, which is supervised by A. E. Eckberg (Eckberg, 1992).

The scheme shown in this figure assumes that the user has a service contract with the ATM network wherein certain QOS parameters have been specified. Upon submitting traffic to the network, the user has the option of identifying individual cells with a certain precedence. This decision by the user allows the ATM network service provider to determine how to treat the traffic in the event of congestion problems.

For traffic that is considered to be essential, the network then assumes the responsibility to assure that this traffic is treated fairly vis-à-vis the type of traffic. As an illustration, if the type of traffic is more tolerant to loss, the network then assumes the responsibility of tagging this traffic as "possible loss traffic", and perhaps discarding this traffic if network congestion becomes a problem.

The major task of the ATM network is to make certain that the total cells presented to the network are consistent with the total cells

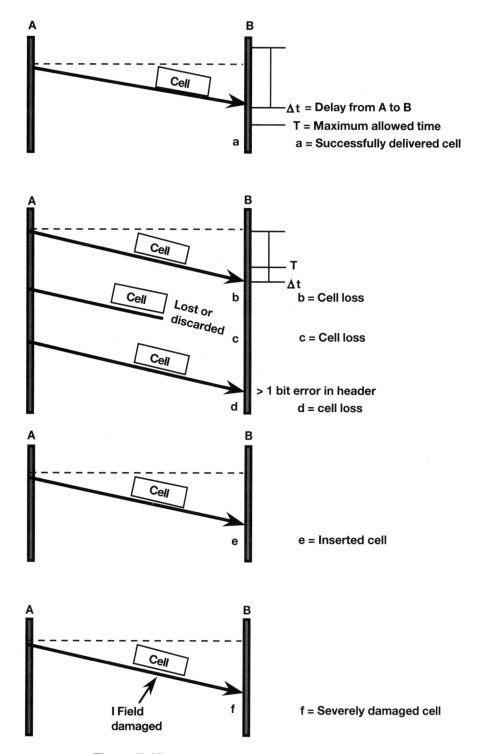

Figure 7–27
Cell transfer performance parameters. [HAND91]

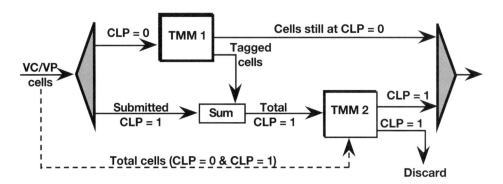

Figure 7–28
Proposed scheme for traffic management. [ECKB 92]

processed by the network. This entails balancing the traffic submitted with cell loss priority of 0 (CLP 0) and CLP 1 to that of the user's service contract.

On the left side of Figure 7–28, the user submits its traffic to the network node with the virtual channel and virtual path identifiers residing in the submitted cell. These identifiers are matched against the user's contract and decisions are made whether or not to grant the user access to the network.

A transmission monitoring machine (TMM1) is responsible for keeping track of all cells submitted by the user with CLP = 0. In the event of unusual problems, or (more likely) if the user violates the contract with the network, the TMM1 can change the CLP of 0 to a 1. The cell that is tagged by the TMM1 is called an excessive traffic tag. TMM1 does not tag all cells. It must pass cells that are within the service contract. These cells come from the user with the CLP bit = 0. This value remains at 0 if traffic conditions are acceptable and the user is within its service contract.

The user may submit traffic to the network with CLP bit = 1 (known as externally tagged cells). The ATM network node will sum these tagged cells with the excessive traffic tagged cells and provide a total of frames with a CLP = 1 to TMM2. TMM2 then makes a decision to (a) discard some of the tagged cells, or (b) permit these cells to enter the network.

Figure 7–29 shows yet another proposed scheme for traffic policing and management. It is based on the idea of a connection-on-demand, wherein a user's service request is matched against the current available bandwidth. Thereafter, resources are "gathered" within the network to support the user payload. The initial service request is submitted to the

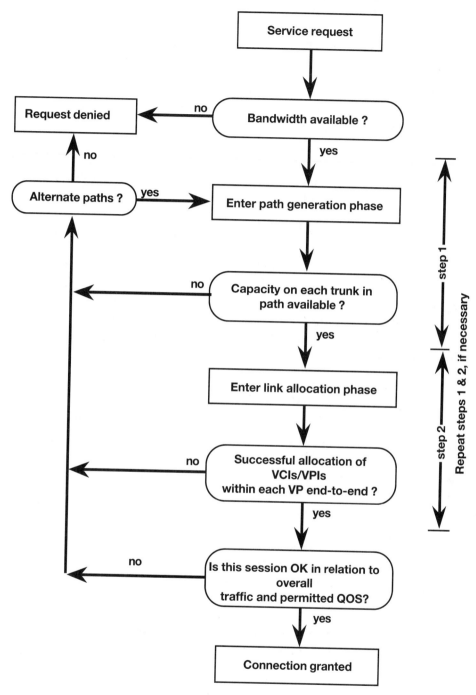

Figure 7–29
Proposed allocation and management scheme. [SYKA92]

ingress node of the network, which makes a determination if sufficient bandwidth is available at the UNI to support the request. If bandwidth is available, the path generation phase is entered and capacity is reserved for each trunk on the virtual path. During the path generation phase, if capacity is not available on any one trunk, the request is denied. If all goes well, the link allocation phase is entered, and VCIs/VPIs are allocated end-to-end. During the link generation phase, if VCI/VPI allocation is not successful on any one trunk, the request is denied. Finally, an analysis is made of the user request in relation to the overall network traffic and the permitted QOS. If this test is passed, a connection is granted.

Will this model work? It might seem preferable to perform the latter overall traffic pattern and QOS test *before* trunk and VCI/VPI allocation operations, since the overhead involved of the allocation procedures would be obviated if this test failed. However, to do so places the allocation process in a Catch-22 situation. If a check is made of the session in relation to overall traffic before the allocation operations, another session might seize the remaining trunk capacity while the check is being performed, which means the service request would have to start over.

ATM Bearer Service Attributes at the UNI

As of this writing, the ATM Forum is studying QOS attributes that will be made available at the UNI. The specific bearer service attributes have been identified, but some of the specific features are not yet complete. Table 7–6 lists the attributes for private and public UNIs, some of which pertain to performance and traffic management. Most of the attributes are self explanatory. A few warrant further explanation.

The subjects of point-to-point and multipoint connections and PVC and SVCs have been covered in earlier parts of this chapter; it is unnecessary to revisit them. The ATM Forum (ATM User-Network Interface Specification, 1993), has defined a set of ATM cell transfer performance parameters to correspond to the ITU-I Recommendation I.350 stipulation for QOS at the UNI. These parameters are summarized in Table 7–7, and the ATM Forum cell rate algorithm is explained in the next section of this chapter. For more detailed information, the reader should refer to Section 3 and Appendix of the ATM User-Network Interface Specification (1993).

A network operator may or may not provide bearer services to its subscribers. However, some of these services must be implemented if the cell relay network is to operate efficiently. As examples, peak rate traffic

Table 7–6 ATM Bearer Service Attributes at the UNI

Support for point-to-point PVCs

Support for point-to-point PCCs

Support for point-to-multipoint PVCs

Support for point-to-multipoint VCCs, SVC

Support for point-to-multipoint VCCs, PVC

Support for PVC

Support for SVC

Support of specified QOS classes

Support of unspecified QOS classes

Multiple bandwidth granularities for ATM connections

Peak cell rate (PCR) enforcement via the usage parameter control (UPC)

Sustainable cell rate (SCR) traffic enforcement via UPC

Traffic shaping

ATM layer fault management

Interim local management interface

---

enforcement, and traffic shaping must be part of the network's ongoing functions in order to keep congestion under control.

ATM Forum and ITU-T Traffic Control and Congestion Control

The ATM Forum and ITU-T have defined algorithms for policing the traffic at the UNI for both CBR and VBR traffic. The traffic parameters employed are: (a) PCR for CBR connections, and (b) PCR, SCR, and maximum burst size (MBS) for VBR traffic. These parameters are provided by the user during the connection establishment with the SETUP message.

An additional parameter is the burst tolerance (BT), which places a restriction on how much traffic can be sent beyond the SCR before it is tagged as excessive traffic. It is calculated as:

$$BT = (MBS - 1) / (1/SCR - 1/PCR)$$

Table 7–7   Cell Transfer Performance Parameters

- The cell error ratio is: errored cells / successfully transferred cells + errored cells

- The severely errored cell block ratio is: severely errored cell blocks / total transmitted cell blocks

  Where: a cell block is a sequence of n cells sent consecutively on a given connection

- Cell loss ratio is: lost cells / total transmitted cells

- Cell misinsertion rate is: misinserted cells / time interval

  Note: calculated as a rate and not as a ratio, since mis-insertion operations are independent of the amount of traffic transmitted

- Mean cell transfer delay is: elapsed time between a cell exit from a measurement point to its entry at another measurement point

- Cell delay variation (CDV) is: the variability of the pattern of cell arrival for a given connection

- Multiple bandwidth granularities are: the provision for bandwidth on demand for each UNI connection

- Peak cell rate (PCR) is: the permitted burst profile of traffic associated with each UNI connection (an upper bound)

- Sustainable cell rate (SCR) is: a permitted upper bound on the average rate for each UNI connection, i.e., an average throughput

- Traffic shaping is: altering the stream of cells emitted into a virtual channel or virtual path connection (cell rate reduction, traffic discarding, etc.)

Generic cell-rate algorithm (GCRA). The generic cell-rate algorithm (GCRA) is employed in traffic policing and is part of the user/network service contract. The GCRA is implemented as a continuous-state leaky bucket algorithm or a virtual scheduling algorithm. The two algorithms serve the same purpose: Make certain cells are conforming (arriving within the bound of an expected arrival time), or nonconforming (arriving sooner than an expected arrival time).

The virtual scheduling algorithm updates a theoretical arrival time (TAT), which is an expected arrival time of a cell. If the arrival is not too early (later than TAT + L, where L is a network-specified limit parameter), then the cell is conforming. Otherwise, it is non-conforming. If cells arrive after a current value of TAT, then TAT is updated to the current time of the arrival of the cell, expressed as $t_a(k)$, where K cell arrives at

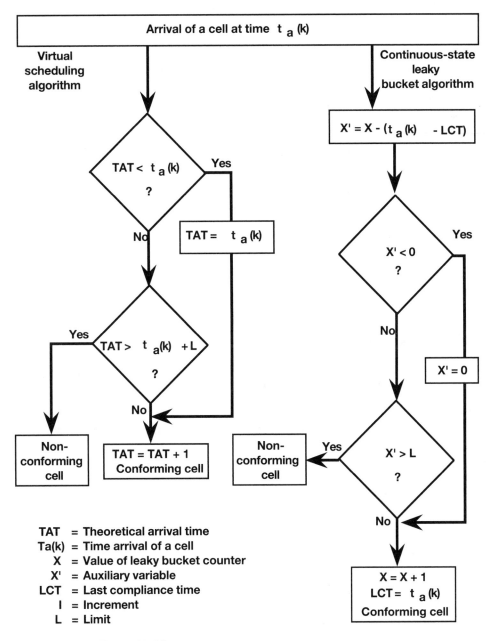

Arrival of a cell at time $t_a(k)$

Virtual scheduling algorithm

$TAT < t_a(k)$?

Yes

No

$TAT = t_a(k)$

$TAT > t_a(k) + L$?

Yes

No

Non-conforming cell

$TAT = TAT + 1$
Conforming cell

Continuous-state leaky bucket algorithm

$X' = X - (t_a(k) - LCT)$

$X' < 0$?

Yes

No

$X' = 0$

$X' > L$?

Yes

No

Non-conforming cell

$X = X + 1$
$LCT = t_a(k)$
Conforming cell

TAT = Theoretical arrival time
Ta(k) = Time arrival of a cell
X = Value of leaky bucket counter
X' = Auxiliary variable
LCT = Last compliance time
I = Increment
L = Limit

Figure 7–30
The generic cell rate algorithm (GCRA). [ATM93a]

time t_a + the increment I. If the cell is nonconforming, TAT is not changed. Other aspects of the virtual scheduling algorithm are shown in Figure 7–30.

The continuous-state leaky bucket algorithm places a bound on the traffic of L + I. The conceptual bucket has a finite capacity; its contents drain out by 1 unit for each cell. Its contents are increased by 1 for each conforming cell. Simply stated, upon the arrival of a cell if the contents of the bucket are less than or equal to the limit L, the cell is conforming. Other aspects of the continuous-state leaky bucket algorithm are also shown in Figure 7–30.

Bandwidth allocation and policing schemes are far from settled, although the ATM Forum and ITU-T approach is supported widely in the industry. Notwithstanding, each month, new papers appear in trade journals, and conferences are held in an attempt to work out this difficult issue. The reader is encouraged to study the IEEE, ACM, and other trade journals for more detail. Indeed, a full book can be written on this issue.

THE ATM BISDN INTERCARRIER INTERFACE (B-ICI)

The ATM internetworking specification is published as the network node interface (NNI) by the ITU-T in Recommendation G.708. Its counterpart, published by the ATM Forum, is called the broadband intercarrier interface (B-ICI) Version 1.0. The principal difference between these two specifications lies at the physical layer. The NNI specifies SDH. The B-ICI specifies SONET or DS3. Of course, other physical layers are technically feasible. Our initial discussions will focus on the B-ICI as published by the ATM Forum in its June 1993 Version 1.0 specification.

Figure 7–31 shows the relationship of the B-ICI to other interfaces and protocols, notably, frame relay, circuit emulation service (CBR traffic), cell relay traffic, and SMDS. All these technologies, as well as ATM, provide for the user to network interface which (as we have learned in this book) is called either a UNI or SNI.

The B-ICI is designed for multi-service operations. The interface supports traffic based on (a) cell relay service (CRS); (b) circuit emulation service (CES), such as DS1 or DS3; (c) frame relay service (FRS); or (d) SMDS. Traffic can be submitted to a network in ATM cells, DS1/DS3 frames, frame relay frames, or SMDSs L2_PDUs. This traffic is sent across the B-ICI in the form of (1) ATM cells over ATM connections; (2) DS1 or DS3 frames (CES), which are encapsulated into AAL type 1

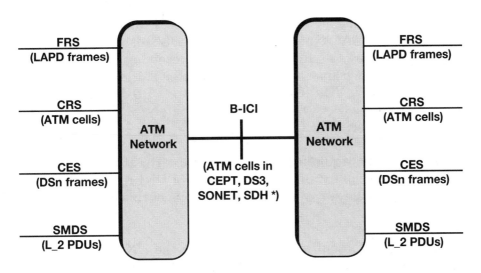

Figure 7–31
ATM B-ICI.
*ATM Forum has defined DS3, SONET

PDUs; (3) frame relay frames, which are encapsulated into AAL type 5 PDUs or mapped into ATM cells; or (4) SMDS L2_PDUs, which are encapsulated into AAL type 3 or type 4 PDUs or mapped directly into ATM cells.

Capacity for each user is assigned based on the type of traffic. Consequently, access classes for SMDS, the committed information rate (CIR), and excess information rate (EIR) for frame relay pertain to this interface. Presently, dynamic bandwidth allocation is not provided.

The ATM B-ICI has considerably more functionality than other ICIs and NNIs discussed thus far in this book. It is (in ITU-T's terms) an interworking unit, which transmits and receives traffic from different systems. The term gateway is sometimes used to describe this function. Other NNIs, such as frame relay, support uni-service traffic—frame relay only. The multi-service aspect of ATM is in keeping with ATM's design to support multimedia traffic. It is good news (and common sense) that the ATM Forum carried this UNI concept through to the B-ICI.

For cell transfer between two ATM networks, the B-ICI specification requires that the two networks retain a relationship between a B-ICI VPC and a UNI VPC. For the transmission of DS1/DS3 frames, the DS1/DS3 VCCs should be translated to unique VCCs at the internetworking interface in order to avoid the use of duplicate VCC values (to those at the UNI). How this is performed is not defined in the standard but is left to the implementation of the carriers. Frame relay frames are encapsulated into AAL type 5 PDUs and the DLCI is sent transparently through the networks and translated at the receiving frame relay interface. Alternately, if frames are mapped to the ATM layer, a DLCI must be translated into a VPI/VCI. Finally, SMDS L2_PDUs can be encapsulated into AAL type 3 or type 4 PDUs. Alternately, SMDS L2_PDUs can be mapped into specific VCC or VPC values.

Physical layer requirements at the B-ICI. The B-ICI physical layer requirements are closely aligned with the ATM UNI requirements. Since these requirements are covered in earlier parts of this chapter, they will not be repeated.

Traffic management at the B-ICI. Like the UNI, the B-ICI stipulates a traffic contract between networks—a carrier-to-carrier service contract. With few modifications, the generic cell rate algorithm (GCRA), described earlier in this chapter, is used at the B-ICI.

For each direction of transmission, the traffic contract consists of several parameters (some optional and some required): (a) a connection traffic descriptor, (b) a requested QOS class, and (c) a definition of a compliant ATM connection. These parameters include: peak cell rate, sustainable cell rate, burst tolerance, cell delay variability tolerance, and conformance definition based on the GCRA. The cell loss priority bit (CLP) may be used at the B-ICI.

Reference traffic loads. The B-ICI specification provides guidance on how the network operator can identify and manage traffic loads in view of complying with the performance objectives of a service contract. The approach is to characterize traffic loads by a utilization factor for a specific type of physical circuit at the B-ICI, and the type of traffic being carried on the circuit. Table 7–8 shows the reference traffic loads for three types of traffic: traffic load type 1 represents CBR traffic (DS1 emulation), and traffic load types 2 and 3 represent VBR traffic. These traffic types are examined with three types of physical links: DS3 with PLCP cell mapping, SONET STS-3c, and SONET STS-12c. Table 7–8 also shows the number of active traffic sources that leads to the link utilization factor.

To illustrate how the reference traffic load is used, consider the requirement for generating reference traffic load type 1. Apply the number of active sources listed in Table 7–8 to each link being tested. The utilization factor is derived from the total traffic from the active sources,

Table 7–8   Reference Traffic Loads for PVC Performance [ATM93a]

Reference Traffic Load Type	Link's Capacity	Number of Active Traffic Sources to Provide Link Utilization Factor	Link Utilization Factor
1	DS3 PLCP	20	0.86
1	STS-3c	73	0.85
1	STS-12c	292	0.85
2	DS3 PLCP	Not applicable	0.85
2	STS-3c	Not applicable	0.85
2	STS-12c	Not applicable	0.85
3	DS3 PLCP	66	0.70
3	STS-3c	242	0.70
3	STS-12c	969	0.70

with each source producing cells at a rate of 4,106 cells a second [(4106 = 1.544 Mbit/s / (47 * 8 bits per cell)]. For CBR traffic, the traffic from the same source must have a uniform distribution over a 244 μsec. interval (244 μsec. = 1/4106 cells/sec.). This source traffic type is compatible with the requirements for AAL type 1, because for its support of 4106 cells/sec. Other reference traffic loads are explained in section 5.1 of the ATM Forum B-ICI documents.

B-ICI layer management operations. The B-ICI layer management operations use the B-ISDN/SONET/SDH F4 and F5 operations, administration and maintenance (OAM) information flows. This topic is covered in Chapter 8 of this book, so it is not repeated here. On another level, the management operations are grouped into the two categories of fault management and performance management. Fault management concerns itself with alarm surveillance, and the verification of ongoing VP and VC connections. Performance management concerns itself with monitoring lost or misinserted cells, cell delay variations, and bit errors. The reader can refer to the section in Chapter 8 titled, "Operation and Maintenance (OAM) Operations" for more information on the F4 and F5 information flows.

OTHER NOTABLE ASPECTS OF ATM

Addressing in an ATM Network

With the addition of SVCs to ATM operations, it is important to have a standardized convention for coding destination and source addresses. Addressing is not an issue with a PVC, since connections and endpoints (destination and source) are defined, and a user need only provide the network with a preallocated VCI/VPI. However, for SVCs, the destination connection can change with each session; therefore, explicit addresses are required. After the call has been mapped between the UNIs, the VCI/VPI values can be used for traffic identification.

The ATM address is modeled on the OSI network service access point (NSAP), which is defined in ISO 8348, and ITU-T X.213, Annex A. A brief explanation of the OSI NSAP follows, and its relationship to ATM addressing (see Figure 7–32).

The ISO and ITU-T describe a hierarchical structure for the NSAP

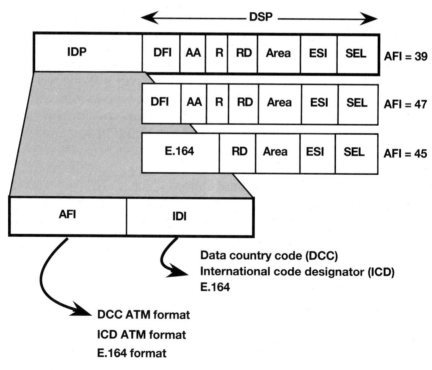

Figure 7–32
The OSI/ATM address formats.

address, and the structure for the NSAP address. It consists of four parts:

1. The *Initial Domain Part* (IDP): Contains the authority format identifier (AFI) and the initial domain identifier (IDI).

2. The *Authority Format Identifier* (AFI): Contains a one-octet field to identify the specific part (DSP). For ATM, AFI field is coded as:

39 = DCC ATM format
47 = ICD ATM format
45 = E.164 format

3. *Initial Domain Identifier* (IDI): Specifies the addressing domain and the network addressing authority for the DSP values. It is interpreted according to the AFI (where AFI = 39, 47, or 45). For ATM, the IDI is coded as (a) a data country code (DCC), in accordance with ISO 3166, (b) an international code designator (ICD), which identifies an international organization, and is maintained by the British Standards Institute, or (c) an E.164 address, which is a telephone number.

4. The *Domain Specific Part* (DSP): Contains the address determined by the network authority. For ATM, the contents vary, depending on value of the AFI. The domain specific part identifier (DFI), specifies the syntax and other aspects of the remainder of the DSP. The administrative authority (AA) is an organization assigned by the ISO that is responsible for the allocation of values in certain fields in the DSP. The R field is a reserved field.

The routing domain identifier (RD) specifies a domain that must be unique either: (a) for AFI = 39, DCC/DFI/AA; (b) for AFI = 47, ICD/DFI/AA, and (c) for AFI = 45, E.164. The area identifies a unique area within a routing domain, and the end system identifier (ESI) identifies an end system (such as a computer) within the area.

The selector (SEL) is not used by an ATM network. It usually identifies the protocol entities in the upper layers of the user machine that are to receive the traffic. Therefore, the SEL could contain upper layer SAPs.

ATM public networks must support the E.164 address and private networks must support all formats.

Network Management

The ATM Forum has defined two aspects of UNI network management: (1) ATM layer management at the M plane, and (2) Interim Local Management Interface (ILMI) Specification. Each of these operations is reviewed in this section; M-plane management is summarized since earlier discussions have explained M-plane operations, and Chapter 8 delves into the subject.

M-Plane management. Most of the functions for ATM M-plane management are performed with the SONET F1, F2, and F3 information flows, which are described in Chapter 8. ATM is concerned with F3 and F4 information flows, which are also covered in Chapter 8. Therefore, discussions are deferred on this topic. The reader can refer to the section in Chapter 8 titled, "Operation and Maintenance (OAM) Operations" for more information on the Fn information flows.

Interim local management interface specification (ILMI). Because the ITU-T and the ANSI have focused on C-plane and U-plane procedures, the ATM Forum has published a interim specification called the ILMI. The major aspects of ILMI are the use of the Simple Network Management Protocol (SNMP) and a management information base (MIB). The MIB is described in the next section.

The ILMI stipulates the following procedures. First, each ATM device supports the ILMI, and one UNI ILMI MIB instance for each UNI. The ILMI communication protocol stack can be SNMP/UDP/IP/AAL over a well-known VPI/VCI value. SNMP is employed to monitor ATM traffic and the UNI VCC/VPC connections based on the ATM MIB with the SNMP Get, Get-Next, Set, and Trap operations.

THE ATM MIB

The ATM Forum has published a management information base (MIB) as part of its ILMI (see Figure 7–33). The ATM MIB is registered under the enterprises node of the standard SMI in accordance with the Internet RFC 1212. The ATM MIB objects are therefore prefixed with 1.3.6.1.4.1.353.

Each physical link (port) at the UNI has a MIB entry that is defined in the atmfPortTable. This table contains: a unique value for each port, an address, the type of port (DS3, SONET, etc.), media type (coaxial cable, fiber, etc.), status of port (in service, out of service, etc.), and other miscellaneous objects.

The atmfAtmLayerTable contains information about the UNI's physical interface. The table contains: the port id, maximum number of

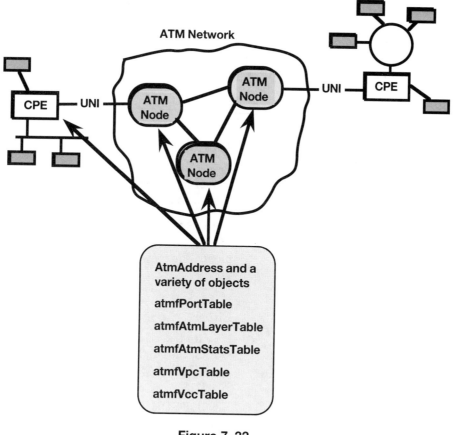

Figure 7–33
The ATM MIB.

VCCs, VPCs supported and configured on this UNI, active VCI/VPI bits on the UNI, and a description of public or private for the UNI.

The atmfAtmStatsTable contains traffic statistics about the ATM layer at the UNI. The table contains: the port id, number of cells received and dropped, number of cells received and not dropped, and number of cells transmitted.

The atmfVpcTable and atmfVccTable contain similar entries for the VPCs and VCCs respectively on the UNI. These tables contain: the port id, VPI or VCI values for each connection, operational status (up, down, etc.), traffic shaping and policing descriptors (to describe the type of traffic management applicable to the traffic), and any applicable QOS that is applicable to the VPI or VCI.

ATM WORKSHEET

The worksheet for ATM is provided in Table 7–9. As suggested in the introduction to this book, it is recommended that the reader attempt to fill out a blank sheet (provided in Appendix D) and then read this section to check your answers.

Table 7–9   ATM Worksheet

Technology name:	Asynchronous Transfer Mode (ATM)
New technology?	Yes, the industry has limited experience on cell-base technology and integrated networks
Targeted applications?	All, it is a multimedia technology
Topology dependent?	Strictly speaking, no, but current implementations are point-to-point
Media dependent?	No, can operate over wire, optical fiber, etc.
LAN/WAN based?	Either
Competes with:	Private leased lines, X.25-based networks, SMDS, Frame relay, and high-speed LAN backbone networks, such as FDDI
Complements:	SONET/SDH as part of the BISDN solution
Cell/Frame based?	Cell based
Connection management?	Yes, PVCs, with SVCs stipulated in later releases by standards groups
Flow control (explicit/implicit)?	Yes, implicit, with flow control field in the ATM header
Payload integrity management?	Depends, ACKs, NAKs, and sequencing are a user responsibility, but the AAL performs sequencing on certain types of traffic
Traffic discard option?	Yes, with the CLP bit
Bandwidth on demand?	Yes, within a contract agreement and network bandwidth availability
Addressing/identification scheme?	Yes, uses labels for identifying connections, which are called VCIs and VPIs

Most of the answers in the worksheet are self-explanatory, but a few comments are in order. ATM and MAN are the technologies discussed in this book that have been designed to support any type of application (voice, video, data, etc.). SONET/SDH certainly supports multimedia traffic, but it is a carrier technology, and does not define the UNI. Frame relay will likely add voice support in the future. Last, ATM is a new technology in that most designers have not had much experience with cell relay networks.

SUMMARY

ATM is a high-speed, low-delay, multiplexing and switching technology. It supports any type of user traffic, such as voice, data, and video applications.

ATM uses small, fixed-length units called cells. Each cell is identified with virtual circuit identifiers that are contained in the cell header. An ATM network uses these identifiers to relay the traffic through high-speed switches.

ATM provides limited error detection operations. It provides no retransmisson services, and few operations are performed on the small header. The intention of this approach is to implement a network that is fast enough to support multi-megabit transfer rates.

ATM has a layer that operates above it, called the ATM adaptation layer (AAL). This layer performs extensive convergence, segmentation, and reassembly operations on different types of traffic.

Synchronous Optical Network (SONET)/Synchronous Digital Hierarchy (SDH)

INTRODUCTION

This chapter examines the Synchronous Optical Network (SONET/SDH)/Synchronous Digital Hierarchy (SDH) systems. It analyzes the advantages and disadvantages of asynchronous and synchronous networks, and explains the digital network synchronization hierarchy, the payloads carried in SONET/SDH envelopes, and multiplexing schemes used in SONET/SDH networks. The chapter examines SONET/SDH equipment such as digital cross connects, add-drop multiplexers, and access nodes (terminals). The operations, administration, maintenance, and provisioning (OAM&P) functions of SONET/SDH are also explained.

The convention in this chapter is to use the terms SONET and SDH interchangeably. SONET/SDH and SDH/SONET mean the same and are used as singular in a sentence. Where they differ, explanations are provided.

PURPOSE OF SONET/SDH

SONET/SDH is an optical-based carrier (transport) network utilizing synchronous operations between the network components. The term SONET is used in North America and SDH is used in Europe. SONET/SDH is:

- A transport technology that provides high availability with self-healing topologies
- A multivendor which allows multivendor connections without conversions between the vendors' systems
- A network that uses synchronous operations with powerful multiplexing and demultiplexing capabilities
- A system that provides extensive OAM&P services to the network user and administrator

SONET/SDH provides a number of attractive features when compared with current technology. First, it is an integrated network standard on which all types of traffic can be transported. Second, the SONET/SDH standard is based on the optical fiber technology which provides superior performance vis-à-vis the older microwave and cable systems. Third, because SONET/SDH is a worldwide standard, it is now possible for different venders to interface their equipment without conversion operations.

Fourth, SONET/SDH efficiently combines, consolidates, and segregates traffic from different locations through one facility. This concept, known as grooming, eliminates back hauling and other inefficient techniques currently being used in carrier networks. Back hauling is a technique in which user payload (say, from user A) is carried past a switch that has a line to A and sent to another endpoint (say, user B). Then, the traffic for B is dropped, and user A's payload is sent back to the switch and relayed back to A. In present configurations, grooming eliminates, but it requires expensive configurations, such as back-to-back multiplexers that are connected with cables, panels, or electronic cross-connect equipment.

Fifth, SONET/SDH eliminates back-to-back multiplexing overhead by using new techniques in the grooming process. These techniques are implemented in a new type of equipment, called an add-drop multiplexer (ADM).

Sixth, the synchronous aspect of SONET/SDH makes for more stable network operations. Later sections in this chapter explain how synchronous networks experience fewer errors than the older asynchronous networks, and provide much better techniques for multiplexing and grooming payloads.

Seventh, SONET/SDH has notably improved OAM&P features relative to current technology. Approximately 5 percent of the bandwidth is devoted to OAM&P.

Eighth, SONET/SDH employs digital transmission schemes. Thus, the traffic is relatively immune to noise and other impairments on the communications channel, and the system can use efficient time division multiplexing (TDM) operations.

Table 8–1 summarizes the main benefits of using SONET/SDH. It is derived from a survey of several vendors.

Equally important, SONET/SDH is backwards compatible, and supports the current transport carriers' systems in North America, Europe, and Japan. This feature is quite important because it allows different digital signals and hierarchies to operate with a common transport system, which is SONET/SDH.

Synchronous Networks

Most digital networks that are in operation today, have been designed to work as asynchronous systems. With this approach, each terminal (device) in the network runs with its own clock. These clocks are not synchronized from a central reference point.

The purpose of the terminal clock is to locate the digital 1s and 0s in

Table 8–1 SONET/SDH Benefits

SONET/SDH Feature	Reduced Capital	Improved Efficiency	Added Revenue
Interconnections	Yes	Yes	
Grooming	Yes	Yes	
Reduced back-to-back muxing	Yes	Yes	
Reduced cabling	Yes		
Better OAM&P		Yes	
Better network management	Yes	Yes	
New services			Yes

Note: This view of SONET/SDH reflects the combined ideas of Northern Telecom Canada, Sprint International, and several PTTs.

the incoming data stream—a very important operation in a digital network. Obviously, if bits are lost in certain payloads, such as data, the traffic may be unintelligible to the receiver. What is more, the loss of bits or the inability to locate them accurately can cause further synchronization problems, so the receiver usually does not even deliver the traffic to the user. It is simpler to discard it and initiate resynchronization efforts.

Because each clock runs independently of others, large variations can occur between the terminal's clock and the rate at which data comes into the terminal. For example, experience has demonstrated that a DS3 signal may experience a variation of up to 1789 bit/s for a 44.736 Mbit/s signal.

Moreover, signals such as DS1s are multiplexed in stages up to DS2, DS3, etc., and extra bits are added to the stream of data (bit stuffing) to account for timing variations in each stream. The lower level signals, such as DS1, are not accessible or visible at the higher rates. Consequently, the original stream of traffic must be demultiplexed if these signals are to be accessed.

SONET/SDH is based on synchronous transmission, meaning the average frequency of all the clocks in the network are the same (synchronous) or nearly the same (plesiochronous). As a result of this approach, the clocks are referenced to a highly stable reference point; so the need to align the data streams or synchronize clocks is unnecessary. Therefore, signals such as DS1/CEPT1 are accessible, and demultiplexing is not needed to access the bitstreams. Also, the signals can be stacked together without bit stuffing. For those situations in which reference frequencies may vary, SONET/SDH uses pointers to allow the streams to "float" within the payload envelope. Indeed, synchronous clocking is the key to pointers; it allows a very flexible allocation and alignment of the payload within the transmission envelope.

This concept of a synchronous system is elegantly simple. By holding specific bits in a silicon memory buffer for a defined and predictable period of time, it is possible to move information from one part of a PDU (a payload envelope) to another part. It also allows a system to know where the bits are located at all times. Of course, this idea is "old hat" to software engineers, but it is a different way of thinking to other designers. As one AT&T engineer put it, "Since the bits are lined up in time, we now know where they are in both time *and* space. So, in a sense, we can now move information in four dimensions, instead of the usual three."

The U.S. implementation of SONET uses a central clocking source; for example, from an end office. This central office must use a highly accurate clocking source known as stratum 3. Stratum 3 clocking requires an accuracy of 1.6 parts in 1 billion elements. A later section returns to this discussion with a more detailed explanation of synchronization, clocking, and pointer operations.

PERTINENT STANDARDS

The SONET/SDH topology is based on standards developed by the American National Standards Institute (ANSI), and the Exchange Carriers Standards Association (ECSA). In addition, Bellcore has been instrumental in the development of these standards. Although SONET has been designed to accommodate the North American DS3 (45 Mbit/s) signal, the ITU-T used SONET for the development and publication of the Synchronous Digital Hierarchy (SDH).

Due to the complexity of implementing a system such as SONET/ SDH, the SONET implementation in the U.S. is divided into three phases. Phase 1 is divided into the implementation of the basic transfer rates, multiplexing scheme, and testing of the frame formats. Phase 2 consists of a number of mapping operations into the optical envelope from other tributaries such as FDDI and ATM. Phase 3 deals with more elaborate implementations to support operations, administration, maintenance, and provisioning (OAM&P).

The SONET/SDH standard has been incorporated into a signaling digital hierarchy standard published by the ITU-T. In addition, Bellcore supports this standard on behalf of the United States Regional Bell Operating Companies (RBOCs). The relevant documents for SONET/SDH are listed in Table 8–2.

TYPICAL SONET/SDH TOPOLOGY

Figure 8–1 shows a typical topology for a SONET/SDH network.[1] End user devices operating on LANs (FDDI, 802.3, 802.5, etc.), and digital transport systems, such as DS1, E1, etc., are attached through a

---

[1] The equipment shown in Figure 8–1 are generic illustrations. Most vendors have designed their equipment in such a way that a machine can be configured to perform the functions of a combination of the services that are offered individually by the machines shown in this figure.

Table 8-2. SONET/SDH Standards

SONET Add-Drop Multiplex Equipment (SONET ADM) Generic Criteria, TR-TSY-000496, Issue 2 (Bellcore, September 1989)

Integrated Digital Loop Carrier System Generic Requirements, Objectives, and Interface, TR-TSY-000303, Issue 1 (Bellcore, September 1986) plus Revisions and Supplements.

Digital Synchronization Network Plan, TA-NPL-000436, Issue 1 (Bellcore, November 1986).

Synchronous Optical Network (SONET) Transport Systems: Common Generic Criteria, TR-TSY-000253, Issue 1 (Bellcore, September 1989). (A module of TSGR, FR-NWT-0000440.)

Transport Systems Generic Requirements (TSGR): Common Requirements, TR-TSY-000499, Issue 3 (Bellcore, December 1989). (A module of TSGR, FR-NWT-000440.)

Synchronous Optical Network (SONET) Transport Systems: Common Generic Criteria, TA-NWT-000253, Issue 6 (Bellcore, September 1990), plus Bulletin No. 1, August 1991.

Generic Reliability Assurance Requirements for Fiber Optic Transport Systems, TA-NWT-000418, Issue 3 (Bellcore, to be issued).

ANSI T1.101 Synchronization Interface Standards for Digital Networks.

ANSI T1.106, Digital Hierarchy—Optical Interface Specifications (Single-mode).

ANSI T1.102, Digital Hierarchy—Electrical Interfaces.

G.700 Framework of the Series G.700, G.800, and G.900 Recommendations

G.701 Vocabulary of Digital Transmission and Multiplexing, and Pulse Code Modulation (PCM) Terms.

G.702 Digital Hierarchy Bit Rates.

G.703 Physical/Electrical Characteristics of Hierarchical Digital Interfaces.

G.704 Synchronous Frame Structures Used at Primary and Secondary Hierarchical Levels.

G.705 Characteristics Required to Terminate Digital Links on a Digital Exchange.

G.706 Frame Alignment and Cyclic Redundancy Check (CRC) Procedures Relating to Basic Frame Structures Defined in Recommendation G.704.

G.707 Synchronous Digital Hierarchy Bit Rates.

G.708 Network Node Interface for the Synchronous Digital Hierarchy.

G.709 Synchronous Multiplexing Structure.

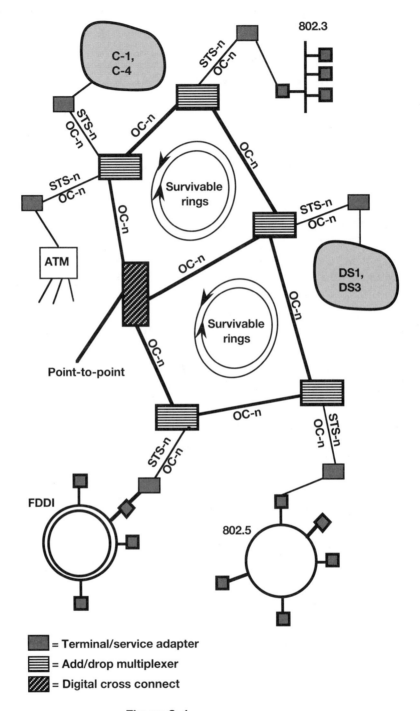

Figure 8–1
SONET/SDH/SDH topology.

SONET/SDH service adapter to the network. This service adapter is also called an access node, a terminal, or a terminal multiplexer. This machine is responsible for supporting the end-user interface by sending and receiving traffic from LANs, DS1, DS3, E1, etc. It is really a concentrator, and at a sending site, it consolidates multiple user traffic into a payload envelope for transport onto the SONET/SDH network. It performs a complementary, yet opposite, service at the receiving site.

The user signals, whatever they are, T1, ATM, E1, etc., are converted (mapped) into a standard format called the synchronous transport signal (STS), which is the basic building block of the SONET/SDH multiplexing hierarchy. The STS signal is an electrical signal, and the notation in Figure 8–1 of "STS-n" means that the service adapter can multiplex the STS signal into higher integer multiples of the base rate. The base rate is 51.84 Mbit/s in North America and 155.520 Mbit/s in Europe. Therefore, from the perspective of a SONET terminal, the SDH base rate in Europe is an STS-3 multiplexed signal (51.84 * 3 = 155.520 Mbit/s).

The terminal/service adapter (access node) shown in Figure 8–1 is implemented as the end-user interface machine, or as an add-drop multiplexer. The latter implementation multiplexes various STS-n input streams onto optical fiber channels, which are now called the optical carrier signal, and designated with the notation OC-n, where n represents the multiplexing integer. OC-n streams can also be multiplexed and demultiplexed with this machine. The term "add-drop" means that the machine can add payload or drop payload onto one of the two channels. Remaining traffic passes straight through the multiplexer without additional processing.

The digital cross-connect (DCS) machine usually acts as a hub in the SONET/SDH network. Not only can it add and drop payload, but it can operate with different carrier rates, such as DS1, OC-n, CEPT1, etc., and can make two-way cross-connections between them. It consolidates and separates different types of traffic. The DCS is designed to eliminate devices called back-to-back multiplexers. As we learned earlier, these devices contain a plethora of cables, jumpers, and intermediate distribution frames. SONET/SDH does not need all these physical components because cross-connection operations are performed by hardware and software.

The topology can be set up as a ring or as point-to-point. In most networks, the ring is a dual ring, operating with two optical fibers. The structure of the dual ring topology permits the network to recover automatically from failures on the channels and in the channel/machine interfaces. This is known as a self-healing ring and is explained later in this chapter.

Figure 8–2 shows another example of a SONET/SDH topology and its multiplexing schemes. Service adapters can accept any signal ranging from DS1/CEPT1 to BISDN. Additionally, sub-DS1 rates (such as DS0) are supported. The purpose of the service adapter is to map these signals into STS-1 envelopes or multiples thereof. In North America, all traffic is initially converted to a synchronous STS-1 signal (51.84 Mbit/s or higher). In Europe, the service adapters convert the payload to an STS-3 signal (155.520 Mbit/s).

Lower-speed signals (such as DS1, and CEPT1) are first multiplexed into virtual tributaries (VTs, a North American term) or virtual

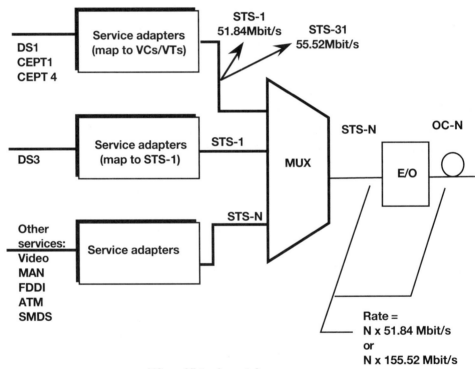

VC = Virtual container
VT = Virtual tributary
STS = Synchronous transport signal
OC = Optical carrier
E/O = Electrical to optical converter

Figure 8–2
SONET/SDH multiplexing.

containers (VCs, a European term), which are sub-STS-1 payloads. Then, several STS-1s are multiplexed together to form an STS-n signal. These signals are sent to an electrical/optical (E/O) converter where a conversion is made to a OC-n optical signal.

Figure 8–3 shows a simplified diagram of a SONET/SDH configuration. Three types of equipment are employed in a SONET/SDH system: (a) path terminating equipment, (b) line terminating equipment, and (c) section terminating equipment. These components are introduced in this section and described in more detail later in the chapter.

The *path terminating equipment* is a terminal or multiplexer, and is responsible for mapping the user payload (DS1, CEPT4, FDDI, etc.) into a standardized format. It must extend to the network elements that assemble and disassemble the payload for the user CPE. The *line terminating equipment* is a hub. It provides services to the path terminating equipment, notably multiplexing, synchronization, and automatic protection switching. It does not extend to the CPE, but operates between network elements. The *section terminating equipment* is a regenerator. It also performs functions similar to HDLC-type protocols: frame alignment, scrambling, and error monitoring. It is responsible for signal reception and signal regeneration. The section terminating equipment may be part of the line terminating equipment.

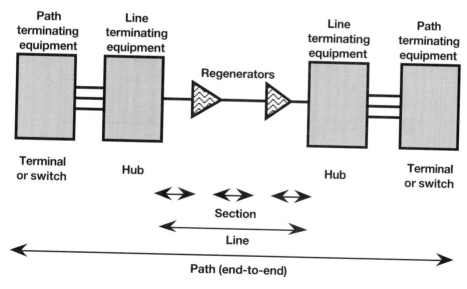

**Figure 8–3
SONET/SDH configuration.**

Each of these components utilizes substantial OAM&P information (overhead). Path level overhead is inserted at the SONET/SDH terminal and carried end-to-end to the receiving terminal. The overhead is added to DS1 signals when they are mapped into virtual tributaries.

Line overhead is used for STS-n signals. This information is created by line terminating equipment such as STS-n multiplexers. The SONET/SDH line concept is important to network robustness, because a line span is protected in case of line or equipment failure, or a deterioration of conditions. Functions operate at the line level to provide for alternate paths—an operation called *protection switching*. Part of the line overhead is used for protection switching.

The *section overhead* is used between adjacent network elements such as the regenerators. It contains fields for the framing of the traffic, the identification of the STS payload, error detection, order wires, and a large variety of network-specific functions.

SONET/SDH LAYERS

The asynchronous transfer mode (ATM) has been designated by the ITU-T to operate with SONET/SDH. Figure 8–4 shows the relationship of the ATM layers and the layers associated with SONET/SDH. The vir-

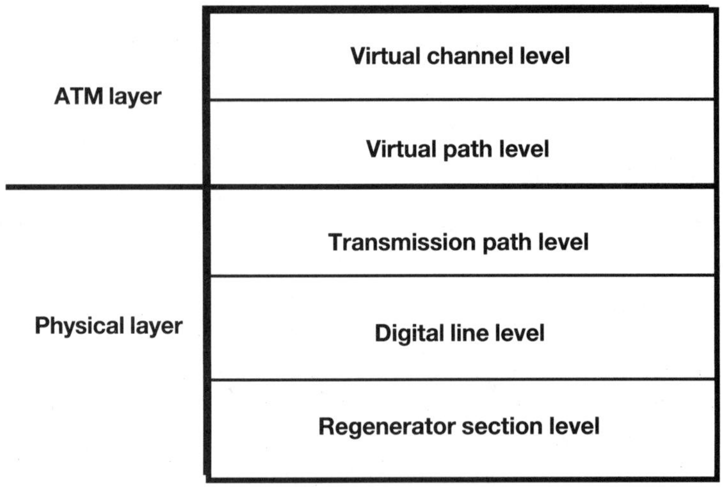

ATM layer	Virtual channel level
	Virtual path level
Physical layer	Transmission path level
	Digital line level
	Regenerator section level

Figure 8–4
The ATM and SDH layers.

tual channel and virtual path layers of ATM run on top of the SONET/SDH physical layer.

The physical layer is modeled on three major entities: transmission path, digital line, and the regenerator section. These layers correspond to the SONET/SDH section, line, and path operations that were introduced in the previous section.

Figure 8–5 shows the relationship of the SONET/SDH and ATM layers with the SONET/SDH channels. The *section* and *photonic* layers comprise the SONET/SDH regenerators. The photonic layer is responsible for converting the electrical signal to an optical signal and then regenerating the optical signal as it is sent through the network. This stack may vary in different implementations. For example, at an ATM switch, the SONET/SDH path layer might not be accessed because it is intended as an end-to-end operation. The manner in which the layers are executed depends on the actual design of the equipment.

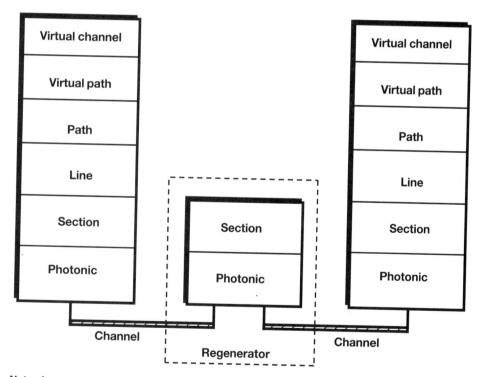

Note: Layer stacks may vary at multiplexers, switches, and other line terminating equipment.

Figure 8–5
A SONET/SDH/ATM Layer Configuration.

SONET/SDH IN MORE DETAIL

Automatic Protection Switching (APS)

One of the more powerful aspects of SONET/SDH is the provision for APS. This feature permits the network to react to failed optical lines and/or interfaces and switch to an alternate facility as illustrated in Figure 8–6. Protection switching operations are initiated for a number of reasons. As examples, the network manager may issue a command to switch the working facility operations to the protection facility for purposes of maintenance, testing, etc. More significantly, APS operations are initiated because of the loss of a connection or the deterioration in the quality of the signal.

APS can be provisioned for a 1:1 or a 1:n facility. With the 1:1

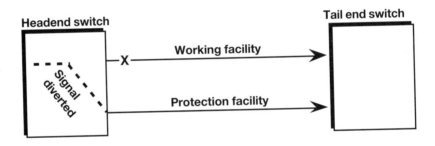

a. **1:1 Protection**

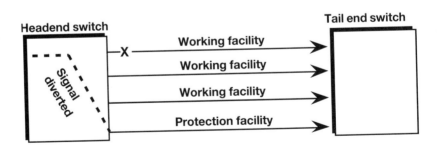

b. **1:n Protection**

Figure 8–6
Protection switching.

option, each working facility (fiber) is backed up by a protection facility (fiber). With a 1:n option, one protection facility may service from one to a maximum of 14 facilities.

As a general practice, the 1:1 operation entails the transmission of traffic on both the working and protection facilities. Both signals are monitored at the receiving end (the tail end) for failures or degradation of signal quality. Based on this analysis, the working or protection facility can be selected and switching operations can be sent back to the sender (headend) to discern which facility is being employed for the transmission of the traffic.

For the 1:n APS, the switching is reverted. That is to say, the traffic is sent across the working facilities and the protection facility is only employed upon the detection of a failure. So the protection facility is not employed until a working facility fails.

THE SDH Multiplexing Structure

Figure 8–7 illustrates the ITU-T SDH multiplexing structure. This structure is quite similar to ANSI's SONET and the ETSI structure, so to simplify matters, one example is shown. The notations in the figure of

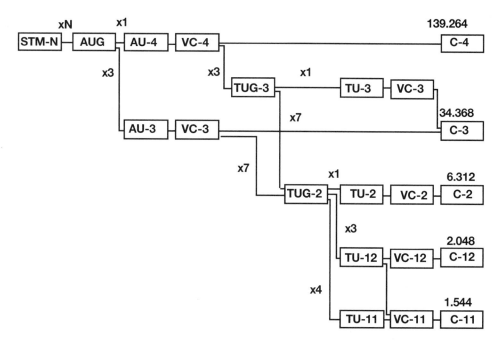

Figure 8–7
The SDH multiplexing structure.

C-1, VC3, etc. are explained in Table 8–3. The reader should refer to this table during this discussion.

At the lowest level, containers (C) are input to virtual containers (VC). The purpose of this function is to create a uniform VC payload envelope. Various containers (ranging from 1.544 Mbit/s to 139.264 Mbit/s) are covered by the SDH hierarchy. Next, VCs are aligned with tributary units (TUs). This alignment entails bit stuffing to bring all inputs to a common bit transfer rate. Next, the VCs are aligned to TUs, where pointer processing operations are implemented.

These initial functions allow the payload to be multiplexed into TU groups (TUGs). As the figure illustrates, the xN indicates the multiplexing integer used to multiplex the TUs to the TUGs. The next step is the multiplexing of the TUGs to higher level VCs, and TUG 2 and 3 are multiplexed into VC 3 and 4. These VCs are aligned with bit stuffing for the administration units (AUs) which are finally multiplexed into the AU

Table 8–3 SDH Multiplexing Structure (See Figure 8–7)

Term	Contents	Use
C-n	$n = 1$ to 4	Contains payload at lowest multiplexing level
VC-n	$n = 1, 2$	Contains single C-n plus VC POH
VC-n	$n = 3, 4$	Contains C-n , TUG-2s, or TU-3s, plus POH for the specific level
TU-n	$n = 1$ to 3	Contains VC plus tributary unit pointer
TUG-2	(TU-n) = 1, 3, 4	Contains various TU-n s
TUG-3	TU-3, 7 TUG-2s	Contains TU-3, 7 TUG-2s
AU-n	$n = 3, 4$	Contains VCs plus AU pointer
STM-1	$n = 1, 3$ AUGs	Contains n synchronously multiplexed STM-1 signals

POH = Path overhead
C   = Container
VC   = Virtual container
TU   = Tributary unit
TUG = Tributary unit group
AU   = Administrative unit
STM = Synchronous transport module

group (AUG). This payload then is multiplexed with an even N integer into the synchronous transport module (STM).

Payloads and Envelopes

Payloads. SONET/SDH is designed to support a wide variety of payloads. Table 8–4 summarizes some typical payloads of existing technologies. The SONET/SDH multiplexer accepts these payloads as sub-STS-1 signals (or VTs).

Table 8–5 shows the relationships of the OC, STS, and SDH levels. The synchronous transport signal-level 1 forms the basis for the optical carrier-level 1 signal. OC-1 is the foundation for the synchronous optical signal hierarchy. The higher-level signals are derived by the multiplexing of the lower level signals. As stated earlier, the high level signals are designated as STS-n and OC-n, where N is an integer number.

As illustrated in Table 8–5, OC transmission systems are multiplexed by the N values of 1, 3, 9, 12, 18, 24, 36, 48 and 192. In the future, multiplexing integrals greater than 192 will be incorporated into the standard. Presently, signal levels OC-3, OC-12, and OC-48 are the most widely supported multiples of OC-1.

Payloads reside in the user payload area of the synchronous payload envelope (SPE). This payload is the revenue-producing traffic that is transported and routed through the SONET/SDH network. The SONET/SDH network is application/service independent. It remains transparent to the user application and vice versa. Indeed, the user payload is transferred intact across the network.

The reader may wish to skip the next sections (and go to "SONET/SDH Equipment") if the details of SONET/SDH (multiplexing and OAM operations) are not of interest.

Table 8–4   Typical SONET/SDH Payloads

Type	Digital Bit Rate	Voice Circuits	T-1	DS3
DSI	1.544 Mbit/s	24	1	–
CEPT1	2.048 Mbit/s	30	–	–
DS1C	3.152 Mbit/s	48	2	–
DS2	6.312 Mbit/s	96	4	–
DS3	44.736 Mbit/s	672	28	1

Table 8–5   SONET/SDH Signal Hierarchy

OC Level	STS Level	Line Rate (Mbit/s)
OC-1*	STS-1	51.840
OC-3*	STS-3	155.520
OC-9	STS-9	466.560
OC-12*	STS-12	622.080
OC-18	STS-18	933.120
OC-24	STS-24	1244.160
OC-36	STS-36	1866.230
OC-48*	STS-48	2488.320
OC-96	STS-96	4876.640
OC-192	STS-192	9953.280

* currently, the more popular implementations

(Note: Certain levels are not used in Europe, North America, and Japan.)

Envelopes.   The basic transmission unit for SONET is the STS-1 envelope (frame). SDH starts at the STS-3 level. All levels are comprised of 8-bit bytes (octets) that are transmitted serially on the optical fiber. For ease of documentation, the payload is depicted as a two-dimensional map (see Figure 8–8). The map is comprised of n rows and m columns. Each entry in this map represents the individual octets of a synchronous payload envelope. (The "F" is a flag and is placed in front of the envelope to identify where the envelope begins.)

The octets are transmitted in sequential order, beginning in the top lefthand corner through the first row, and then through the second row, until the last octet is transmitted—to the last row and last column.

The envelopes are sent contiguously and without interruption, and the payload is inserted into the envelope under stringent timing rules. Notwithstanding, a user payload may be inserted into more than one envelope, which means the payload need not be inserted at the exact beginning of the part of the envelope that is reserved for this traffic. It can be placed in any part of this area, and a pointer is created to indicate where it begins. This approach allows the network to operate synchronously, yet accept asynchronous traffic.

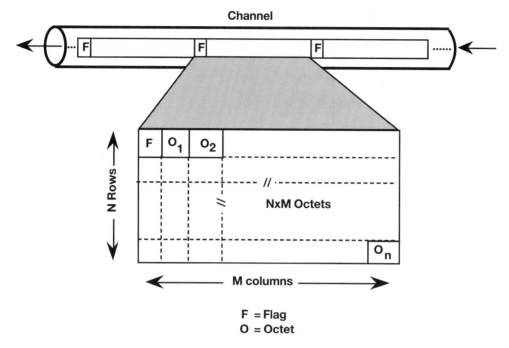

Figure 8–8
The synchronous payload envelope.

Figure 8–9 depicts the SONET STS-1 envelope. It consists of 90 columns and 9 rows of 8-bit bytes (octets), and carries 810-bytes or 6480-bits. SONET/SDH transmits at 8000 frames/second. Therefore, the frame length is 125 microseconds (μsec.). This approach translates into a transfer rate of 51.840 Mbit/s (6480 × 8000 = 51,840,000).

The first three columns of the frame contain transport overhead, which is divided into 27-bytes, with 9-bytes allocated for *section overhead* and 18 bytes allocated for *line overhead*. The other 87 columns comprise the STS-1 SPE (although the first column of the envelope capacity is reserved for STS path overhead). The section overhead in this envelope is also known as region overhead in certain parts of the world and the line overhead is also known as the multiplex section. The frame consists of two distinct parts: the *user SPE part* and the *transport part*.

Since the user payload consists of 86 columns or 774-bytes, it operates at 49.536 Mbit/s (774 * 8 bits per byte * 8000 = 49,536,000). Therefore, the user payload can support VTs up to the DS3 rate (44.736 Mbit/s).

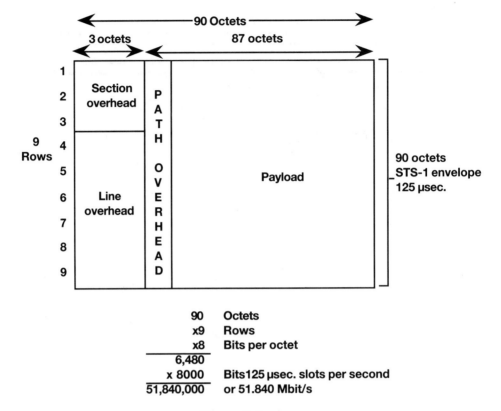

	Octets
90	Octets
x9	Rows
x8	Bits per octet
6,480	
x 8000	Bits125 µsec. slots per second
51,840,000	or 51.840 Mbit/s

Figure 8–9
STS-1 envelope.

The SDH envelope begins at STS-3. As shown in Figure 8–10, it consists of three STS-1 envelopes, and operates at a bit rate of 155.52 Mbit/s ($51.840 \times 3 = 155.52$ Mbit/s). The STS-3 SPE has sufficient capacity to carry a broadband ISDN H4 channel.

The original SONET standard published by Bellcore had no provision for the European rate of 140 Mbit/s. Moreover, it was inefficient in how it dealt with the European 2.048 Mbit/s system. Bellcore and the T1 committee accommodated European requests and accepted the basic rate for SDH at 155.52 Mbit/s, and other higher multiplexing rates were also approved. All parties worked closely to accommodate the different needs of the various administrations and countries, which resulted in a world-wide multiplexing structure that operates with North American, European, and Japanese carrier systems.

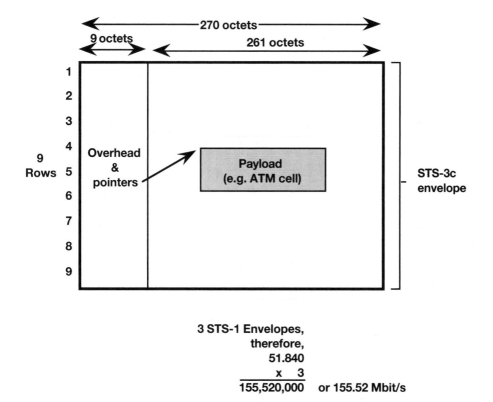

Figure 8–10
STS-3 envelope.

Payload Pointers

SONET/SDH uses a new concept called pointers to deal with timing variations in a network. The purpose of pointers is to allow the payload to "float" within the payload. As Figure 8–11 shows, the SPE can occupy more than one frame. The pointer is an offset value that shows the relative position of the first byte of the payload. During the transmission across the network, if any variations occur in the timing, the pointer need only be increased or decreased to compensate for the variation. While SONET/SDH is designed to be a synchronous network, different networks may operate with different clocks at slightly different rates. So, the pointers and the floating payload allow the network to make adjustments to these variations. In effect, payload pointers allow the existence of asynchronous operations within a synchronous network.

Traffic must be synchronized in a SONET/SDH network equipment

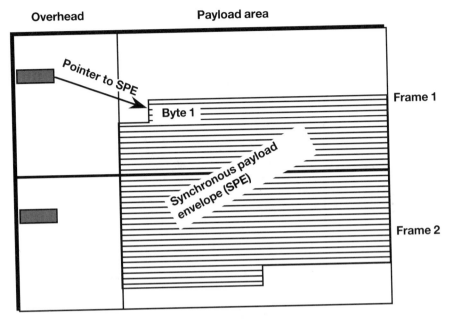

Figure 8–11
Floating payloads.

before it is multiplexed. For example, as individual transport signals arrive at a multiplexer, they may be misaligned due to timing and bit rate differences. The bit rate variation could occur because of asynchronous operations between other equipment. The SONET/SDH equipment will synchronize the traffic such that: (a) the individual transport overhead bytes are aligned and (b) payload pointers have been changed to adjust the user payload within the envelope. Therefore, two types of timing adjustments take place. One timing adjustment takes place within the SPE with pointers, and the other takes place with buffer adjustments at the receiver.

Examples of Payload Mapping

This section gives the reader an overview of how several user payloads are mapped into the SPE. During this overview, keep in mind that SONET/SDH can carry today's carrier payloads. However, the reverse case is not true; today's carriers cannot carry SONET/SDH payloads.

SONET/SDH is designed to be "backward compatible" with the carrier technologies in Europe, North America, and Japan. Therefore, the

DS and CEPT payloads are supported and carried in the SONET/SDH envelope. These payloads range from the basic DS-0/CEPT0 of 64 kbit/s up to the higher-speed rates of DS-4 and CEPT5.

In addition, cells are also carried in the SONET/SDH envelope, which results in the technology being both backward compatible (supporting current technology) and "forward compatible" (supporting cell technology).

SONET/SDH supports a concept called virtual tributaries (VT) or virtual containers (VC)—the former phrase is used in SONET, and the latter phrase is used in SDH. Figure 8–12 shows the major VT support provided in the OC-1 payload of 51.84 Mbit/s.

Through the use of pointers and offset values, VTs/VCs such as DS1, DS3, CEPT1, etc. can be carried in the SONET/SDH envelope. The standard provides strict and concise rules on how various VTs/VCs are mapped into the SONET/SDH envelope.

VTs/VCs are used to support sub-STS-1 levels, which are lower-speed signals. To support different mixes of VTs/VCs, the STS-1 SPE can be divided into seven groups. As seen in Figure 8–13, each group occupies columns and rows of the SPE and may actually contain 4, 3, 2, or 1 VTs/VCs. For example, a VT group may contain one VT 6, two VT 3s, three VT 2s or four VT 1.5s. Each VT/VC group must contain one size of VTs/VCs, but different groups can be mixed in one STS-1 SPE. The four

OC-1 = 51.84 Mbit/s

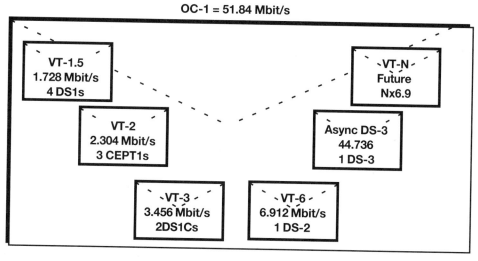

Figure 8–12
Virtual tributary payloads in the 51.84 Mbit/s envelope.

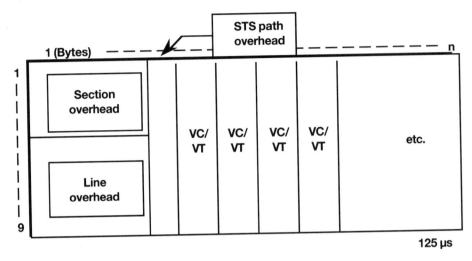

Figure 8–13
Virtual containers (VCs)/Virtual tributaries (VTs).

sizes of the VT supported by SONET are: VT 1.5 = 1.728 Mbit/s, VT 2 = 2.304 Mbit/s, VT 3 = 3.456 Mbit/s, VT 6 = 6.912 Mbit/s.

As stated earlier, the various administrations and standards groups from Japan, North America, and Europe worked closely together to accommodate the three different regional signaling standards. The initial SONET standards published in the United States in 1984 were reviewed by Japan and the European PTTs to see if these requirements would meet their needs. During this time, the ANSI T1 committee had become involved with Bellcore in the development of SONET.

Discussions continued through the European Telecommunications Standard Institute (ETSI), and agreement was reached on a subset of the multiplexing schemes of the three regions. ETSI stressed the importance of the intermediate rates of 8 and 34 Mbit/s in contrast to the ease of doing international networking. Reason prevailed, compromises were reached, and the importance of international internetworking came to the fore with multiplexing schemes based on 1.5, 2.48, and 6.312 Mbit/s capabilities.

Figure 8–14 shows the structure of the virtual tributaries (VT) and virtual containers (VC). VT1.5 is called VC1-11 in Europe and it accommodates the T1 rate. VT2 is called VC-12 in Europe. It accommodates the Europe CEPT1 rate of 2.048 Mbit/s. VT3 is not employed in Europe. It is used in North America to optimize multiplexing DS1c transport sig-

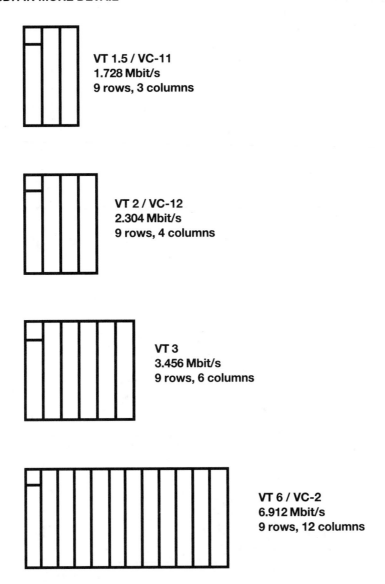

VT 1.5 / VC-11
1.728 Mbit/s
9 rows, 3 columns

VT 2 / VC-12
2.304 Mbit/s
9 rows, 4 columns

VT 3
3.456 Mbit/s
9 rows, 6 columns

VT 6 / VC-2
6.912 Mbit/s
9 rows, 12 columns

Figure 8–14
The tributaries and containers to accommodate regional multiplexing schemes.

nals. VT6 is called VC-2 in Europe and it accommodates the 6.912 Mbit/s rate from all three regions.

Figure 8–15 shows the envelope of a VT 1.5 group. The 1.5 Mbytes occupy three columns and nine rows. The user traffic consists of 24-bytes in accordance with a T1 24-slot frame. The remaining three-bytes are

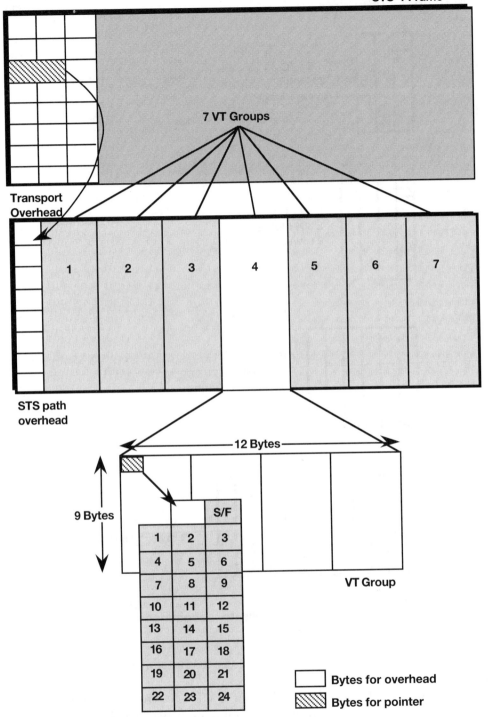

Figure 8–15
The VT 1.5 group.

used for SONET control. A VT 1.5 group supports four VT 1.5 transmissions to occupy the full 12 columns of the VT structure. The VT is organized as a super frame structure and spread over multiple frames. Each frame contains VT payload pointers, which locate the payload in the envelope, as well as specify the size of the VT. The pointers also have error checking capabilities as well as some bits that provide for path status information. These bits have the same functions of the bytes as other parts in the payload envelope (B3, C2, and G1), and will be discussed later.

Two options dictate how the VT payload is mapped into the frame. One option is called the floating mode, for obvious reasons. Floating mode provides a convenient means to cross-connect transport signals in a network. Floating mode allows each VT SPE to float with respect to the complete envelope. This approach also obviates the use of slip buffers, which have been used in the past to phase-align the individual multiplexed signals as required. While the use of the slip buffers allows the system to repeat or delete a frame to correct for frequency variations, they should be avoided, if possible, because they impose additional complexity and may further impair the system. Payload pointers eliminate the need for slip buffers.

Another option is called the locked mode. With this approach, the pointers are not used and the payload is fixed within the frame. It cannot float. Locked mode is simpler, but it requires that timing be maintained throughout the network, but since all signals have a common orientation, the processing of the traffic is performed efficiently. However, slip buffers must be employed to adjust to any timing and synchronization differences that may be present in the system.

Mapping and Multiplexing Operations

This section expands the explanations in the previous sections, and shows more examples of the SONET/SDH mapping and multiplexing functions. One idea should be kept in mind when reading this section: One principal function of these operations is to create a payload format and syntax that is the same for any input (DS1, CEPT1, etc.), after the initial multiplexing and pointer processing is complete. Therefore, the T1 and CEPT1 rates of 1.544 Mbit/s and 2.048 Mbit/s respectively are mapped and multiplexed into a 6.912 Mbit/s frame. Then, these signals are multiplexed further into higher levels.

This section does not show all the mapping and multiplexing possibilities, but concentrates on the VT 1.5 group (1.544 Mbit/s) and the VC-12 group (2.048 Mbit/s). Figure 8–16 shows the VT 1.5 which multiplexes

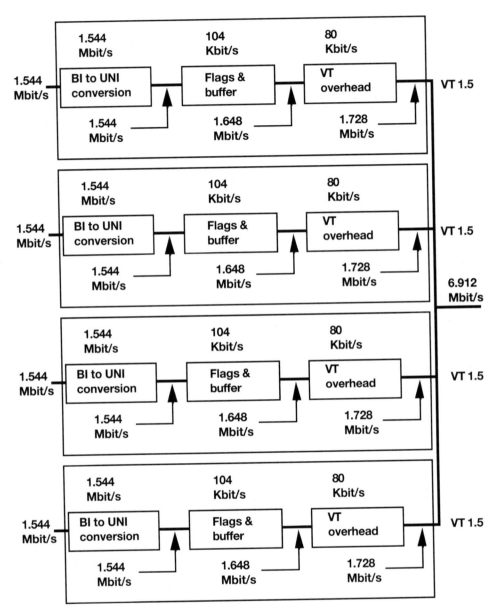

Figure 8–16
Virtual tributary group (VTG) 1.5.

314

four DS1 systems. Each 1.544 Mbit/s DS1 signal is converted to a 1.728 Mbit/s virtual tributary. The additional bits are created to provide flags, buffering bits, conversion bits, and VT headers.

The four DS1s are multiplexed to equal a 6.912 Mbit/s VT output. Then the 6.912 Mbit/s output is input into additional multiplexing functions in which each VT has path, line, and section overhead bits added. In addition, pointer bits are added to align the VT payload in the SONET/SDH envelope. The result of all these operations is a 51.840 Mbit/s STS-1 signal, which is shown later in Figure 8–18.

The European CEPT1 is converted to a VT2 (called VC-12 in Europe) signal of 6.912 Mbit/s (see Figure 8–17). It can be seen that the

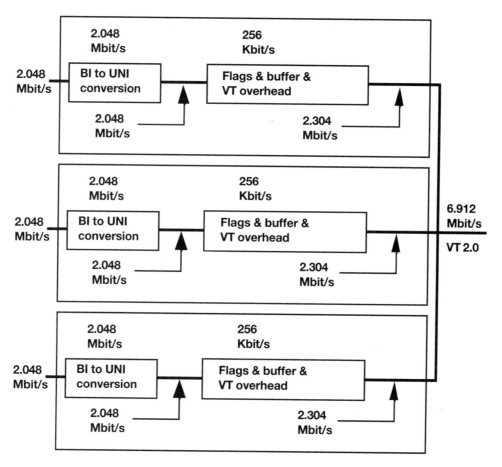

Figure 8–17
Virtual tributary 2 (VT 2) or tributary unit group 2 (TUG-2).

approach of the multiplexing scheme is to provide a preliminary payload output of 6.912 Mbit/s for all input streams.

For the CEPT1 conversion, 3 CEPT transmissions of 2.048 Mbit/s are input into the SONET/SDH conversion operation. The operation adds flags, buffering bits, and VT overhead bits which equal 256 kbit/s. Therefore, each CEPT signal is converted to a 2.304 Mbit/s signal. 3 CEPT1 signals at 2.304 equal the desired 6.912 Mbit/s output stream. This stream is called the transmission unit group 2 (TUG-2) in the SDH standards.

Figure 8–18 shows (a) how 28 DS1s are mapped into the DS1 payload, and (b) how a DS3 transmission is mapped into the STS-1 payload. The two mappings in Figure 8–17 are taken from two separate operations on two separate input streams. They are shown together to illustrate how both input streams are mapped first to 48.384 Mbit/s, second to 50.112 Mbit/s, and finally to 51.840 Mbit/s.

The purpose of the initial multiplexing and mapping is to create an intermediate stream of 48.384 Mbit/s; thereafter, both DS1, CEPT, and DS3 transmissions are treated the same. All these transmissions have path, line, and section overhead bits added as well as the STS-1 pointer. The result is shown on the right hand side of the figure as the 51.840 Mbit/s STS-1 envelope.

Error Checking, Diagnostics, and Restoration

The enhanced network management attributes of SONET/SDH are one of the reasons carriers are migrating to this technology. AT&T Network Systems International (in London, England) compared to the time required to restore operations due to a failure in a network node for a current digital service path operating at 140 Mbit/s and for an SDH network. The comparison was broken down into five major operations: failure identification, route selection, route implementation, continuity testing, and authorization and transfer (see Figure 8–19).

A *failure identification* determines the point of failure in the span or the path level. This step may be performed manually or with software, and it is always necessary. Part of this operation is to determine the order in which the digital paths are to be restored. *Route selection*, as its name implies, entails the analysis of available alternate routes and the capacity of these routes to handle more traffic. *Route implementation* involves the connection of the various intermediary parts such as digital cross connects that were identified during the route selection process.

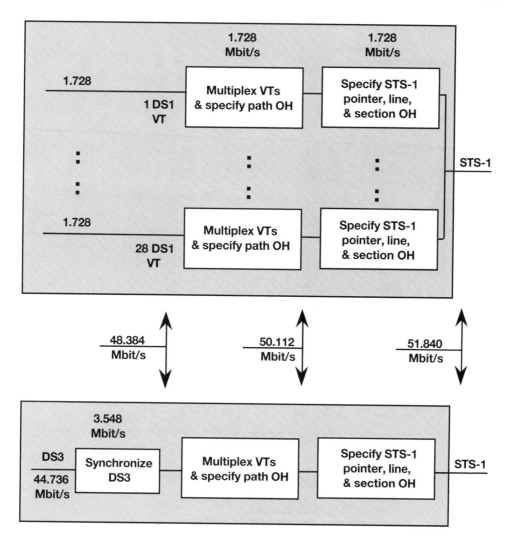

Figure 8–18
The STS-1 payload.

Continuity testing is used to determine if the path is fully connected and of sufficient quality to be used. Finally, *authorization* and *transfer* completes the final cross connect operations that are needed to reestablish the service to the user.

Figure 8–19 compares the relative times needed by a current network and an SONET/SDH network to complete these five operations.

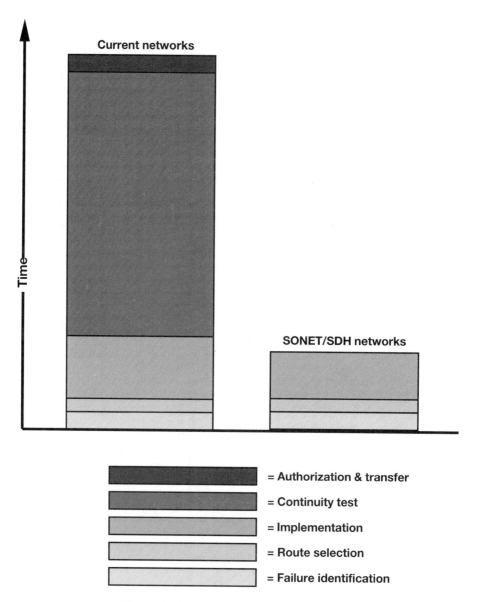

Figure 8–19
Comparison of network restoration times.

Absolute times are not used because each incident of failure and recovery entails different parts of a network. Nonetheless, SONET/SDH eliminates some of these operations completely, and reduces the time of others.

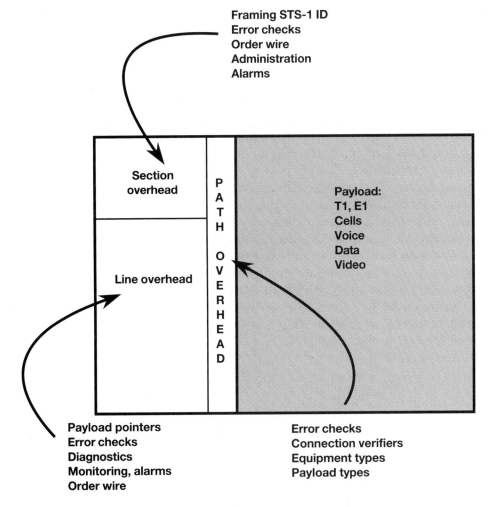

Framing STS-1 ID
Error checks
Order wire
Administration
Alarms

Section
overhead

P
A
T
H

O
V
E
R
H
E
A
D

Line overhead

Payload:
T1, E1
Cells
Voice
Data
Video

Payload pointers
Error checks
Diagnostics
Monitoring, alarms
Order wire

Error checks
Connection verifiers
Equipment types
Payload types

Figure 8–20
SONET/SDH control fields.

The Control Headers and Fields

Figure 8–20 provides a general view of the control fields which are used by the SONET/SDH equipment for control and signaling purposes. The section overhead and the line overhead make up the transport overhead, which consumes nine rows of the first three columns of each STS-1 payload. This equals 27-bytes which is allocated to the transport overhead. Nine bytes are allocated for the section overhead, and 18-bytes are allocated for the line overhead. These fields are explained in the next section. On a more general note, headers are used to provide signaling control, alarms, equipment type, framing operations, error checking operations, etc.

Section overhead. Figure 8–21 shows the section overhead fields which are used by the SONET/SDH equipment for control and signaling purposes. Each box in this figure is one byte of information. The A1 and A2 bytes are the framing bytes. They are provided with all STS-1 and STS-n signals. The bit pattern is always 1111011000101000 in binary, or F628 in hexadecimal. The purpose of these octets is to identify the beginning of each STS-1 frame. The receiver initially operates in a search mode and examines bits until the candidate A1 and A2 pattern is detected. Afterwards, the receiver changes to the maintenance mode, which correlates the received A1 and A2 values with the expected values. If this mode detects the loss of synchronization, the search mode is then executed to once again detect the framing bits.

The C1 bit is used for STS-1 identification. It is a unique number that is assigned to each STS-1 of an STS-n signal. The C1 byte in the STS-1 is set to a number that corresponds to its order in the STS-n frame. The C1 value is assigned to each STS-1 signal before the signal is byte-interleafed into an STS-n frame. The C1 byte is simply incremented from zero through n to indicate the first, second, third, through n STS-1 signals to appear in the STS-n signal. The identifying number does not change in the STS-1 signal until byte-deinterleaving occurs.

The B1 byte is the bit interleaved parity (BIP-8) bit. SONET/SDH performs a parity check on the previously sent STS-1 frame and places the answer in the current frame. The bit-8 byte checks for transmission errors over a section. It is defined only for the first STS-1 of the STS-n signal.

The E1 byte is an orderwire byte. It is a 64 kbit/s voice path which can be used for maintenance communications between terminals, hubs, and regenerators.

The F1 byte is set aside for the network provider to use in any manner deemed appropriate, but it is used at the section terminating equipment within a line. Finally, the D1, D2, and D3 bytes are for data communications channels and are part of 192 kbit/s operations that are used for signaling control, administrative alarms, and other OAM.

Line overhead. The line overhead occupies the bottom 6-octets of the first three columns in the SONET/SDH frame (see Figure 8–22). It is processed by all equipment except for the regenerators. The first two

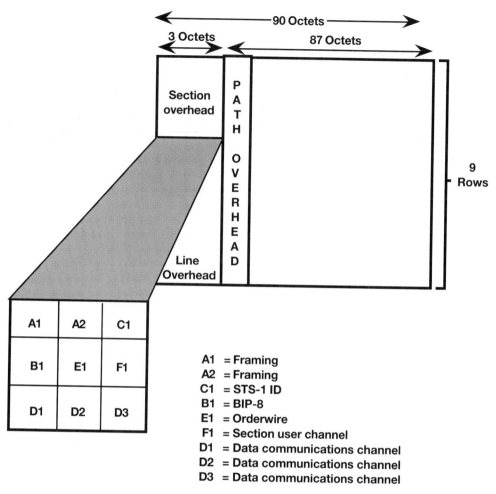

A1 = Framing
A2 = Framing
C1 = STS-1 ID
B1 = BIP-8
E1 = Orderwire
F1 = Section user channel
D1 = Data communications channel
D2 = Data communications channel
D3 = Data communications channel

Figure 8–21
Section overhead.

bytes (labeled H1 and H2) are pointers that indicate the offset in bytes between the pointer and the first byte of the SPE. This pointer allows the SPE to be allocated anywhere within the SONET/SDH envelope, as long as capacity is available. These bytes also are coded to indicate if any new data are residing in the envelope.

The pointer action (H3) byte is used to frequency justify the SPE. It is used only if negative justification is performed. The B2 byte is a BIP-8 parity code which is calculated for all bits of the line overhead.

Octets K1 and K2 are the automatic protection switching (APS)

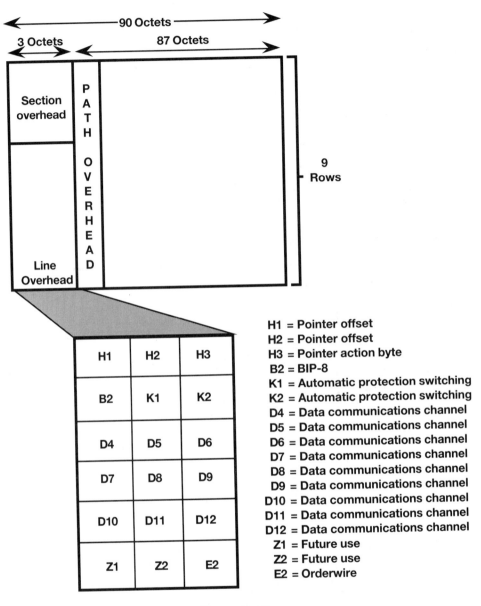

Figure 8–22
Line overhead.

322

bytes. They are used for detecting problems with line terminating equipment for bidirectional traffic and for alarms and signaling failures. They are used for network recovery as well.

All the data communications channel octets (D4-D12) are used for line communication and are part of a 576 kbit/s message to be used for maintenance control, monitoring, alarms, etc. The Z1 and Z2 bytes had been reserved for further growth, and are now partially defined. The Z2 byte is now defined to support a line layer for n block error operation on a broadband ISDN UNI. We shall see this operation later in this chapter and also in the ATM chapter. Finally, the E2 byte is an orderwire byte.

Path overhead.   Figure 8–23 illustrates the path overhead (POH). The path overhead remains with the payload until the payload is demultiplexed finally at the end MUX (the STS-1 terminating equipment). The path overhead bytes are processed at all points of the SONET/SDH system. SONET/SDH defines four classes provided by path overhead. The classes are summarized as follows:

Class A   Payload independent functions (required)

Class B   Mapping dependent functions (not required for all payload types)

Class C   User specific overhead functions

Class D   Future use functions

All path terminating equipment must process class A functions. Specific and appropriate equipment also processes class B and class C functions.

For class A functions, the path trace byte (J1) is used to repetitively transmit a 64-byte fixed length string in order for the recipient path terminating equipment to verify a connection to the sending device.

The BIP-8 (B3) is also a class A field. Its function is the same as that of the line and section BIP-8 fields, to perform a BIP-8 parity check calculated on all bits in the path overhead.

The path signal label (C2) is used to indicate the construction of the STS payload envelope. The path signal label can be used to inform the network that different types of systems are being used, such as SMDS, FDDI, etc.—something like a protocol identifier for upper layer protocols. Currently it is coded to indicate if a path terminating equipment is not sending traffic; that is to say, that the originating equipment is inten-

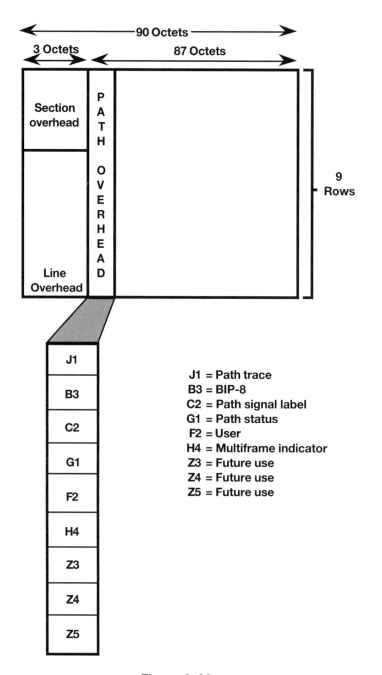

Figure 8–23
Path overhead.

324

tionally not sending traffic. This signal prevents the receiving equipment from generating alarms.

The path status byte (G1) carries maintenance and diagnostic signals such as an indication for block errors, etc., for class A functions. For class B functions, a multiframe indicator byte (H4) allows certain payloads to be identified within the frame. It is used, for example, for VTs to signal the beginning of frames. It also can be used to show a DS0 signaling bit, or as a pointer to an ATM cell.

For class C functions, the 1 byte F2 is used for the network provider. For class D functions, the growth bytes of Z3 through Z5 are available.

SONET/SDH EQUIPMENT

Figure 8–24 shows some of the major equipment that is used in SONET/SDH networks. The terminal multiplexer is used to package incoming T1, E1, and other signals into STS payloads for network use. The architecture of the terminal multiplexer consists of a controller, which is software driven, a transceiver which is used to provide access for lower speed channels, and a time slot interchanger (TSI) which feeds signals into higher-speed interfaces.

The add drop multiplexer (ADM) replaces the conventional back-to-back M13 devices in DS1 cross connections. ADM is actually a synchronous multiplexer which is used to add and drop DS1 signals onto the SONET/SDH ring. The ADM is also used for ring healing in the event of a failure in one of the rings. The add drop multiplexer can be reconfigured for continuous operations.

The ADM must terminate both OC-n connections as well as conventional electrical connections. The ADM can accept traffic from incoming OC-n and insert it into an outgoing OC-n. ADMs can also provide groom and fill operations, although this capability is not defined in the current Bellcore standards manuals.

The ADM multiplexers are required to convey the DS-n signals as they are—without alteration. They operate bidirectionally, which means they can add drop DS1, E1, etc. signals from either direction. ADMs support both locked and floating mode signals. And, the ADM uses both electrical and optical interfaces, which are specified in great detail in the ITU-T and ANSI/Bellcore documents.

It is conceivable that ADMs will become protocol converters as the technology matures. This means that instead of only providing simple

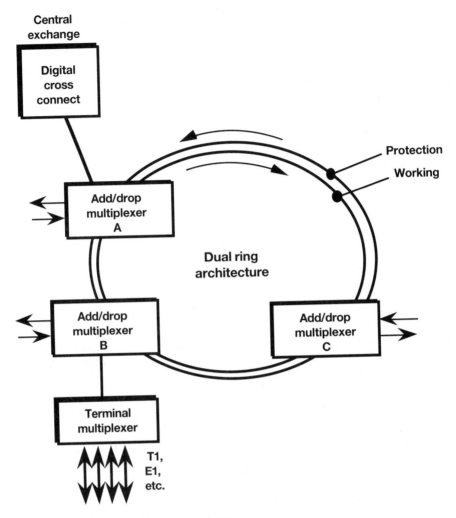

Figure 8–24
SONET/SDH equipment.

multiplexing and bridging functions, they may also perform protocol conversion functions by internetworking SONET/SDH with LANs, SMDS, frame relay, etc.

The topology for the ring can take several forms. Figure 8–24 shows the simplest arrangement, known as a unidirectional self-healing ring (USHR). Two fibers are used in this example; one is a working fiber, and the other is a protection fiber. In the event of a failure on a fiber or at an interface to a node, the ring will take corrective action (self-heal) and cut out the problem area. An example of this operation is provided shortly.

Although not shown in Figure 8–24, the topology can also be established to include four fibers and operate with a second arrangement, known as bidirectional SHR (BSHR), in which case traffic shares the working and protection fibers between two nodes, and is routed over the shortest path between nodes.

Figure 8–25 shows payload coming into an ADM from the ring and from a local source (or even another ring). Part of the traffic is "dropped," part of the traffic is "inserted", and part of the traffic is passed unaltered to the next ADM.

Path protection switching (PPS) is achieved by using fields in the overhead headers. Figure 8–26 shows how the operations occur. During normal operations (Figure 8–26a), STS-1 signals are placed on both fibers, so the protection fiber carries a duplicate copy of the payload, but in a different direction, and as long as the signals are received at each node on these fibers, it is assumed all is well. When a problem occurs, such as a fiber cut between nodes A and B, the network changes from a ring (loopback) network to a linear network (no loopbacks). In this example, node A detects a break in the fiber, and sends an alarm to the other nodes on the working fiber. The effect of the signal is to notify node B of the problem. Since node B is not receiving traffic on the protection fiber from node A, it diverts its traffic on to the fiber, as shown in Figure 8–26b.

The digital cross-connect (DCSs) systems are used to cross-connect VT (see Figure 8–27). One of their principal jobs is to process certain of the transport and path overhead signals and map various types of tributaries to others; in essence they provide a central point for grooming and consolidation of user payload. The DCS is also tasked with trouble isolation, loopback testing, and diagnostic requirements. It must respond to alarms and failure notifications. The DCS performs switching at the VT level, and the tributaries are accessible without demultiplexing. It can segregate high bandwidth traffic from low bandwidth traffic and send them out to different ports.

Figure 8–27 shows the traffic on one ring (the outer ring of the two dual rings). Some traffic is relayed around the same ring and other traffic is diverted and cross-connected to the other ring.

For some applications, it may be necessary to provide extra capacity in the system. For others, it may be necessary to ensure survivability of the network in the event of problems. In either case, path protection

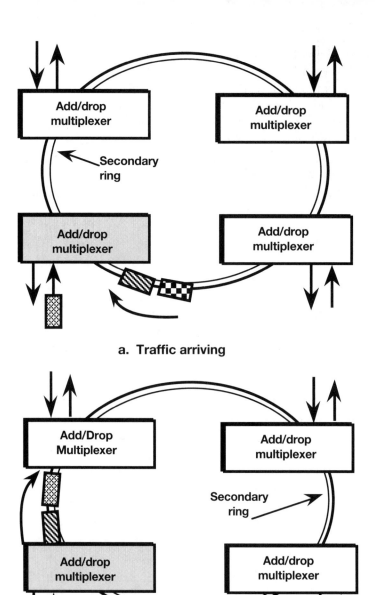

a. Traffic arriving

b. Results of traffic operations

 Protocol data units

Figure 8–25
ADM operations.

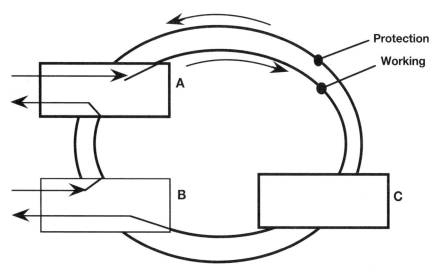

a. USHR, normal operations

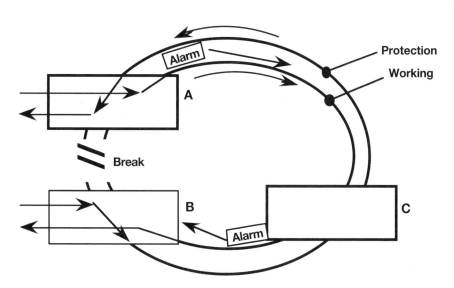

b. USHR, ring recovery operations

Figure 8–26
Normal operations and ring recovery operations.

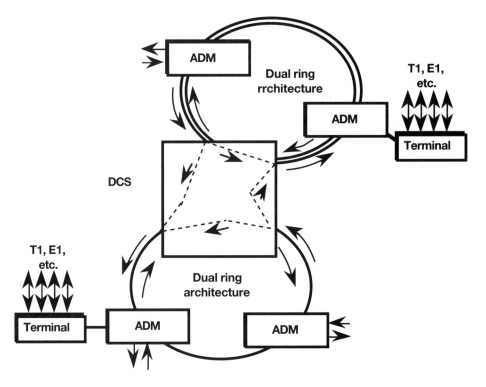

ADM = Add/drop multiplexer

Figure 8–27
Digital cross connects (DCS) systems.

switching (PPS), and self-healing rings (SHRs) are employed, and in some instances multiple PPS rings may be employed. An SHR is a collection of nodes joined together by a duplex channel. The latter topology is illustrated in Figure 8–28.

This arrangement is quite flexible. The rings are connected together, but they are independent of each other, and can operate at different speeds, and the rings can be expanded by adding other ADMs, DCSs, and PPS rings to the existing topology.

The ADMs in Figure 8–28 are called serving nodes. Any traffic that is passed between the rings is protected from a failure in either of the serving nodes. The small boxes within the ADM depict selectors which can pass the signals onto the same ring, or to the other ring.

Figure 8–29 shows how SONET/SDH is being deployed. Figure

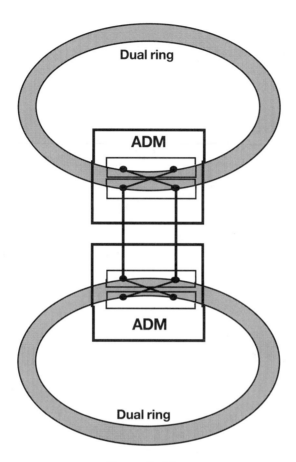

Figure 8–28
Multiple rings.

8–29a is a typical non-SONET/SDH access network using the current optical fiber technology. Most of the connections are point-to-point between high-density metropolitan areas that employ optical fiber terminals at each end, and, if necessary, repeaters on the line. Figure 8–29b shows how SONET/SDH is being used to improve and modernize the network. SONET/SDH is deployed between cities A, B, and C. This deployment occurs for a number of reasons: (a) growth of the subscriber base between these cities, (b) the current network's capacity is exhausted, or (c) the desire to modernize the current T1/E1 system. Eventually, a SONET/SDH network could connect cities E and F.

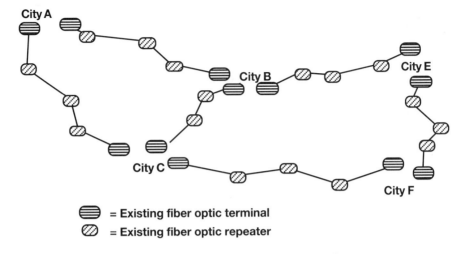

= Existing fiber optic terminal

= Existing fiber optic repeater

a. Typical structure of the network

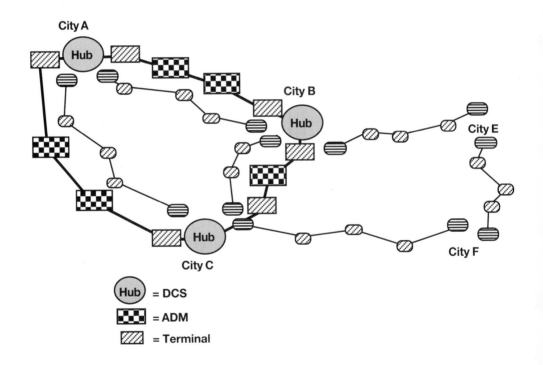

Hub = DCS

= ADM

= Terminal

b. SONET/SDH deployment

Figure 8–29
Typical network structure versus SONET/SDH structure.

332

OTHER NOTABLE ASPECTS OF SONET/SDH

Operation, Administration, and Maintenance (OAM) Operations

The operation, administration, and maintenance (OAM) functions are associated with the hierarchical, layered design of SONET/SDH. Figure 8–30 shows the five levels of the corresponding OAM operations, which are labeled F1, F2, F3, F4, and F5. F1, F2, and F3 functions reside at the physical layer; F4 and F5 functions reside at the ATM layer.

The Fn tags depict where the OAM information flows between two points, (see Figure 8–31 for one of several possibilities). The five OAM flows occur as follows:

F5 OAM information flows between network elements performing VC functions. From the perspective of a BISDN configuration, F5 OAM operations are conducted between B-NT2/B-NT1 endpoints. F5 deals with degraded VC performance, such as late arriving cells, lost cells, cell insertion problems, etc.

F4 OAM information flows between network elements performing VP functions. From the perspective of a BISDN configuration, F4 OAM flows between B-NT2 and ET. F4 OAM reports on an unavailable path or a virtual path (VP) that cannot be guaranteed.

F3 OAM information flows between elements that perform the assembling and disassembling of payload, header and control (HEC) operations, and cell delineation. From the perspective of a BISDN configuration, F3 OAM flows between B-NT2 and VP cross connect and ET.

F2 OAM information flows between elements that terminate section endpoints. It detects and reports on loss of frame synchronization and degraded error performance. From the perspective of a BISDN configuration, F2 OAM flows between B-NT2, B-NT1, and LT, as well as from LT to LT.

F1 OAM information flows between regenerator sections. It detects and reports on loss of frame and degraded error performance. From the perspective of a BISDN, F1 OAM flows between LT and regenerators.

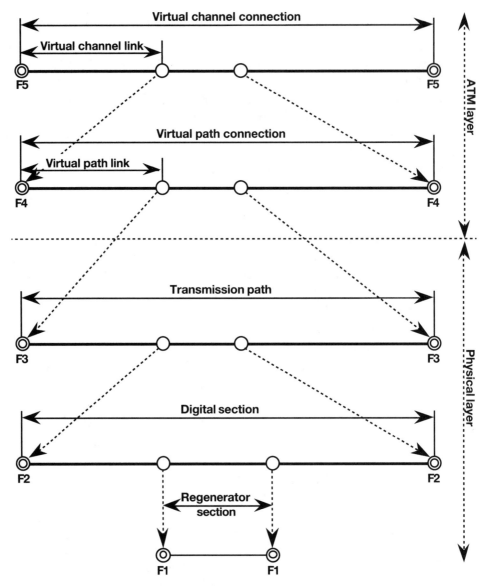

Figure 8–30
The hierarchical relationship.

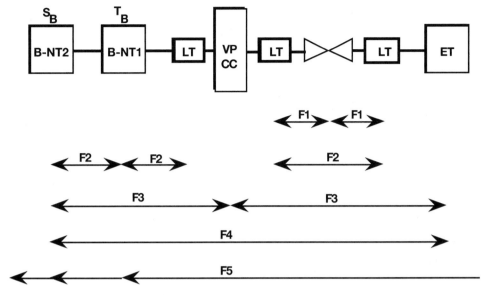

Figure 8–31
Information flows in the BISDN.

Progress in SONET/SDH Penetration

SONET and SDH installations are used mostly for systems that are expanding or new. It is expected that older carrier systems will be retired and replaced with SONET/SDH systems. Nonetheless, most estimates are that T1 and E1 will still be quite prevalent through the end of this century. Only time will tell, but recent announcements and deployments of SONET/SDH technology (as well as the drop in prices) lead this writer to think that most of the older technology will be replaced by SONET/SDH by the year 2000.

SONET/SDH WORKSHEET

The worksheet for SONET/SDH is provided in Table 8–6. As suggested in the introduction to this book, it is recommended that the reader attempt to fill out a blank sheet (provided in Appendix D), and then read this section to check your answers. This section amplifies the answers in the table that might not be self-evident.

Table 8–6   SONET/SDH Worksheet

Technology name:	SONET/SDH
New technology?	Somewhat, synchronous networks are not common, and extensive OAM in carrier networks is new
Targeted applications?	SONET/SDH are application indepedent
Topology dependent?	No, but dual rings provide the best alternative
Media dependent?	No, but optical fiber is needed for the higher bit rates
LAN/WAN based?	WAN-based
Competes with:	Current carrier transport systems
Complements:	Current carrier transport systems
Cell/Frame based?	Neither, uses concept of an envelope, but actually resembles a frame
Connection management?	No
Flow control (explicit/implicit)?	No
Payload integrity management?	No
Traffic discard option?	No
Bandwidth on demand?	No
Addressing/identification scheme?	No

SONET/SDH is not designed to support any particular type of application; it has no convergence functions and is not aware of the type of traffic that resides inside the SPE. Other protocols, such as ATM, MANs, and SMDS must deal with the application-specific issues.

Strictly speaking, SONET/SDH is not topology dependent, nor must it operate on a specific type of media. However, the higher transmission rates must use optical fiber, and dual ring topologies are preferred for their ability to self-heal.

Most of the other answers to the questions in the worksheet are "No," because the emerging technologies or user protocols utilized above SONET/SDH are designed to provide these other services.

SUMMARY

Modern telecommunications and applications need increased carrier capacity for wide area transport service. Broadband WANs provide the answer, and SONET/SDH is being positioned to provide this high-speed transport service. From the perspective of the ITU-T and other standards groups, ATM will provide the switching operations for a BISDN with the underlying physical operations of the ATM traffic supported by the SONET/SDH operations.

Mobile Communications Technologies

INTRODUCTION

This chapter is devoted to mobile-wireless communications systems and is organized into two major sections: (a) cellular systems, and (b) cordless systems. This organization is somewhat arbitrary, because cordless systems also use cells (albeit, small cells), and the two technologies are sure to converge in the future. However, for the present, this approach will ease the task of explaining them, because the two technologies use different protocols and standards.

The first part of the chapter introduces the major features of mobile communications systems, and subsequent parts explain cellular and cordless systems.

THE PURPOSE OF MOBILE COMMUNICATIONS SYSTEMS

The purpose of a mobile communications system can be inferred from the name of the technology: the provision for telecommunications services between mobile stations and fixed land stations, or between two mobile stations. This simple statement is quite modest in view of the impact that mobile communications is having on our society. Mobile communications is no longer a niche market; it is now a widespread technology that is of keen interest to the average consumer. Of all the technologies discussed in this book, mobile communications is the fastest growing and shall be the one that touches us all. In the not too distant future, it will pose a major challenge to fixed networks and hardwire user-to-network interfaces (UNIs), such as frame relay and SMDS.

It should prove helpful to distinguish (as best as one can with similar and complementary technologies) between two forms of mobile communications: cellular and cordless. Perhaps the best approach is to examine their major attributes and compare them with each other. First, a cellular system usually has a completely defined network which includes protocols for setting up and clearing calls and tracking mobile units through a wide geographical area. So, in a sense, it defines a UNI and an NNI. With cordless personal communications, the focus is on access methods vis-à-vis a closely located transceiver—usually within a building. That is to say, it defines a rather geographically limited UNI.

Cellular systems operate with higher power than do the cordless personal communications systems. Therefore, cellular systems can communicate within large cells with a radius in the km range. In contrast, cordless cellular communication cells are quite small, usually in the order of 100 meters.

With this introduction in mind, the remainder of the chapter provides an overview of (a) cellular systems, and (b) cordless systems.

TYPICAL CELLULAR SYSTEMS TOPOLOGY

Figure 9–1 depicts a typical topology for a cellular radio system. The principal components of this system are the mobile telephone switching office (MTSO), the cell, its base transceiver station (BTS), and a mobile unit.

The MTSO is the control element for cellular systems. It is responsible for switching the calls to the cells, providing for backup, interfacing with telephone networks, monitoring traffic for charging activities, performing testing and diagnostic services, and overall network management.

The mobile unit is the mobile transceiver, such as an automobile, truck, etc., which contains a frequency-agile modem that allows it to synchronize on a particular frequency, which is designated by the MTSO.

The cell site with the BTS is the interface between the mobile unit and the MTSO. Receiving signals and directions from the MTSO, it

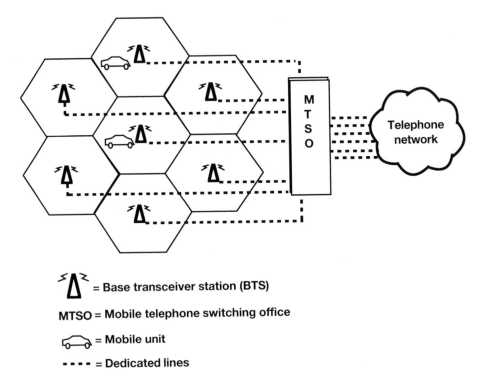

= Base transceiver station (BTS)

MTSO = Mobile telephone switching office

= Mobile unit

▪ ▪ ▪ ▪ = Dedicated lines

Figure 9–1
A typical cellular system topology.

sends and receives traffic to/from the mobile unit. As Figure 9–1 shows, the hexagonal pattern of the cell is used to determine how to position the BTS and how to reuse frequencies. The cells can range from about a mile to 25 miles. Of course, the clearly defined hexagonal pattern only approximates the radio waves.

In certain cells, such as the highly populated urban areas, increased local traffic can use up available radio channels. However, some of the system's capacity can be increased by the continual decrease of the cell size with the associated decrease in the transmitted power of the base stations. The reduction in cell size radius allows the available bands to be reused in noncontiguous cells. The approach allows the cellular carrier provider to decrease and increase the cell sizes to accommodate to growth or shrinking populations of this mobile subscriber base.

Figure 9–2 shows one way of increasing channel capacity in cells by splitting them into smaller sizes by decreasing the cell radius. As the fig-

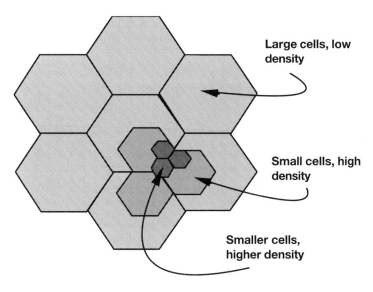

Large cells, low density

Small cells, high density

Smaller cells, higher density

Figure 9–2
Cell splitting.

ure suggests, the larger cell can be split into smaller cells, which can be split further into even smaller cells by decreasing the power of the transmitters. Therefore, large cells (perhaps up to 20 miles) can be employed for suburban areas while very small cells can be employed in high density areas. However, it should be emphasized that cell splitting requires careful design when the system is being initially set up to try to decrease the amount of adjustments that have to be made in the system. Also, small cells require more frequent hand-offs (as the mobile unit moves across cells), which places additional overhead on the network.

CELLULAR SYSTEMS OPERATIONS IN MORE DETAIL

Instead of using a generic example for this section, a specific implementation of a cellular system is examined. The Groupe Speciale Mobile/Global System for Mobile Communications (GSM) forms the basis for what is called today as digital cellular system 1800 (DCS1800). It is highlighted in this chapter because it is the first cellular system that will be implemented worldwide (instead of each country or geographical region implementing a closed system). Before GSM is examined, an overview is provided of other cellular systems and their market penetration.

Cellular System Types and Market Penetration

All cellular systems employ (in one fashion or another) the techniques described in the previous section of this chapter. Their principal differences are their use of the frequency spectrum and the spacing between the channels. Table 9–1 summarizes the original, major systems and their principal attributes (Lee, 1989). Obviously, these systems are not compatible, thus the reason for the implementation of a common cellular technology, GSM. Table 9–1 also contains information on the use of cellular systems, including both the original and newer systems (Steward, 1992). The Advanced Mobile Phone System (AMPS) is implemented in over 40 countries, with the main subscriber base in the United States. The Total Access Communications System (TACS) has its largest subscriber base in the United Kingdom, and operates in 20 other countries. The Nordic Mobile Telephone system (NMT) is implemented as NMT450 and NMT900 and found mostly in the Nordic countries, although it has presence in 31 countries. The C450 system is installed mainly in Germany.

The last part of Table 9–1 shows the penetration rates of cellular systems in the top ten countries (Steward, 1992). Obviously, the Nordic countries lead the world in the population percentage that uses cellular systems.

GSM

Initial work on GSM began in 1982 by a group within the European Telecommunications Standards Institute (ETSI). Originally, it was called Groupe Speciale Mobile, which led to the acronym GSM.

The GSM/DCS was initiated by the European Commission (EC) by adapting, in June 1987, a Recommendation and Directive for the Council of Ministers. The aim of this document was to end the incompatibility of systems in the mobile communications area and to create a European-wide communications system structure. The aim was also to accelerate the efforts of the individual countries to develop a European-wide digital cellular system. Initially, 13 countries in Europe signed onto the agreement. As of this writing, 18 countries now are participating.

GSM/DCS is targeted to include a wide variety of services including (naturally) speech transmissions and message handling services (such as

Table 9–1   Summary of Cellular Systems

A. Early Cellular Systems in kHz and km [LEE89]

	AMPS	TACS	NMT	T (450C)
Base station	870–890	935–960	463–467.5	461.3–465.74
Mobile station	825–845	870–915	453–457.5	451.3–455.74
Spacing	45	45	10	10
Coverage radius	2–20	2–20	1.8–40	5–30
Modulation	FM	FM	FM	FM

B. Market Penetration [STEW92]

	Subscribers	% of Subscribers	Number of Countries
AMPS	8,174,448	59.4	42
TACS	2,214,860	16.1	21
NMT	1,799,250	13.0	33
450C	711,190	5.2	11
Others	860,960	6.3	1
	3,760,708	100	108

C. Penetration Rates [STEW92]

Country	Subscribers	% of Subscribers	Years in Service
Sweden	564,060	6.7	11
Finland	276,110	5.8	10
Norway	227,240	5.4	10.5
Iceland	12,240	5.1	6
Denmark	168,900	3.2	10.5
Hong Kong	168,420	3.0	8.5
Switzerland	164,080	2.9	4.5
Faeroe Islands	1,330	2.8	3.5
United States	7,557,148	2.6	8.5
Singapore	66,00	2.4	4

X.400, facsimile transmission, emergency call, and different types of data transmission services) between mobile units (or any portable unit).

The initial goal for GSM was to have commercial services available in 1991, but that goal was not achieved. Tests have been performed on the technical feasibility of GSM, and commercial implementations are being placed into the industry today.

Figure 9–3 illustrates the major components of GSM and a typical GSM topology. GSM provides a cell size up to 20-30 km. The interface

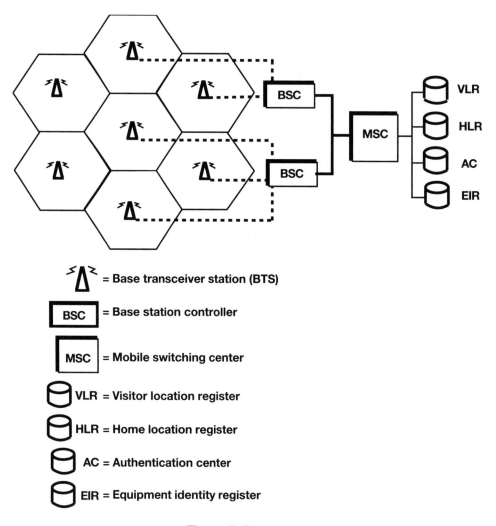

Figure 9–3
DCS 1800/GSM900.

with the mobile station (MS) is provided through the base transceiver station (BTS). These two components operate with a range of radio channels across an air interface. The BTSs are controlled by the base station controller (BSC), which is a new cellular network element that was introduced by GSM. It is responsible for the hand over operations of the calls as well as for controlling the power signals between the BTSs and MSs—thus relieving the switching center of several tasks.

The mobile MSC is the heart of GSM and is responsible for setting up, managing, and clearing connections as well as routing the calls to the proper cell. It provides the interface to the telephone system as well as provisioning for charging and accounting services.

GSM requires the use of two databases called the home location register (HLR) and visitor location register (VLR). These databases store information about each GSM subscriber. The HLR provides information on the user, its home subscription base, and the supplementary services provided. The VLR stores information about subscribers in a particular area. It contains information on whether mobile stations are switched on or off, and if any of the supplementary services have been activated or deactivated.

In addition, two other major components are part of GSM. The authentication center (AC or AUC) is used to protect each subscriber from unauthorized access or from a subscription number being used by unauthorized personnel. It operates closely with the HLR.

Finally, the equipment identity register (EIR) is used for the registration of the type of equipment that exists at the mobile station. It can also provide for security features such as blocking calls that have been determined to emanate from stolen mobile stations, as well as preventing certain stations that have not been approved by the network vendor from using the network.

GSM interfaces.   Four interfaces are defined in the GSM structure (see Figure 9–4). Two mandatory interfaces are the U_m *interface* and the *A interface*. The U_m interface is the air interface between the mobile station and the BTS. It is based on TDMA operations and the ISDN Q.931 protocol. The A interface exists between the mobile services switching center and the base station system. It is based on a modified Q.931 that runs on top of PCM-30, SS7 message transfer part (MTP), and signaling connection control part (SSCP).

A third interface, called *Abis*, defines operations between the BSC and the BTS. It is based on 2 Mbit/s PCM-30 transmission link, and LAPD.

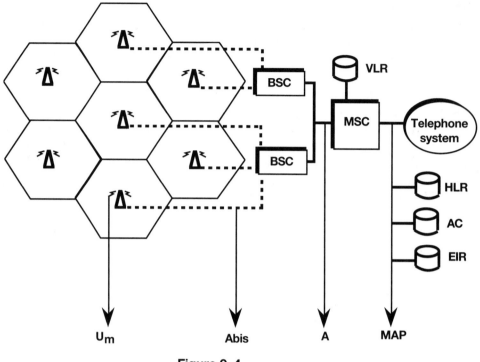

Figure 9–4
The GSM interfaces.

Finally, the mobile *application part* (MAP) defines the operations between the MSC and the telephone network as well as the MSC, the HLR, the VLR, and the EIR. MAP is implemented on top of SS7.

GSM is designed to permit *functional partitioning*. The major partitioning occurs at the A interface. One side of the interface deals with the MSC, HLR, and VLR operations and the other side of the interface deals with the BSS and air operations.

The air interface (the U_m interface) is usually of the most interest to individuals. This interface, shown in Figure 9–5, uses a combination of frequency division multiplexing (FDM) and time division multiplexing (TDM) techniques. The original GSM system operated in the 900 MHz range with 890 to 915 MHz allocated for the mobile station-to-base station transmissions and the 935 to 960 MHz allocated for the base station-to-mobile station transmissions. The DCS1800 now uses channel spectrum space from 1710 to 1785 MHz and 1805 to 1880 MHz, therefore increasing the capacity in bandwidth of DCS1800 over that of GSM900.

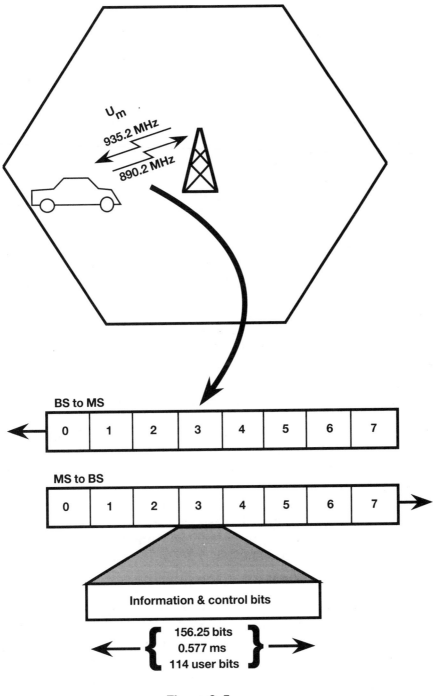

Figure 9–5
The U$_m$ interface.

124 channel pairs operate at full duplex with the uplink and downlink allocated different carrier frequencies. In this example, one channel is allocated to the 935.2 carrier and another channel is allocated to the 890.2 carrier. Thereafter, these frequency division multiplexed channels are time division multiplexed. TDM slots are allocated with 8 slots to a frame. Information and control bits operate in these slots. Each individual slot comprises 156.25 bits with a bit time of .544 ms. However, the user only receives 114-bits of this slot. The remaining bits are used for synchronization and other control functions.

The instance of one particular time slot (such as slot number three) in each frame makes up one physical channel. This means that the physical channel uses one slot every 4.615 ms.

Note from Figure 9–5 that the same structure exists both on the uplink and downlink channels. In addition, the carriers are separated into 124 channel pairs with a 200 kHz spacing to prevent channel interference.

Call routing.   Figure 9–6 shows an example of GSM call routing. In step 1, a telephone user places a call through the public telephone network to the mobile unit. The call is routed to a gateway MSC (step 2), which examines the dialed digits and determines that it cannot route the call further. Therefore, in step 3, it interrogates the called user's home location register (HLR) through the SS7 TCAP. The HLR interrogates the visitor location register (VLR) that is currently serving the user (step 4). In step 5, the VLR returns a routing number to the HLR, which passes it back to the gateway MSC. Based on this routing number, the gateway MSC routes the call to the terminating MSC (step 6). The terminating MSC then queries the VLR to match the incoming call with the identity of the receiving subscriber (steps 7 and 8). In step 9, the BSS[1] receives a paging request from the terminating MSC and sends out a paging signal, upon return of the user signal the call is completed (step 10).

Location updating.   Figure 9–7 shows an example of how a cellular subscriber can roam from cell to cell and how the system keeps track of the subscriber's location. When a mobile station crosses a cell boundary, the mobile unit automatically sends its location update request (which also contains its identification) to the BSS. The message is routed to the

---

[1] Some vendors use the term base station system (BSS) for BSC.

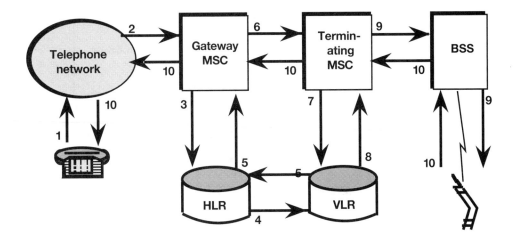

1 = Call made to mobile unit

2 = Telephone network recognizes number and gives to gateway MSC

3 = MSC can't route further, interrogates user's HLR

4 = Interrogates VLR currently serving user (roaming number request)

5 = Routing number returned to HLR and then to gateway MSC

6 = Call routed to terminating MSC

7 = MSC asks VLR to correlate call to the subscriber

8 = VLR complies

9 = Mobile unit is paged

10 = Mobile unit responds; MSCs convey information back to telephone

Figure 9–6
Example of GSM call management. From: Bell Northern Research, Global Systems for Mobile Communications, *Telesis, 94,* **1992.**

new cell's MSC which checks its VLR (new VLR in Figure 9–7). If the new VLR has no information about the message identity for this user (because the user has moved recently to this area), it sends a location update request message to the user's home location register (event 2). This message includes the identity of the user as well as the identity of the VLR that is sending the message. In event 3, the HLR stores the subscribers new location as the new VLR and then downline loads the user's subscription database to the new VLR. Upon receiving this information, the new VLR sends the acknowledgment of the location update

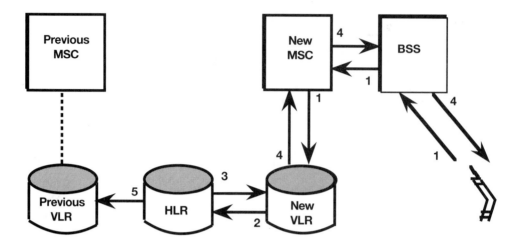

1 = Location update request

2 = Location update message

3 = Subscription data return

4 = Location update ACK

5 = Location cancellation message

Figure 9–7
Location updating.

through the new MSC to the BSS, and back to the originating mobile user (event 4). Finally, in event 5, the HLR sends a location cancellation message to the old VLR to clear the subscriber's data from its database.

The mobile subscriber must only be known to one VLR at a time. In this example, when the subscriber has roamed to another area (another cell) the VLR has had to be updated. It can be seen that the HLR is the master of the subscriber databases and therefore coordinates changes to the VLRs and MSCs as the subscriber roams through the cells.

Other notable aspects of GSM. Standards groups took almost seven years to develop GSM, and it has taken vendors about five years to develop their equipment and software for implementation. All these efforts are beginning to bear fruit, as the initial systems start to be installed. As examples, the first international GSM traffic is now underway between Europe and Asia, under the Nordic Group of NMT-GSM and the Hong Kong Telecom CSL. Vodafone will install a GSM network that covers approximately 90 percent of the U.K. population and now has roaming agreements covering Germany, Denmark, Finland, Sweden.

In a nutshell, GSM is now penetrating Europe, and will be installed in other countries as well. It is the first internationally recognized standard for cellular mobile systems. The European Commission (EC) projects 16 million GSM subscribers by the year 2000 (Williamson, 1992).

CORDLESS SYSTEMS OPERATIONS IN MORE DETAIL

Cordless systems (also called cordless telephones) are undergoing rapid technological changes. At the same time, their growth is almost exponential in some countries. Presently, countries use different protocols and standards, but this situation will change as vendors migrate to common approaches. These approaches are discussed further below.

The recently coined term used to describe the new cordless telephony technologies is personal communications systems (PCS). Definitions vary on PCS, but it is generally acknowledged that PCS represents a concept, and not a technology. It uses the cellular and cordless technologies, and each subscriber is assigned a personal identification number which is independent of any geographical restriction. Through the use of a small, lightweight handset, a person can be reached at any time. The PCS concept is to provide low-cost mobile services to a pedestrian in a variety of indoor and outdoor environments—usually in a small area, due to the need to use the small, portable handsets (with small batteries and low power capabilities).

The next section focuses on two prevalent cordless systems used throughout the world, CT2 and CEPT.

CT2

The CT2 technology (cordless telephone or cordless communications) is a continuation of work that was done in the mid-1980s on cordless standards. While CT2 was not envisioned to be a long-running standard, it has been widely supported because it is an evolution of ongoing technology.

In the United Kingdom, the government granted four licenses for public systems in 1989, with the requirement that the operators had to support the interworking between systems. The U.K.'s Department of Trade and Industry (DTI) did not stipulate a standard interface between the operators, but cooperatively, these operators developed a common air interface (CAI) between their systems which was accepted later as a standard by the ETSI.

The CT2 technology operates on time division multiplexing (TDM) concepts that support a speech rate of 32 kbit/s using ADPCM for speech coding. Operating frequencies at this time range between 8.461 and 868.1 MHz. Carrier spacing is 100 kHz. The transmitter data rate is 72 kbit/s.

During the past few years, the CT2 CAI standards were accepted in the industry (after being reviewed and approved by a number of industry organizations and several European PTTs). As a consequence of this acceptance, several manufacturers decided to abandon their proprietary plans and migrate forthwith to the standard (by using the standard in their products). Some vendors such as Ferranti's ZonePhone and Hayes Communications Forum Phone chose to stay with proprietary standards for the time being.

CT2 CAI exists as an organized European standard along with Digital European Cordless Telecommunications (DECT). It has received support throughout the world, from countries such as Canada, Japan, and other Asian countries. CT2 is capable of supporting voice and data traffic. The work that enhanced CT2 to support data was largely through the efforts of the Canadian vendors in their research and trials.

For CT2 data capabilities, four types of services have been defined. These services operate on ISDN B channels at 32 kbit/s in a circuit-mode environment. These services are:

Asynchronous services are supported for full duplex operations with speeds ranging from 300 to 19,100 bit/s. Flow control operations are provided with this service as well as automatic request for repeat (ARQ) and forward error control (FEC) provisions. *Transparent synchronous services* are also provided for full duplex channels. These services operate on the 32 kbit/s B channel, although lower rates can be supported as well. With these implementations, FEC is provided but ARQ is not. The third type of service is the conventional *X.25 service*, and the fourth type of service is conventional *ITU-T group 3 fax service*.

DECT

The DECT standard was initiated by the CEPT in the mid-1980s, as a method to move to a second generation cordless telephone system. It was viewed also as a suitable technology to support high capacity PABX systems (Grillo, MacNamee, & Rashidzadoh, 1993). Several alternatives were evaluated by the CEPT in 1987 and eventually a technology known as Digital European Cordless Telephony was chosen. Later the name was retitled to the Digital European Cordless Telecommunications standard to reflect that it supported not only voice traffic, but also data traffic. With the formation of ETSI in 1988, the responsibility for DECT standardization was passed to this organization. In 1992, DECT became a formal European standard known as the European Telecommunications Standard, ETS 300-175. It is now "enforced" as a standard by the European Commission Directive 91/287.

DECT operates on frequencies ranging from 1880 to 1990 MHz. Like CT2, it uses ADPCM for speech coding, in accordance with ITU-T G.721. Also, like CT2, the speech rate is 32 kbit/s using TDMA as an access method.

DECT does not restrict itself to voice transmission only. As depicted in Figure 9–8, it is also designed to support data applications. For data transmission, DECT uses the layered protocol stack based on the OSI Model and the IEEE 802 standards. The conventional physical medium access control (MAC) and logical link control (LLC) layers are employed. A gateway examines the addresses contained in packets and routes the packets accordingly. This routing entails the relaying of the traffic to attached LANs, other gateways, or attached cordless devices (such as workstations, printers, fax machines, servers, etc.).

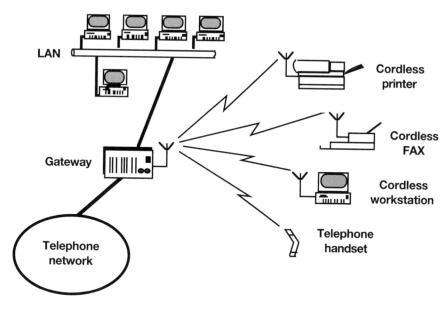

Figure 9–8
Data support with DECT.

It is envisioned that most traffic will be carried with the connection-oriented procedure based on the DECT standard, although (technically) no reason exists for traffic to be sent in a connectionless mode. In addition, the gateway provides connections for voice traffic between workstations, telephones, and the telephone network.

One of the more widely known implementations of DECT with data is Olivetti's cordless LAN. This product was demonstrated in 1991 at the ITU Telecom Exhibition in Geneva, Switzerland. The Olivetti products were rolled out in 1992.

Bellcore View of PCS

The Bellcore view of PCS is based on five access services provided between the Bellcore client company (BCC) network and the PCS wireless provider (PWP) network (see Figure 9–9).

The PCS Access Service for Networks (PASN) is a connection service for the provision of transport and signaling services to and from the PCS service provider (PSP). The PASN provides translation functions of a universal personal telecommunication number (UPT#) to a routing number (RN). PASN also ensures that calls are delivered to the appropriate PCS as well as PWP.

The PCS Access Service for Controllers (PASC) is designed for use with the PWP across radio channels and some type of automatic link transfer (ALT) capability. (ALT refers to the processing of signaling information during an active call without disrupting the call—in other words, an out-of-band signaling function.) The PASC performs all the functions of PASN in addition to dialtone functions and call delivery operations.

The PCS Access Service for Ports (PASP) is designed to interface into a PWP that has very little functionality. In this situation, the service provider would own the license and the radio ports, but the BCC would handle most of the functions. In this type of service, the PCS provides the functions in PASN and PASC.

The PCS Service for Data (PCSD) is, as its name implies, for services for transport of database information. It is an important component in PCS because it contains the information needed to establish and manage connections.

Finally, the PCS Access Service for External Service Providers (PASE) is used to accommodate other PCS support services such as specialized voice mail, paging, alerting, etc. for other service providers.

The Bellcore vision for the PCS entails the use of the *universal personal telecommunication* (UPT) identification based on work done by the

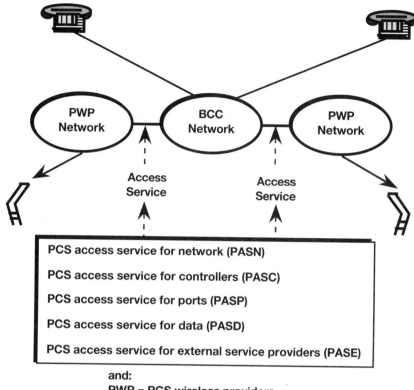

Figure 9-9
Bellcore view of PCS.

ITU-T. UPT includes the ability of a caller to reach another subscriber at any location by simply dialing a single *UPT number* (UPT#). This number is also used to identify unambiguously any user to the network. Presently, it is envisioned that UPT#s will be implemented with the E.164 numbering plan. In addition to the UPT, there will also be in place a *terminal identifier* (TID) which is a code actually stored in the *user terminal* (UT). The TID allows BCC networks and PWP networks to identify a specific piece of equipment. The format for the TID includes a country code, the provider of the terminal, and the terminal ID itself.

One other identifier, which is important to the Bellcore PCS, is the *routing number* (RN). This is used to deliver calls through the networks to specific locations. It is used in conjunction with the UPT# to provide a complete service to the user. The RN is also an E.164 number.

Other Standardization Efforts for PCS

Other standards organizations are working (as of this writing) on defining standards for PCS networks. The ANSI T1 technical subcommittee T1P1 has been working on PCS standards for a number of years. While the work is still in the initial stages of development, it is similar in many ways to the Bellcore model explained above. The T1P1 subcommittee is also working with the ITU-T and the CCIR with the aim of harmonizing the standards from different groups.

The ITU-T is also undertaking studies on PCS, which is published under the document I.39. The service aspects of the ITU-T are similar to some of the Bellcore concepts.

The PCS technology is maturing rapidly and it is hoped that the standards groups will continue to work closely with each other—ideally, to align the PCS standards close to each other throughout the world. This is not an impossible task; its feasibility has been demonstrated with the efforts of the standards groups in developing worldwide standards for the ATM, SONET/SDH, and GSM technologies.

OTHER NOTABLE ASPECTS OF THE CORDLESS AND PCS TECHNOLOGIES

The U.S. market for personal communications systems (PCS) is more fragmented than in other countries, and is far from settled. To illustrate, many issues dealing with technical, procedural, political, and marketing matters are not resolved. As of this writing, the implementation of coherent PCS products in the U.S. is still on the horizon.

Unlike other countries, the U.S. market allows for more diversification, more noncompatible implementations, and more testing. This seemingly confused approach often leads to superior products—all due to the wide market in the U.S., *and* the relative lack of regulation.

Notwithstanding, the Federal Communications Commission (FCC) has published rules by which PCS licensees will be expected to operate. These rules are not final yet and are really nothing more than proposals to allow comment and future experimentation.

Many unresolved issues remain. One of the biggest is the allocation of the spectrum for PCS. Invariably taking spectrum away from any user end group will create controversy and resistance. For example, the proposed allocation of some of the spectrum space for PCS is met with fierce opposition from public utilities who have been using this spectrum space for their microwave systems.

Then there is the problem of how to license the service providers.

How many providers should exist within the country? Should they be allocated to geographical areas? Who should be allowed to apply? These are all questions that have yet to be answered.

Indeed, it is expected that many of these issues will not be resolved for several years.

Canada has made noteworthy progress in the PCS arena. The efforts have been coordinated closely by the Department of Communications (DOC) and have resulted in a fairly substantial implementation of PCS.

The effort began in November 1989, when the DOC issued guidance for applications and trials. The purpose of the trials was to determine technical feasibility of the system (and spectrum requirements), as well as to reach a consensus on what the standards would be for the actual implementation.

After a review of the applications, in May 1990, 22 experimental licenses were issued and 10 trials were initiated throughout Canada. One noteworthy trial was conducted in Ottawa by Cantel Personal Communications. This was the first two-way PCS trial in the world. It consisted of approximately 200 participants with 300 telephones. Communications occurred by individuals carrying their handsets into shopping centers, business offices, streets, etc. The trial was quite satisfactory and CT2 performed well.

After the trials were evaluated, the DOC and several advisory committees chose CT2 for the Canadian standard. This decision was made in December 1990. Later, in September 1991, the spectrum space of 944 to 948 MHz was chosen for the system.

In early 1993, DOC will award the commercial licenses. Licenses are granted based on the ability of companies to provide coast-to-coast service. Therefore, it is expected that some of the smaller companies will form consortiums in their bid for a license. Several telephone companies have formed these consortiums, one consists of twelve telephone companies called Mobility Canada.

The Cellular Digital Data Packet System Specification (CDPD)

Several public telecommunications carriers have published a specification for a wireless extension to existing data networks. It is named the Cellular Digital Data Packet System Specification (CDPD). Its goal is to provide a wireless packet data connectivity to mobile communications users ("Cellular Digital Packet Data System Specification," 1993). The companies that produced this specification are listed in Table 9–2.

Table 9–2   Companies that Developed CDPD

Ameritech Mobile Communications, Inc.

Bell Atlantic Mobile Systems

Contel Cellular, Inc.

GTE Mobile Communications, Inc.

McCaw Cellular Communications, Inc.

NYNEX Mobile Communications, Inc.

PacTel Cellular

Southwestern Bell Mobile Systems

The intent of the CDPD developers is to utilize the unused capacity ot the existing Advanced Mobile Phone Systems (AMPS), and existing data communications protocols, such as the Connectionless Network Layer Protocol (CLNP), the Internet Protocol (IP), the OSI transport layer, the Transmission Control Protocol (TCP), etc. In effect, existing protocols running at layers 3 and above are run on top of CDPD, which operates at the lower two layers. CDPD also specifies a wide variety of upper-layer protocols for directory managment, electronic messaging, home location management, etc. Many of these services are OSI- and Internet-based, such as X.500, X.400, the Domain Name System (DNS), etc.

CDPD's architecture is OSI-based, and is derived from ISO 7498 and the CCITT's OSI X.200 Recommendations. The developers have used OSI concepts and terminology wherever possible. The key components and interface are illustrated in Figure 9–10. Each CDPD service provider supports three interfaces: (1) the airlink interface (A): the interface between the service provider and the mobile subscriber, (2) the external interface (E): the interface between the service provider and external networks, and (3) the interservice provider interface (I): the interface between cooperating CDPD service providers.

Two basic network entities exist in this architecture. The end system (ES) is a user device, which is called a host in Internet terminology. Each ES must be identified with at least one globally unique network entity identifer (NEI). The intermediate system (IS) is an internetworking unit, which is called a router in Internet terminology. In addition to supporting conventional protocols, such as TCP/IP, the ES may also run a CDPD-defined operation called the Mobile Network Location Protocol

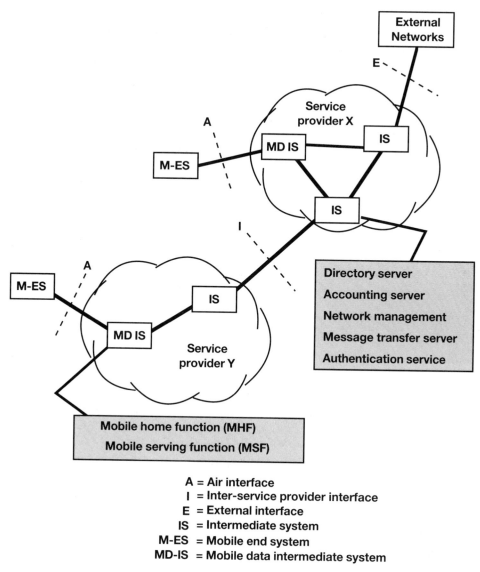

A = Air interface
I = Inter-service provider interface
E = External interface
IS = Intermediate system
M-ES = Mobile end system
MD-IS = Mobile data intermediate system

Figure 9–10
The Cellular Digital Packet Specification (CDPD).

(MNLP), which provides location information in the system. The latter device is called a Mobile Data Intermediate System (MD-IS).

The MD-IS is the only entity that has any knowledge of the mobility of the ESs, and it provides two routing services. The Mobile Home Function (MHF) provides a packet forwarding function, and supports a database for its serving area (its "home" ESs and ISs). Each ES must belong to a fixed home area, and MHF keeps track of this information. The Mobile Serving Function (MSF) handles the packet transfer services for visiting MSs. Like most mobile systems, a visiting ES must register with the serving MD-IS, which notifies the visiting ES's MD-IS of its current location.

CDPD services and servers. A CDPD service provider must provide a number of "servers" for its domain. These servers reside in an IS and are called fixed end systems (F-ESs). The servers perform the following operations:

- Directory server:

  Provides X.500 directory services for name-to-address translation, information on subscribers, and other yellow and white page functions. Lower layer access to a directory server must adhere to a specific CDPD protocol, and ISO's Connection Oriented Transport Service (COTS), and optionally, ISO's Connectionless Network Service (CLNS), X.25, PVC frame relay, point-to-point, or a LAN profile.

- Accounting server:

  Provides a repository for accounting information, including the collection and distribution of this information to interested parties. Lower layer access to an accounting server must adhere to the same conventions as the directory server.

- Network management:

  Provides network management services through the use of either common management information protocol (CMIP), or simple network management protocol, version 2 (SNMPv2). The CCITT/OSI System Management Functional areas (CCITT M.3400) must be supported (fault management, configuration management, accounting manage-

- Message transfer server: Provides an electronic messaging service through the use of CCITT's X.400 Recommendation. Lower layer access to the message transfer server must adhere to the same conventions as the directory server, except ISO's Connectionless Network Service (CLNS) is not allowed.

- Authentication service: Provides an authentication service for a home MD-IS to verify a M-ES's credential by the user of secret key encryption and decryption. The Authentication Protocol must run on top of OSI's Association Control Service Element (ACSE, X.217) and the remote operations service element (ROSE, X.219)

It is too early to know the impact of CDPD. As of this writing the vendors are writing the software for the system.

Third Generation Mobile Systems

Even though the second generation cordless systems are not yet firmly in place, work is underway for a third generation system. This work is spearheaded, once again, in Europe and sponsored by the European Commission, the CCIR and the ETSI. The European Commission sponsors this work through the Research on Advanced Communications for Europe (RACE). The goal of RACE is to create a third generation mobile system by the year 2000. This system is called the Universal Mobile Telecommunications System (UMTS).

Significant progress has been made on RACE with the support of the World Administrative Radio Congress (WARC) in allocating 230 MHz of spectrum space for the UMTS efforts. Representatives from 127 countries approved the use of this spectrum space at the International Telecommunication Union (ITU) World Administrative Radio Conference

1992, held in Torremolinos, Spain. This important action allows the CCIR to move forward with an effort called the Future Public LAN Mobile Telecommunications System (FPLMTS). The goals of CCIR are the development of standards for the implementation of a seamless world-wide mobile networks, encompassing PCS and cellular technologies.

Some concluding thoughts.   The likely topology for these new systems will be what is called a mixed-cell architecture; variable-sized cells will be implemented that are tailored to meet specific geographical areas and traffic demands. Figure 9–11 shows this topology.

The small picocells likely will be an enhanced version of the current cordless technology, exhibiting a small, lightweight handset; the larger cells probably will use enhanced GSM features. However, changes must be made for this new generation of mobile systems.

The challenge for the technology is to develop a handset that can be used seamlessly across all the cells. Obviously, it will be quite awkward if different handsets must be employed by the picocell, microcell, and macrocell.

The issues are far from settled and while they are being addressed,

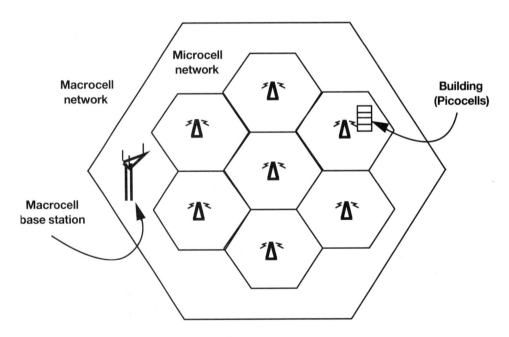

Figure 9–11
Future mobile systems.

some of them may seem intractable. For example, how can the mobile system support both (a) high and low power requirements, (b) short and long distance requirements, (c) different frequency spectrums, and (d) multiple protocols—all in the same handset?

In conclusion, it is a good idea to put things in perspective. What seemed to be intractable problems in this industry a few short years ago are no longer problems. In fact, solutions to these problems are now implemented in everyday products. As an example, as late as the mid-1980s, forward error correction (FEC) was considered to be so esoteric and expensive that it was used in only very sophisticated applications. Today, FEC runs in instruments as simple as personal computers. The "intractable problems" of the past seem rather simple, when viewed from the present, and as we learned in Chapter 1, technical issues that seem like intractable problems are only minor roadblocks.

MOBILE COMMUNICATIONS SYSTEMS WORKSHEET

The worksheet for mobile communications technologies is provided in Table 9–3. As suggested in the introduction to this book, it is recommended that the reader attempt to fill out a blank sheet (provided in Appendix D) and then read this section to check your answers.

The mobile communications technologies are somewhat comparable to a carrier transport technology, since they are designed to operate at the lower layers of the OSI Model, with emphasis on the physical layer. However, unlike T1/E1 and SONET/SDH, mobile communications have UNIs and some have NNIs. Nonetheless, many of the other answers to the questions in the worksheet are "No", because the other emerging technologies or user protocols are designed to provide many of the UNI services. Of course, their most distinguishing features are their use of the radio spectrum, and the geographical cell.

MOBILE COMMUNICATIONS SUMMARY

Wireless mobile communications are maturing rapidly throughout the world. Notable progress has been made in Europe due to the coordination efforts of the European Commission, ETSI, and other standards bodies.

GSM technology has become the most prevalent implementation for cellular systems with high bandwidth and wide area requirements. CT2 and DECT have assumed prominent positions for wireless communications for more restricted geographical areas.

Table 9–3   Mobile Communications Worksheet

Technology name:	Mobile communications
New technology?	No, technology is ten years old
Targeted applications?	Mobile telephone (voice), with increasing use for data
Topology dependent?	Yes, technology is quite dependent on cell topology
Media dependent?	Yes, relies on wireless, radio media
LAN/WAN based?	Both, cellular uses larger cells over metropolitan area, and cordless uses small cells within a building
Competes with:	Hard-wired systems
Complements:	Any application requiring transient nodes
Cell/Frame based?	If digital, frame-based, usually with TDM slots; otherwise, analog based
Connection management?	Yes, similar to telephone calls
Flow control (explicit/implicit)?	No, upper layers provide
Payload integrity management?	No, upper layers provide
Traffic discard option?	No, upper layers provide
Bandwidth on demand?	No
Addressing/identification scheme?	Yes, uses telephone numbers, user IDs and OSI-based NSAPs

Europe has fostered many of the mobile communications standards, and is deploying GSM, CT2 throughout EC and non-EC countries. Canada has provided leadership in implementing personal computer systems (PCS) with CT2 technology. The U.S. efforts have been fragmented due to the nature of the regulatory process and the competitive environment.

Standards are leading rapidly to products and further research efforts for third generation systems. Again, the focus of these efforts is in Europe.

A Tutorial on Communications Networks

INTRODUCTION

The purpose of this appendix is to provide a basic tutorial on communications networks. An experienced reader need not read this appendix, if subjects such as modems, multiplexers, analog-to-digital conversion, and packet switching are understood.

DATA COMMUNICATIONS NETWORKS

Classifying Networks

Data communications networks consist of data communications components, such as user terminals, host computers, modems, multiplexers, and switches. Typically, these components are called a network when several to many of them operate together to exchange data and share resources. The information is exchanged between these components through switches or other means of transporting traffic across the media, such as broadcasting the traffic to all stations in a network.

A good analogy to the data network is the telephone network, because the telephone network performs a service to a telephone user in the same manner the data network performs a service for the data communications user.

Networks may be classified as *private* or *public*. Private networks, as their name implies, are owned and managed by an enterprise and are used by no one except that enterprise. In contrast, a public network (while owned by an enterprise) "rents" its network to the public. Again, the analogy of the telephone system is relevant. As an example, AT&T, Sprint, and MCI are public telephone interexchange carriers who rent their services to whomever is willing to pay the fees charged for the services.

Networks can be classified as a *broadcast* or *switched*. Broadcast networks are distinguished by one-to-many transmission characteristics. This means that one communications device transmits to more than one device. This idea is found in commercial TV and radio broadcasting, where one station sends to many receivers. Broadcast networks are widely available in networks in which the machines are in close proximity because it is relatively easy to send the signal to all stations. In addition, broadcast is quite popular in satellite transmissions, wherein the satellite station transmits traffic to (potentially) thousands of receivers.

In contrast to broadcast networks, a switched network is not designed to transmit on a one-to-many relationship. An individual transmission is sent to the physical device (called a switch) wherein the switch determines how to route the data further. This approach does not mean that a switched network could not use broadcast topology (which indeed it can). Rather, it may not be economically or technically feasible to send traffic to all parties in a switched network. Broadcasting onto a switched network means that each user must be physically connected to the network to receive the message.

Wide Area and Local Area Networks

Until recently, it has been relatively easy to define *wide area networks* (WANs) and *local area networks* (LANs) and to point out their differences. It is not so easy today, because the term wide area and local area do not have the meaning they once had. For example, a LAN in the 1980s was generally confined to a building or a campus where the components were no more than a few hundred or a few thousand feet from each other. Today LANs may span scores of miles.

Nonetheless, certain characteristics are unique to each of these networks. A WAN is usually furnished by a third party. For example, many WANs are termed public networks because the telephone company or a public data network owns and manages the resources and sells these services to users. In contrast, a LAN is usually a private network because it is privately owned. The cables and components are purchased and managed by an enterprise.

LANs and WANs can also be contrasted by their transmission capacity (in bit/s). Most WANs generally operate in the kbit/s range, whereas LANs operate in the Mbit/s range.

Another principal feature that distinguishes these networks is the error rate. WANs are generally more error prone than LANs, due to the wide geographical terrain over which the media must be laid. In contrast, LANs operate in a relatively benign environment because the data communications components are housed in buildings where humidity, heat, and electrical controls are managed.

Network Components

The major network components must employ a *communications medium*; for example, a telephone communications line or a LAN channel. Thereafter, the other components vary, depending on the needs of the organization.

Figure A–1 shows a typical configuration of a data communications system. Certainly, computers are part of the data communications network, and most networks also utilize LANs as well as long distance communications lines.

Some computer systems contain a *front-end processor*. The purpose of this machine is to off-load communications tasks from the host computer, thus freeing the host computer to devote its CPU cycles to user applications and database operations. This approach was pioneered by IBM, with its development of the IBM 37×5 family of front-end processors. While Figure A–1 shows the use of a front-end processor, many organizations choose not to install these machines.

Figure A–1 also shows the *databases* that are attached to the central computer (host). Since the purpose of a data communications system is to transport data between users and user devices (data terminal equipment, DTEs), the databases are a significant component in the process. For example, the amount and frequency of the data transferred between user applications affect the "load" on the data communications components.

Most user organizations have installed a LAN or several LANs. The LAN can be attached to a front-end processor or directly to the host com-

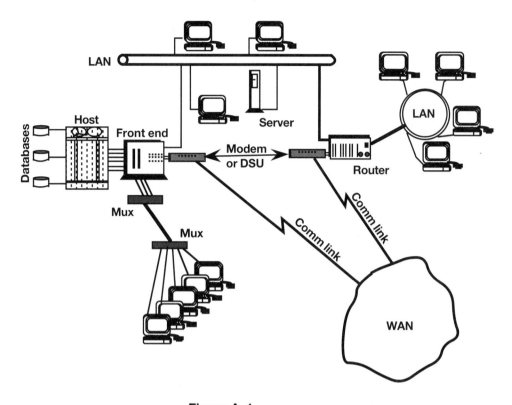

Figure A–1
Network components.

puter. In most instances, the front-end processor is not aware of many of the activities occurring on the LAN.

LAN topologies vary widely in the industry, with Ethernet and token ring-type networks dominating the field. This example shows an Ethernet-type LAN, wherein the LAN devices are attached to a single channel (such as a coaxial cable or a twisted-wire pair); and a token ring LAN, wherein the devices are attached to a cable that forms a logical ring.

Typically, functions are further off-loaded from the host computer with the use of database servers, terminal servers, and other support machines. Consequently, communications between terminals and servers on a network occur without requiring the use of the host computer resources.

Servers are used in many networks, and many communications processes today are organized around a client-server model. This model is based on the notion that personal computers and workstations (actually, any type of machine) operate as clients. These clients communicate over a network to a server. The client's responsibility is to formulate a simple request for a service that the server performs and sends back as a reply to the client. This request/reply concept is imbedded into most client/server networks that provide services such as file services, database services, electronic mail services, terminal services, etc.

The client/server approach is widely used in LANs in which diskless workstations access servers to obtain application software, files, and electronic mail. This approach always uses a request/reply dialogue and the request is always initiated by the client.

If the workstations and other computers are located more than a few hundred feet from the front-end or the host computer, modems are used to terminate the communications line with the host computer and remote workstation. In some installations, modems are not used but modem-like devices (such as line drivers) will provide the signaling interface. A modem is used if the communications line uses analog signaling. A line driver is used if the line uses digital signaling.

In many installations, the user devices do not utilize the full capacity of the communications line. Consequently, multiplexers (MUXs) are installed to permit users to share the line. Although not shown in Figure A–1, the MUX may also use modems if the communications line is analog. Otherwise, the MUX uses a line driver (or a similar device) to achieve digital transmission and reception.

As organizations have increased their use of data communications systems and LANs, it has become necessary to attach some of the LANs together to share the resources of the servers and the host computers.

For the attachment of these LANs, an additional component is needed. Its generic name is an *internetworking unit* (IWU). This component relays the traffic between the networks and may also provide protocol conversion functions if the networks' protocols differ. It may not be necessary to install an IWU because many organizations choose to attach each LAN to the host computer or the front-end processor and then use these two devices to relay the traffic between the LANs. The host/front-end processor is performing the IWU functions.

The IWU is capable of performing a variety of functions. Depending on the function performed, the IWU could be termed a bridge, router, gateway, or even (when bridge and router functions overlap) a brouter. Whatever the term used, the component still is involved in relaying traffic between networks.

For widely dispersed organizations, the WAN becomes a very important component in the communications system. WANs can be interfaced to the host computer in a number of ways. One common approach is to dedicate a connection into a front-end processor (or the host directly) for the communications line to the WAN. Another approach is to use one of the interfaces of a LAN router to provide the interface into the WAN.

The communications lines between the host computer or the LAN IWU can be dedicated or dial-up lines. In addition, they may be either digital or analog lines.

The WAN interface is usually provided with a switch. The switch interfaces to the host computer or the organization's IWU. For longer distances between the network and the host system, the switch also requires modems or DSUs for its connection into the line. The number of lines between the switch and the user organization's computer is usually determined by the amount of throughput required and reliability requirements. It is not unusual to have more than one line attached for purposes of backup (in case problems occur on one of the communications links).

Within the network itself, other switches are employed to provide for the relay of traffic through the network to the final destination. The communications lines inside this network are typically high capacity digital lines operating in the 56 kbit/s to 1.544 Mbit/s range. The section in this appendix titled, "Circuit, Message, Packet, and Cell Switching" explains the operations within the network.

VOICE NETWORKS

Figure A–2 illustrates a typical telephone network. The customers, either in homes or offices, connect through the telephone system into the class 5 office, which is also called the central office (CO), local exchange, or end office (shown in the figure as switching office). Thousands of these offices may be installed around a country. Connection is provided to the CO through a pair of wires or four wires called the local loop or subscriber loop, respectively.

Connections between COs are also provided by a facility called a tandem center (also called a tandem switch or toll center). The tandem center interconnects COs that do not have direct connections with each other. Four classes of tandem centers exist in the U.S. and Canada: toll centers, primary centers, sectional centers, and regional centers. The regional center (ten in the U.S.) has the largest area to serve, with each lesser class serving a smaller area. The components are arranged in a

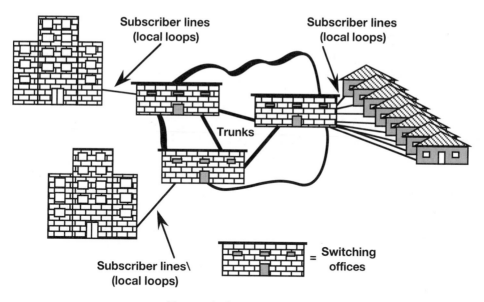

Figure A–2
The telephone network.

hierarchy starting with the customer location at the bottom of the hierarchy, and the regional center at the top of the hierarchy in this fashion:

> **Top of hierarchy**
> Regional center
> Sectional center
> Primary center
> Toll center
> Central office
> Subscriber telephone

The use of direct or indirect connections between offices with tandem trunks or other tandem switching systems depends on several factors: distances between offices, the traffic volume between offices, and the potential for sharing facilities among the customers within the geographical area. In the case of intermediate traffic volumes or longer distances, the telephone system generally establishes a combination of direct and tandem lengths.

The system is designed for each switching center to be connected to an office of a higher level, except at the highest level. The top-level offices are shortest path and/or the fewest number of switches. This design approach reduces the delay of establishing the connection with the other telephone set, and the fewer number of intermediate switches reduces the expense to the telephone company. As the tandem path becomes longer, it must go through more components, incurring more delay and additional expense.

The system is built around high-usage trunks (or high-volume trunks), which carry the bulk of the traffic. High-usage trunks are established when the volume of calls warrants the installation of high-capacity channels between two offices. Consequently, trunk configurations vary depending on traffic volume between centers. The system attempts to switch the call down into the hierarchy, across the hierarchy, or, as a last resort, up into the hierarchy. Routing the call up usually entails more intermediate switching, thereby increasing the connection delay and the telephone company's cost to obtain the connection.

Nonhierarchical Routing

Most telephone companies now use a routing concept called *nonhierarchical routing* (NHR). This technology is not constrained by a fixed hierarchical structure, but allows a choice of path based on heavy overflow traffic from the fixed topology. NHR means fewer busy signals and faster connections to the end-user. It has been demonstrated that nonhierarchical routing can provide up to a 15 percent cost reduction over the traditional routing schemes of the telephone system.

The telephone companies are using a variety of approaches in their nonhierarchical schemes. In the U.S., AT&T uses a *decentralized* approach, each node involved in the route stores a predetermined series of alternate paths. The sequence of these paths and their number will vary depending on the time of day and the day of the week. Bell Canada uses a *centralized dynamic* technique. The routes through the Bell Canada network change as the traffic conditions change. Bell Canada maintains a central routing processor which receives switching information. The information contains the load on the units and descriptions of idle trunks that are available. The processor then selects the routes and sends this information back to the distributed switches. Many of the European systems do *source routing*. With this technique, the originating office chooses the routes.

Voice-oriented applications obviously use the telephone network. In addition, many data communications systems use the telephone network for transmission and reception of data. The process is workable, but expensive and awkward, because the majority of telephone local loops to our homes and offices are designed and constructed to transport human speech signals. These signals exhibit the well-known analog waveform characteristics.

However, computers, printers, terminals and other data communications devices are digital devices: numbers and other symbols are represented by discrete signals. Typically, these symbols take the form of binary (1 and 0) values by abrupt changes in voltage levels, current flow, or the pulsing of a laser/light emitting diode on an optical fiber system. Therefore, the interworking between voice and data systems requires rather cumbersome and expensive conversion systems.

History of and Inherent Problems with Coexistence of Analog and Digital Systems

Since the early 1960s, telephone companies, interexchange carriers, and network vendors have been implementing systems with digital technology. Today, many transmission components use digital transmission schemes, including such diverse devices as PBXs, multiplexers, and switches.

The process of digitization was developed in the 1960s to overcome some of the limitations of analog transmission and analog image recording. Several problems arise regarding the analog signal and how it is transmitted across the channel. First, the signal is relayed through amplifiers and other transducers. These devices are designed to perform the relaying function in a linear fashion; i.e., the waveform representing the signal maintains its characteristics from one end of the channel to the other. A deviation from this linearity creates a distortion of the waveform. All analog signals exhibit some form of nonlinearity (therefore, a distortion). Unfortunately, the intervening components to strengthen the signal, such as amplifiers, also increase the nonlinearity of the signal.

Second, all signals (digital and analog) are weakened (or attenuated) during transmission through the medium. The decay can make the signal so weak that it is unintelligible at the receiver. A high-quality wire cable with a large diameter certainly mitigates decay, but it cannot be eliminated.

Third, the current practice of using a mix of analog and digital systems in a tandem link creates significant engineering problems. For example, high-speed modems use analog phase shift modulation. These modems are particularly sensitive to phase shift. Digital systems occasionally create a phase shift when resynchronizing a digital link. This resynchronization creates serious problems for the voice band modems.

Fourth, voice and data systems have evolved using different signaling and control systems. The telephone signaling systems do not interface with data systems very well. However, it is accepted today that one digitally based control system can be used for voice and data networks, and systems have been designed to accomplish this feat. This approach reduces dramatically many interface and engineering problems.

Analog-to-Digital Conversion

Digital systems overcome many of these problems by representing the transmitted data with digital and binary images. The analog signal is converted to a series of digital numbers and transmitted through the communications channel as binary data. The digital numbers represent samples of the waveform.

Of course, digital signals are subject to the same kinds of imperfections and problems as the analog signal—decay and noise. However, the digital signal is discrete: The binary samples of the analog waveform are represented by specific levels of voltages, in contrast to the nondiscrete levels of an analog signal. Indeed, an analog signal has almost infinite variability. As the digital signal traverses the channel, it is only necessary to sample the absence or presence of a digital binary pulse—not its degree, as in the analog signal.

The mere absence or presence of a signal pulse can be more easily recognized than the magnitude or degree of an analog signal. If the digital signals are sampled at an acceptable rate and at an acceptable voltage level, the signals can then be completely reconstituted before they deteriorate below a minimum threshold. Consequently, noise and attenuation can be completely eliminated from the reconstructed signal. Thus, the digital signal can tolerate the problems of noise and attenuation much better than the analog signal.

The periodic sampling and regeneration process is performed by regenerative repeaters. The repeaters are placed on a channel at defined intervals. The spacing depends on the quality and size of the conductor, the amount of noise on the conductor, its bandwidth, and the bit rate of the transmission. The early digital systems required a repeater to be placed on the line at 6000 feet intervals. Today's optical fiber repeaters are spaced scores of miles apart from each other.

Several methods are used to change an analog signal into a representative string of digital binary images. Even though these methods entail many processes, they are generally described in three steps: *sampling*, *quantizing*, and *encoding*.

The devices performing the digitizing process are called channel banks or primary multiplexers. They have two basic functions: (1) converting analog signals to digital signals (and vise versa at the other end); and (2) combining the digital signals into a single time division multiplexed (TDM) data stream (and demultiplexing them at the other end). Multiplexing is discussed later in this appendix.

Analog-to-digital conversion is based on Nyquist sampling theory (see Figure A–3). This theory states that if a signal is sampled instantaneously at regular intervals and at a rate at least twice the highest frequency in the channel, the samples will contain sufficient information to allow an accurate reconstruction of the signal.

The accepted sampling rate in the industry is 8000 samples per second. Based on Nyquist sampling theory, this rate allows the accurate reproduction of a 4 kHz channel, which is used as the bandwidth for a voice-grade channel. The 8000 samples are more than sufficient to capture the signals in a telephone line if certain techniques (discussed shortly) are used.

The samples are stored and collected at the predetermined rate of 8000 times per second. The idea of the sampling process is to modulate a pulse carrier (see Figure A–3) with the sample of the analog signal. The pulse carrier is a sampling pulse clock, and can be modulated in a number of ways; pulse amplitude modulation (PAM) is the system used for this analysis.

With PAM, the pulse carrier amplitude is varied with the value of the analog waveform. As Figure A–3 shows, the pulses are fixed with respect to duration and position. PAM is classified as a modulation technique because each instantaneous sample of the wave is used to modulate the amplitude of the sampling pulse.

The 8 kHz sampling rate results in a sample pulse train of signals with a 125 microseconds (µsec.) time period between the pulses (1 second/8000 = .000125). Each pulse occupies 5.2 µsec. of this time period. Consequently, it is possible to interleave sampled pulses from other signals within the 125 µsec. period. The most common approach in North America utilizes 24 interleaved channels, which effectively fills the 125 µsec. time period (.0000052 * 24 = .000125). The samples are then multiplexed using TDM and put into a digital TDM frame. TDM provides an efficient and economical means of combining multiple signals for trans-

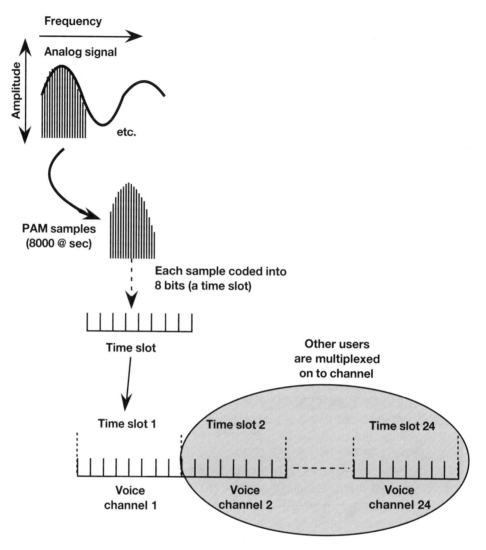

Figure A–3
Analog-to-digital conversion.

mission on a common facility, and is examined in a later section of this appendix.

In our efforts to understand the emerging technologies, it is important to understand that the 125 µsec. time period remains in most of these systems. It is used in FDDI II, ATM, MAN, SDH, and SONET.

Importance of the conversion process. We have now explained the method to convert analog signals into digital signals. While this example explained a voice-to-digital conversion, similar procedures allow video, music, and TV signals to be converted to digital signals, as well. As we shall see, all the emerging technologies discussed in this book (with some exceptions in the cellular and cordless arena) assume the user payload is all-digital. This approach is a significant step toward integrated networks, because all traffic is digitized and can be accommodated with one transport system.

Data Images over Voice Channels

Before proceeding further, one other conversion process must be examined: sending discrete digital bits over the telephone line. The telephone line has been designed for the transmission of analog voice signals. As discussed earlier, the shape and nature of the signals are quite different, and almost all local loops into homes and offices are analog. Therefore, even though it is quite possible to use all digital images inside a sophisticated integrated voice/data/video network, it is still necessary (and will be for many years to come) to work with analog signaling on local lines.

It is possible to redesign the analog link so that it will carry the digital signals, but such a redesign is an expensive process. Moreover, as just stated, the telephone local loops into homes and most offices are built for analog transmissions. Can you imagine how great a task it would be to replace all the telephone voice systems with digital systems?

In order to use the analog signal for data transmission (digital transmission), the industry developed a very clever and relatively simple device called a modem. The term is a derivative of the words *mod*ulator and *dem*odulator.

Three basic methods of digital-to-analog modulation are employed by modems (see Figure A–4). Some modems use more than one of the methods. Each method impresses the data on an analog *carrier* signal, which is altered to carry the properties of the digital data stream. The three methods are called *amplitude modulation, frequency modulation,* and *phase modulation*. They work as follows. Amplitude modulation alters the amplitude of the signal in accordance with the digital bitstream; frequency modulation alters the frequency of the signal in accordance with the digital bitstream; phase modulation alters the phase of the signal in accordance with the digital bitstream.

So, it has been demonstrated that digital signals can be sent over

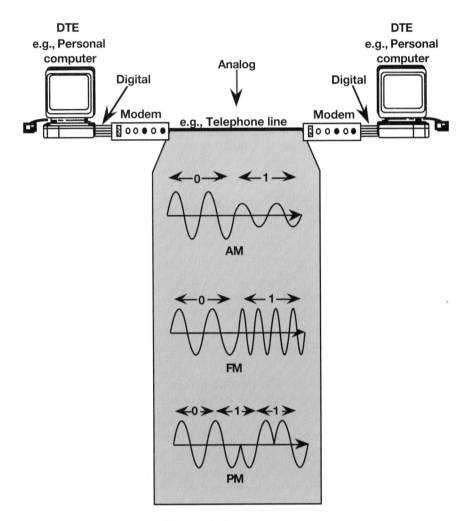

Figure A–4
Modulation techniques.

analog facilities, and these techniques have allowed the use of the telephone system to interconnect computers—in effect, providing a worldwide network for data communications. This capability has had the
effect of revolutionizing the world of information processing, and was the
fundamental component of the "wired society." It is difficult to overesti-
mate the effect this simple machine has had on our society.

Yet, it should also be emphasized that this technique still uses analog waveforms. Ultimately, these systems will be replaced with an all-digital technology. And as stated previously, most of the emerging technologies discussed in this book operate on the supposition that all transmissions are digital images.

FDM, TDM, and STDM

Most of the emerging technologies use some form of multiplexing. As discussed earlier in this appendix, multiplexers (MUXs) accept lower-speed voice or data signals from terminals, telephones and user applications and combine them into one high-speed stream for transmission onto a link. A receiving multiplexer demultiplexers and converts the combined stream into the original multiple lower-speed signals. Since several separate transmissions are sent over the same line, the efficiency ratio of the path is improved.

In the previous decade, the most widely used multiplexing technique was frequency-division multiplexing (FDM). While it has largely disappeared from end-user equipment, it is still widely used in telephone, microwave, and satellite carrier systems. As shown is Figure A–5a, this approach divides the transmission frequency range (the bandwidth) into narrower bands (called subchannels). The subchannels are lower-frequency bands, and each band is capable of carrying a separate voice or data signal. Consequently, FDM is used in a variety of applications such as telephone systems, radio systems, and the familiar cable television (CATV) that many people have installed in their homes. CATV provides a separate FDM band for each TV channel.

FDM decreases the total bandwidth available to each user, but the narrower bandwidth is usually sufficient for the low-speed devices. The transmissions from the multiple users are sent simultaneously across the path. Each user is allocated a fixed portion of the frequency spectrum.

Time-division multiplexing (TDM) provides a user the full channel capacity but divides the channel usage into time slots (see Figure A–5b). Each user is given a slot and the slots are rotated among the attached users. The TDM cyclically scans the input signals (incoming data) from the multiple incoming points. Bits, bytes, or blocks of data are separated and interleaved together into frames on a single high-speed communications line. TDMs are discrete signal devices and will not accept analog data. If a device has no traffic, its slot is empty, because the slots are preassigned to the devices. This approach works well for constant bit rate (CBR) applications, but leads to wasted capacity for variable bit rate

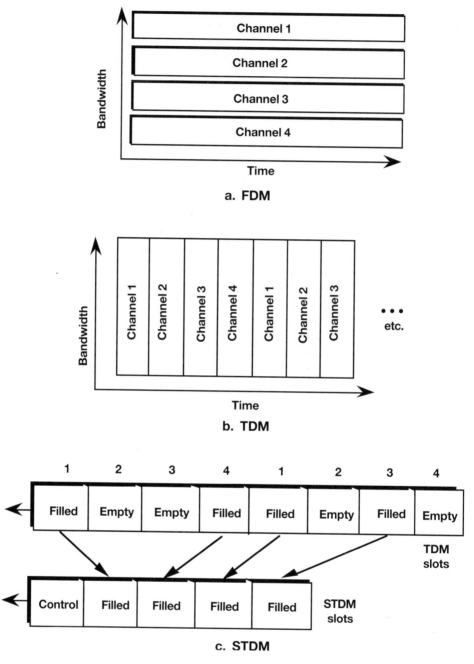

a. FDM

b. TDM

c. STDM

Figure A–5
Multiplexing operations.

(VBR) applications. The conventional TDM wastes the bandwidth of the communications line for certain applications when the time slots are not used. Vacant slots occur when an idle terminal has nothing to transmit in its slot or a conversation experiences periods of silence between the conversing parties.

Statistical TDMs (STDMs) dynamically allocate the time slots among active terminals (see Figure A–5c). Dedicated time slots (TDMs) are not provided for each port on a STDM. Consequently, idle terminal time does not waste the line's capacity. It is not unusual for two to five times as much traffic to be accommodated on lines using STDMs in comparison to a TDM.

However, it must be emphasized that statistical multiplexing performs well with "bursty" input streams. This term describes a system in which transmissions are sporadic, occurring at irregular intervals, i.e., VBR applications. The STDM takes advantage of this characteristic and assigns slots to devices that need the slots. Statistical multiplexers will not improve a system such as digitized voice, or remote job entry (RJE), which exhibits continuous, nonbursty characteristics.

Circuit, Message, Packet, and Cell Switching

Voice, video, and data signals are relayed in a network from one user to another through switches. This section provides an overview of prevalent switching technologies.

Figure A–6 compares and contrasts four types of switching technologies. The first is called *circuit switching* (Figure A–6a), which provides a direct connection between two components. The direct connection of a circuit switch serves as an open "pipeline", permitting the two end-users to utilize the facility as they see fit—within bandwidth and tariff limitations. Many telephone networks use circuit switching systems.

Circuit switching only provides a path for the sessions between data communications components. Error checking, session establishment, frame flow control, frame formatting, selection of codes, and protocols are the responsibility of the users. Little or no care of the data traffic is provided in the circuit switching arrangement. Consequently, the telephone network is often used as the basic foundation for a data communications network, and additional facilities are added by the value-added carrier, network vendor, or user organization.

Today, most circuit switching systems do not provide a direct connection through the switches. Rather, the traffic is stored in fast queues

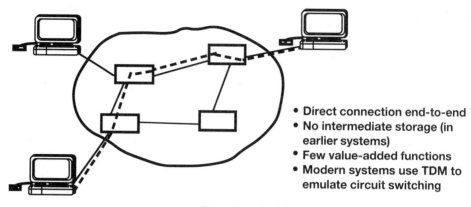

- Direct connection end-to-end
- No intermediate storage (in earlier systems)
- Few value-added functions
- Modern systems use TDM to emulate circuit switching

a. Circuit switching

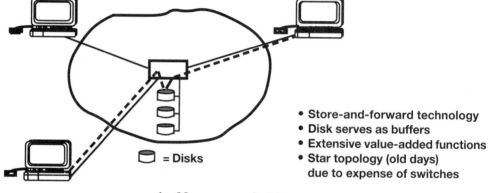

◯ = Disks

- Store-and-forward technology
- Disk serves as buffers
- Extensive value-added functions
- Star topology (old days) due to expense of switches

b. Message switching

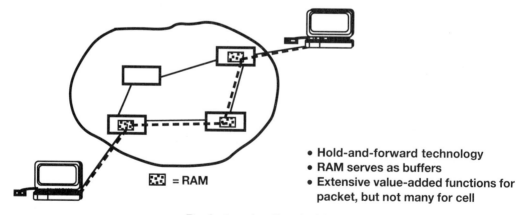

▦ = RAM

- Hold-and-forward technology
- RAM serves as buffers
- Extensive value-added functions for packet, but not many for cell

c. Packet and cell switching

Figure A–6
Comparison of circuit, message, packet, and cell switching.

in the switch and switched on to an appropriate output line with TDM techniques—a technique known as circuit emulation switching (CES).

In the 1960s and 1970s, the pervasive method for switching data communications traffic was *message switching*. The technology is still widely used in certain applications, such as electronic mail, but it is not employed in a backbone network (the primary switches are not message switches). The switch is typically a specialized computer. It is responsible for accepting traffic from attached terminals and computers. It examines the address in the header of the message and switches (routes) the traffic to the receiving station.

Message switching is a store-and-forward technology: The messages are stored temporarily on disk units at the switches. The traffic is not considered to be interactive or real-time. However, selected traffic can be sent through a message switch at very high speeds by establishing levels of priority for different types of traffic. In the 1960s and 1970s, switches were so expensive that most organizations built their networks with a star topology as shown in Figure A–6b, wherein one message switch routed traffic between all end-points. Of course, this often led to bottlenecks at the switch during heavy traffic loads, and the network was inoperable if the one switch went down.

Because of the problems with message switching, the industry began to move toward a different data communications switching structure in the 1970s: *packet switching* as shown in Figure A–6c. Packet switching distributes the risk to more than one switch, reduces vulnerability to network failure, and provides better utilization of the lines than does message switching.

Packet switching is so named because user data (for example, messages) are divided into smaller pieces. These pieces, or packets, have protocol control information (headers) placed around them and are routed through the network as independent entities.

Packet switching has become the prevalent switching technique for data communications networks. It is used in such diverse systems as private branch exchanges (PBXs), LANs and even within multiplexers.

Each packet only occupies a transmission line for the duration of the transmission; the line is then made available for another user's packet. Packet switching employs statistical multiplexing techniques and takes advantage of the intermittent (bursty) transmission characteristics that are common to many data applications and workstations.

The idea of using a small packet is to limit its size so it does not occupy the line for an extended period. Therefore, bursty (asynchronous) traffic is supported quite well in a packet network environment. Notwithstanding, the modern packet network is designed to support continuous, high-volume traffic as well as asynchronous, low-volume traffic, and each user is given a variable transmission capability based on the user's throughput and response time requirements.

The concepts of *cell switching* and packet switching are similar; each attempts to process the traffic in memory as quickly as possible. However, as discussed in Chapter 1, cell switching is distinguished by the use of a small protocol data unit (PDU) relative to packet switching. For example, the ATM switch uses a cell size of 53-octets, while the X.25-based packet switches default to packet sizes of 128-octets with extensions to 4096-octets. The principal difference is that packet networks process a variable size PDU and cell networks process a fixed size PDU. Both networks accept variable size traffic, but an ATM network segments this traffic into small, fixed length cells *before* the traffic enters the network. Cell switching is discussed in considerable detail in Chapter 7, so we need not elaborate any further in this tutorial.

Network Routing Operations

The majority of WANs are based on the following criteria: (a) how control is maintained in the network, (b) how routing directories or tables are managed, and (c) how routing decisions are made (route determination).

The following discussion examines each of these criteria, and Figure A–7 provides a reference to be used during this discussion. It should be understood during this analysis that many vendors use various combinations of the various methods discussed here. As we shall see, some vendors use a combination of (for example) partial and full directories in their large networks. Additionally, some vendors use a combination of source and non-source routing as well.

Network routing control is usually categorized as *centralized* or *distributed* routing. The centralized routing network uses a network control center to determine the routing of the packets. The packet switches are limited in their functions. They are not very "intelligent," which usually translates into decreased costs in operating the switches. However, centralized control suffers from the vulnerability of one central site to possible failure. Consequently, network control centers (NCCs) are usually duplicated (duplexed). Centralized routing is also vulnerable to bottlenecks at the central site; therefore considerable care must be taken to ensure that the central site has the capability to handle all routing tasks.

Distributed routing requires more intelligent switches, thereby providing a more resilient network, because each node makes its own routing decisions without regard to a centralized control center. Distributed routing is also more complex, but its advantages over the centralized approach has made it the preferred routing method in most networks today.

The manner in which a network stores its routing information varies. Typically, routing information is stored in a software table (also known as a directory). This table contains a list of destination nodes. These destination nodes are identified with some type of network address. Along with the network address (or some type of label, such as a virtual circuit identifier) there is an entry describing how the switch/router is to relay the traffic. In most implementations, this entry simply

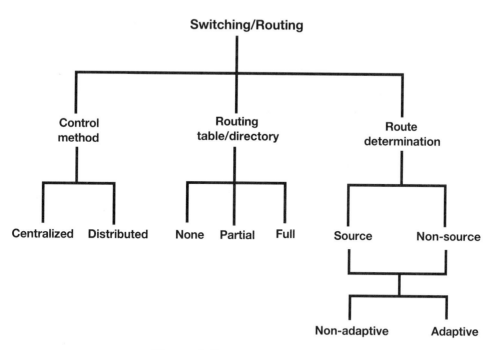

Figure A–7
Switching/routing operations.

lists the next node that is to receive the traffic in order to relay it to its destination.

Small networks typically provide a full routing table at each routing node. With this approach, the switch has full knowledge of the topology of its network and has reachability information to all nodes within the network. An update to the table occurs with each node sending routing information to every other node.

For larger networks, full directories are cumbersome. They require too many entries and are expensive and time-consuming to update. Moreover, the messages that are exchanged between the routers for a large network using full directories can reduce the available bandwidth for end-user traffic. Consequently, these larger networks are often divided into smaller subnetworks (also called areas). In this manner, a designated router within each area can advertise its area's reachability to other area gateways or routers. This diminishes the amount of traffic that must be exchanged within the full network and also reduces the size of the routing tables.

Broadcast networks contain no routing directories. Their approach is to send the traffic to all devices. The devices examine the destination address and, if appropriate, copy the traffic and send it to the appropriate entity in the machine. Additionally, fully meshed networks use no routing tables because they already connect to every node in the network.

The method in which internetworking PDUs (i.e., frames, cells, datagrams, or packets) are routed between networks is sometimes a source of confusion. The two major methods to perform routing are *source routing* and *non-source routing*.

Source routing derives its name from the fact that the transmitting device (the source) dictates the route of the PDU through a network or several networks. The source (host) machine places the addresses of the "hops" (the intermediate networks or routers) in the PDU. Such an approach means that the IWUs need not perform address maintenance; they simply use an address in the PDU to determine where to route the frame.

In contrast, non-source routing makes decisions about the route and does not rely on the PDU to contain information about the route. Non-source routing is usually associated with bridges and is quite prevalent in LANs.

We learned that most communications networks perform routing by the a routing directory or table. The directory contains directions that instruct the switches to transmit a packet to one of several possible output links at the switch. Typically, packet network directories are organized around these approaches.

Fixed (static) directory. This directory changes at system generation time. It is established when users subscribe to the network, and remains static for an individual session, unless problems occur in the network

Session-oriented directory (logon). This directory changes with each user session—each time a user logs on to the network. It remains static for an individual session once the session is established.

Adaptive (dynamic) directory (during session). This routing table may change during the user session. The change can occur if conditions in the network change, such as the loss of a link between switches. In such an instance, the network adjusts its routing tables to reflect the change, and traffic is routed "around" the problem area.

THE CHALLENGE OF INTEGRATING VOICE, DATA, AND VIDEO APPLICATIONS

At first glance, it might appear that the integration of voice, data, and video is a simple matter. After all, once the analog signals have been converted to digital images, all transmissions can be treated as data. However, if we examine the transmission requirements of voice and data, we find that they are quite different. Figure A–8 acts as a reference point for this discussion.

Voice and lower quality video transmissions exhibit a high tolerance for errors. If an occasional PDU is distorted, the fidelity of the voice or video reproduction is not severely affected. In contrast, data PDUs have a low tolerance for errors. One bit corrupted likely changes the meaning of the data. Also, high-quality video has less tolerance for errors as well.

Yet another difference between voice, data, and video transmissions deals with network delay. For the packetized voice to be translated back to an analog signal in a real-time mode, the network delay for voice packets must be constant and generally must be low—usually less than 200 milliseconds. For data packets, the network delay can vary considerably. Indeed, the data packets can be transmitted asynchronously through the network, without regard to timing arrangements between the sender and the receiver. In contrast, video transmissions must have a timing relationship between the sender and the receiver.

Voice and video packets can afford (on occasion) to be lost or dis-

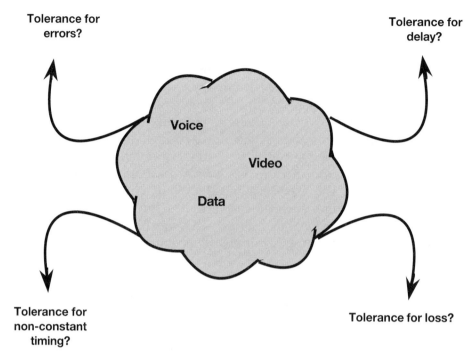

Tolerance for errors?

Tolerance for delay?

Tolerance for non-constant timing?

Tolerance for loss?

Figure A–8
Integrating voice, data, and video.

carded. In the event of excessive delays in the network, the packets may be discarded because they are of no use. Again, the loss does not severely affect voice fidelity if the lost packets are less than 1 percent of the total packets transmitted. As discussed before, data packets can ill afford to be lost or discarded.

Finally, voice and video transmissions require a short queue length at the network nodes in order to reduce delay, or at least to make the delay more predictable. The short voice packet queue lengths can experience overflow occasionally, with resulting packet loss. However, data packets require the queue lengths to be longer to prevent packet loss in overflow conditions.

Notwithstanding these problems, the integration of voice and data is an idea whose time has come. As we shall see in this section, the concept of packet switching plays a key role in integrated voice/data packet networks. Two concepts that have gained wide attention are *fast packet switching (FPS)* and *hybrid switching*. These terms are used in deference to industry practice.

Some of the emerging technologies, such as FDDI II, MAN, and ATM use a variety of the techniques explained in the next section.

Fast Packet Switching (FPS)

FPS employs many of the conventional packet switching conventions, and also the use of very high-speed switches and high-capacity trunks. The technique relies on a session-oriented protocol that uses the same physical path through the network for the duration of the connection, unless severe problems occur, such as damaged lines, failed switches, etc.

The network route is determined at the source node by the node accessing a network topology database. The database contains information on the current load on the switches and trunks in the network. Periodically, the database is updated at each node to reflect the capacities and delays within the network.

When a route for a call is first calculated, the source node also calculates and stores several backup routes. In the event the connection request is not successful, the source selects the next best route and reinitiates the call. During a session, a failure of a node or link will cause the source node to choose a backup path. The destination node also follows suit and uses the same physical path.

Fast packet switching is an attractive combination of adaptive and non-adaptive packet switching. It chooses a path before a session begins (adaptive) and keeps the same path during the session (non-adaptive). The ongoing updates to the topology database increase the speed of the call setup since the route is predetermined. Moreover, since each node stores the topology data base, it is not necessary to route the call setup request to a control center.

The FPS network must ensure that video packets and a certain ratio of the voice packets experience as little delay as possible. One approach to manage delay is through the use of a total delay (T) parameter that is periodically adjusted. The source node initializes a packet lifetime field in the packet header to zero, and each node increments the field by its own delay value. Since each node has a knowledge of the network topology, a reasonably accurate time can be calculated of a probable one-way delay. This one-way delay factor (C) is added to a variable amount of delay (V) and the accumulated delay (L) in order to equal T ($T = L + C + V$). On occasion, when $L + C + V$ exceed T, the packet is discarded because it has lost its value.

T is recalculated to keep the end-to-end delay low, as well as to keep discarded packets at a low percentage to the total packets transmitted. V can be changed to keep the delay constant. If T exceeds a maximum threshold, voice calls will be blocked. Moreover, if network congestion occurs, some schemes "dynamically trim" the 32 kbit/s adaptive differential pulse code modulation signal (ADPCM) to a reduced rate of 24 or 16 kbit/s.

Hybrid Switching

Hybrid switching uses a combination of circuit switching for voice connections and packet switching for data connections. The actual integration of the two technologies can be achieved in a number of ways. Generally, the integration is accomplished in the switch.

Hybrid switching uses the conventional concepts of TDM and framing. The bitstream on each trunk arrives in slots and the hybrid switch assigns these slots to different connections. The system is designed for each frame to be the standard T1 carrier length of 125 μsec.

A frame consists of N slots (for example, in the T1 carrier, the frame is 24 slots). A specific number of these N slots are reserved for data packets and they cannot be used for voice transmissions. If multiple frames were examined, one data packet could be spread out over several contiguous and successive frames.

In hybrid switching, as in fast packet switching, the circuit switched connections have priority over the remaining slots in the frame (N–M). However, the packet traffic is allowed to use the N–M area of the frame, if the circuit switched traffic leaves any of these areas unused. This provides for temporary additional packet capacity for the data.

The voice packets in a hybrid system are managed through an additional feature called digital speech interpolation (DSI). During the intervals in which two individuals do not speak, with periods of silence, the time intervals are filled with other transmissions. If circuit capacity is available and not needed for the DSI, the packet traffic will be allowed to "grab" some of the additional capacity. The attractive feature of DSI is that it is often possible to concentrate as many as twice the number of incoming circuits to the outgoing circuits, since more than half of a typical voice conversation is silence.

The call setup for hybrid switching is the same as the call setup structure for fast packet switching. The data packet handling procedures for hybrid switching is also the same as that in FPS.

Layered Protocols, OSI, and TCP/IP

INTRODUCTION

This appendix introduces the architecture of the Open Systems Interconnection (OSI) standards, the architecture of the Internet protocols, and the Transmission Control Protocol (TCP)/Internet Protocol (IP). The purpose of the appendix is to provide the reader with enough information on these subjects to understand their relationships and roles in the emerging communications technologies.

PROTOCOLS AND THE OSI MODEL

Machines communicate through established conventions called protocols. Since computer systems provide many functions to users, more than one protocol is required to support these functions. A convention is also needed to define how the different protocols of the systems interact with each other to support the end user. This convention is referred to by several names: network architecture, communications architecture, or computer-communications architecture. Whatever the term used, most systems are implemented with layered protocols.

The Open Systems Interconnection (OSI) Model was developed in the early 1980s by several standards organizations, principally led by ITU-T and ISO. It is now a widely used layered model for describing how communications are standardized between different vendors' equipment and for the design and implementation of computer networks (even though the OSI protocols have seen only limited use).

The ITU-T publishes its OSI Model specifications in the X.200– X.290 Recommendations. The X.200 documents contain slightly over 1100 pages. The ISO publishes its OSI Model in several documents, but does not use a common numbering scheme.

The Model is organized into seven layers. Each layer contains several to many protocols that are invoked based on the specific needs of the user. Each protocol in a layer need not be invoked, and the Model provides a means for two users to negotiate the specific protocols they desired for the session.

As suggested in Figure B–1, each layer is responsible for performing specific functions to support the end user application. One should not think of a layer as monolithic code, rather each layer is divided into smaller operational entities. These entities are then invoked by the end user to obtain the services defined in the Model.

An end user is permitted to negotiate services within (a) layers in its own machine, or (b) layers at the remote machine. This capability allows, for example, a relatively low-function machine to indicate to a relatively high-capability machine that it may not support all the operational entities supported by the high-level machine. Consequently, the machines will still be able to communicate with each other, albeit at a lesser mode of service. Conversely, if two large-scale computers, each with the full OSI stack, wish to exchange traffic with a rich functional environment, they may do so by negotiating the desired services.

OSI Layer Operations

The layers of the OSI Model and the layers of vendors' models (such as IBM's SNA) contain communications functions at the lower three or four layers. From the OSI perspective, as demonstrated in Figure B–2, it is intended that the upper four layers reside in the host computers.

This does not mean that the lower three layers reside only in the network. The hardware and software implemented in the lower three layers also exist at the host machine. End-to-end communications, however, occurs between the hosts by invoking the upper four layers, and between the hosts and the network by invoking the lower three layers.

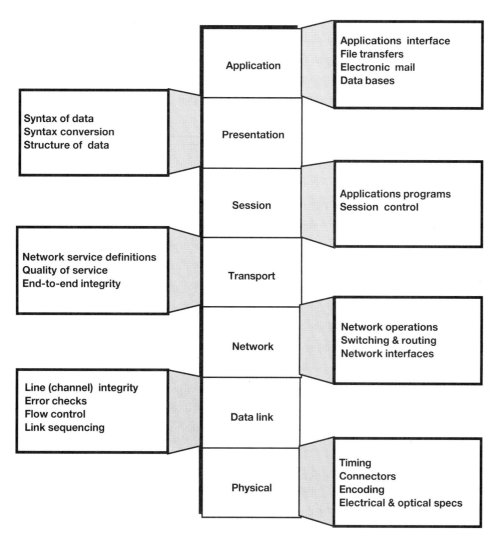

Figure B-1
Functions of layers.

This concept is shown in this figure with the arrows drawn between the layers in the hosts and the network. Additionally, the upper four layers may also reside in the network, for the network components to communicate with each other and obtain the services of these layers.

The end user rests on top (figuratively speaking) of the application layer. Therefore, the user obtains all the services of the seven layers of the OSI Model.

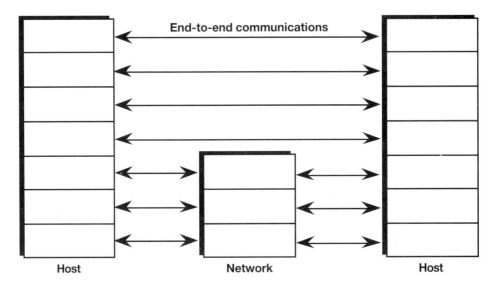

Figure B–2
Position of host and network layers.

In the OSI Model, a layer is considered to be a *service provider* to the layer above it (see Figure B–3). This upper layer is considered to be a *service user* to its lower layer. The service user avails itself of the services of the service provider by sending a *transaction* to the provider. This transaction informs the provider as to the nature of the service that is to be provided (at least, requested).

In so far as possible, the service provider does provide the service. It may also send a transaction to its user to inform it about what is going on.

At the other machine (B in this figure), the operation at A may manifest itself by service provider B's accepting the traffic from service provider A, providing some type of service, and informing user B about the operation. User B may be allowed to send a transaction back to service provider B, which may then forward traffic back to service provider A. In turn, service provider A may send a transaction to user A about the nature of the operations at site B.

The OSI Model provides several variations of this general scenario.

In accordance with the rules of the Model, a layer cannot be bypassed. Even if an end user does not wish to use the services of a particular layer, the user must still "pass through" the layer on the way to

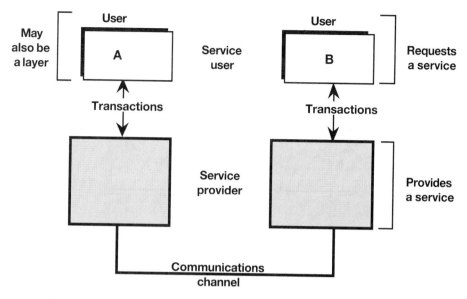

Figure B–3
The layer as a service provider.

the next adjacent layer. This pass-through may only entail the invocation of a small set of code, but it still translates to overhead.

However, every function in each layer need not be invoked. A minimum subset of functions may be all that is necessary to "conform" to the standard.

Layered network protocols allow interaction between functionally paired layers in different locations without affecting other layers. This concept aids in distributing the functions to the layers. In the majority of layered protocols, the data unit, such as a message or packet, passed from one layer to another is usually not altered, although the data unit contents may be examined and used to append additional data (trailers/headers) to the existing unit.

Each layer contains entities that exchange data and provide functions (horizontal communications) with peer entities at other computers. For example, in Figure B–4, layer N in machine A communicates logically with layer N in machine B, and the N+1 layers in the two machines follow the same procedure. Entities in adjacent layers in the same computer interact through the common upper and lower boundaries (vertical communications) by passing parameters to define the interactions.

Typically, each layer at a transmitting station (except the lowest in

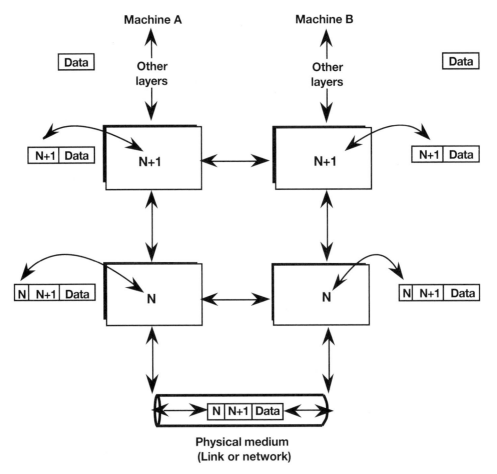

Figure B–4
Adding header information for horizontal communications.

most systems) adds "header" information to data. The headers are used to establish peer-to-peer sessions across nodes, and some layer implementations use headers to invoke functions and services at the N+1 or N adjacent layers. The important point to understand is that, at the receiving site, the layer entities use the headers created by the *peer entity* at the transmitting site to implement actions.

Figure B–5 shows an example of how machine A sends data to machine B. Data is passed from the upper layers or the user application to layer N+1. This layer adds a header to the data (labeled N+1 in the

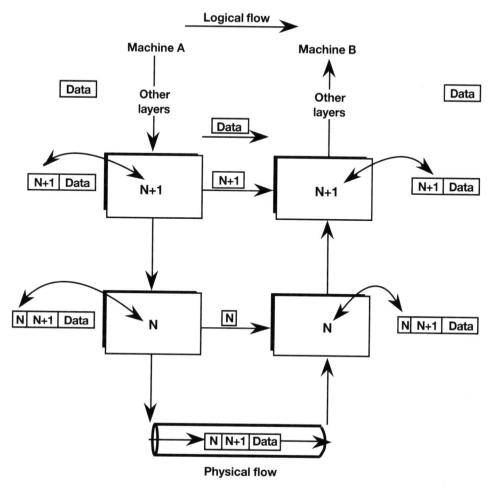

Figure B–5
Machine A sends data to machine B.

figure). Layer N+1 also performs actions based on the information in the transaction that accompanied the data from the upper layer.

Layer N+1 then passes the data unit with its N+1 header to layer N. Layer N performs the requested actions, based on the information in the transaction, and adds its header N to the N+1 traffic. This appended traffic is then passed across the communications line (or through a network) to the receiving machine B.

At B, the process is reversed. The headers that were created at A are used by the *peer layers* at B to determine that actions are to be taken. As the traffic is sent up the layers, the respective layer "removes" its header, performs defined actions, and passes the traffic on up to the next layer.

The user application at site B is presented only with user data—which was created by the sending user application at site A. These user applications are unaware (one hopes) of the many operations in each OSI layer that were invoked to support the end user data transfer.

The headers created and used at peer layers are not to be altered by any non-peer layer. As a general rule, the headers from one layer are treated as transparent "data" by any other layer.

There are some necessary exceptions to this rule. As examples, data may be altered by a non-peer layer for the purposes of compression, encryption, or other forms of syntax changing. This type of operation is permissible, as long as the data are restored to the original syntax when presented to the receiving peer layer.

As an exception to the exception, the presentation layer may alter the syntax of the data permanently, when the receiving application layer has requested the data in a different syntax (such as ASCII instead of BIT STRING).

Protocol entities.   One should not think that an OSI layer is represented by one large monolithic block of software code. While the Model does not dictate how the layers are coded, it does establish the architecture whereby a layer's functions can be structured and partitioned into smaller and more manageable modules. These modules are called *entities*.

The idea of the Model is for peer entities in peer layers to communicate with each other. Entities may be active or inactive. An entity can be software or hardware. Typically, entities are functions or subroutines in a program.

A user is able to "tailor" the universal OSI services by invoking selected entities through the parameters in the transactions passed to the service provider, although vendors vary on how the entities are actually designed and invoked.

Service access points (SAPs).   SAPs are OSI addresses and identifiers. The OSI Model states that: *an (N+1)-entity requests (N)-services via an (N)-service access point (SAP) which permits the (N+1)-entity to interact with an (N)-entity.*

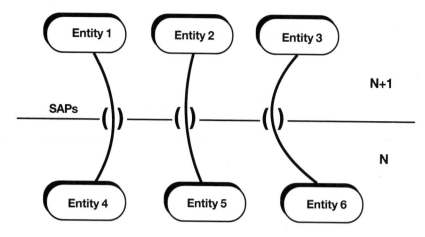

Figure B–6
OSI service success points (SAPs).

Perhaps the best way to think of a SAP is that it is a software port (an identifier) that allows the two adjacent layers in the same machine to communicate with each other (see Figure B–6). SAPs may also be exchanged across machines in order to identify a process in the machine. In the OSI Model, the SAP can identify a protocol entity that resides in a layer. For example, a SAP value could be reserved for E-mail, while a different SAP value could identify file server software. The reader may know about a UNIX socket—a concept that is similar to the OSI SAP.

THE INTERNET PROTOCOLS (TCP/IP)

In the early 1970s, several groups around the world began to address the problem of network and application compatibility. At that time, the term "internetworking," which means the interconnecting of computers and/or networks, was coined. The concepts of internetworking were pioneered by the ITU-T, the ISO, and especially the original designers of the ARPANET. (The term "ARPA" refers to the Advanced Research Projects Agency, which is a U.S. Department of Defense [DOD] organization.)

Perhaps one of the most significant developments in these standardization efforts was ARPA's decision to implement the Transmission Control Protocol (TCP) and the Internet Protocol (IP) around the UNIX operating system. Of equal importance, the University of California at Berkeley was selected to distribute the TCP/IP code. Because the TCP/IP code was nonproprietary, it spread rapidly among universities, private companies, and research centers. Indeed, it has become the standard suite of data communications protocols for UNIX-based computers.

In order to grasp the operations of TCP/IP, several terms and concepts must first be understood. The Internet uses the term *gateway* or *router* to describe a machine that performs relaying functions between networks, which are often called subnetworks. The term does not mean that they provide fewer functions than a conventional network. Rather, it means that the two networks consist of a full logical network with the subnetworks contributing to the overall operations for internetworking. Stated another way, the subnetworks comprise an *internetwork* or an *internet*.

An internetworking gateway is designed to remain transparent to the end user application. Indeed, the end user application resides in the host machines connected to the networks; rarely are user applications placed in the gateway. This approach is attractive from several standpoints. First, the gateway need not burden itself with application layer protocols. Since they are not invoked at the gateway, the gateway can dedicate itself to fewer tasks, such as managing the traffic between networks. It is not concerned with application level functions such as database access, electronic mail, and file management.

Second, this approach allows the gateway to support any type of application because the gateway considers the application message as nothing more than a transparent protocol data unit (PDU).

In addition to application layer transparency, most designers attempt to keep the gateway transparent to the subnetworks and vice versa. That is to say, the gateway does not care what type of network is attached to it. The principal purpose of the gateway is to receive a message that contains adequate addressing information which then enables the gateway to route the message to its final destination or to the next gateway. This feature is also attractive because it makes the gateway somewhat modular; it can be used on different types of networks.

The Internet Layers

Figure B–7 shows the relationship of subnetworks and gateways to layered protocols. In this figure it is assumed that the user application in host A sends an application PDU to an application layer protocol in host B, such as a file transfer system. The file transfer software performs a

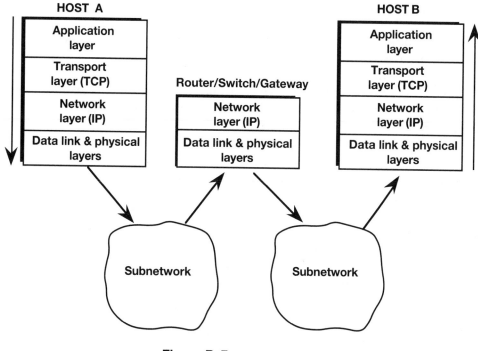

Figure B–7
Internet layer operations.

variety of functions and appends a file transfer header to the user data. In many systems, the operations at host B are known as *server* operations, and the operations at host A are known as *client* operations.

As indicated with the arrows going down in the protocol stack at host A, this PDU is passed to the transport layer protocol, TCP. This layer performs a variety of operations and adds a header to the PDU passed to it. The PDU is now called a *segment*. The traffic from the upper layers is considered to be data to the transport layer.

Next, TCP passes the segment to the network layer, also called the IP layer, which again performs specific services and appends a header. This unit (now called a *datagram* in internet terms) is passed down to the lower layers. Here, the data link layer adds its header as well as a trailer, and the data unit (now called a *frame*) is launched into the network by the physical layer. Of course, if host B sends data to host A, the process is reversed and the direction of the arrows is changed.

The Internet protocols are unaware of what goes on inside the network. The network manager is free to manipulate and manage the datagram portion of the frame in any manner necessary. However, in most instances the datagram remains unchanged as it is transmitted through the subnet. The frame passes through the subnetwork until its arrival at the gateway where it is processed through the lower layers and passed to the IP (network) layer (where it is again called a datagram). Here, routing decisions are made based on the addresses provided by the host computer.

After these routing decisions have been made, the datagram is passed to the communications link that is connected to the appropriate subnetwork (consisting of the lower layers). The datagram is re-encapsulated into the data link layer PDU (usually called a frame) and passed to the next subnetwork. As before, this unit is passed through the subnetwork transparently (usually), where it finally arrives at the host B.

The host B receives the traffic through its lower layers and reverses the process that transpired at host A. That is to say, it decapsulates the headers by stripping them off in the appropriate layer. The header is used by the layer to determine the actions it is to take; the header governs the layer's operations.

IP Functions

IP is an example of a connectionless service. It permits the exchange of traffic between two host computers without any prior call setup. However, these two computers usually share a common connection-oriented transport protocol. Since IP is connectionless, it is possible that the datagrams could be lost between the two end user's stations. For example, the IP gateway enforces a maximum queue length size, and if this queue length is violated, the buffers will overflow. In this situation, the additional datagrams are discarded in the network. For this reason, a higher level transport layer protocol (such as TCP) is essential to recover from these problems.

IP hides the underlying subnetwork from the end user. In this context, it creates a virtual network to that end user. This aspect of IP is quite attractive because it allows different types of networks to attach to an IP gateway. As a result, IP is reasonably simple to install and, because of its connectionless design, it is quite robust.

Since IP is a best effort datagram-type protocol, it has no reliability mechanisms. It provides no error recovery for the underlying subnetworks. It has no flow-control mechanisms. The user data (datagrams) may be lost, duplicated, or even arrive out of order. It is not the job of IP to deal with these problems. As we shall see later, most of the problems are passed to the next higher layer, TCP.

IP supports fragmentation operations. The term "fragmentation" refers to an operation wherein a PDU is divided or segmented into smaller units. This feature can be quite useful because all networks do not use the same size PDU.

For example, X.25-based WANs typically employ a PDU (called a packet in X.25) with a data field of 128-octets. Some networks allow negotiations to a smaller or larger PDU size. The Ethernet standard limits the size of a PDU to 1500-octets. Conversely, proNET-10 stipulates a PDU of 2000 octets.

Without the use of fragmentation, a gateway would be tasked with trying to resolve incompatible PDU sizes between networks. IP solves the problem by establishing the rules for fragmentation at the gateway and reassembly at the receiving host.

IP is designed to rest on top of the underlying subnetwork—insofar as possible in a transparent manner. This means that IP assumes little about the characteristics of the underlying network or networks. As stated earlier, from the design standpoint this is quite attractive to engineers, because it keeps the subnetworks relatively independent of IP.

As the reader might expect, the transparency is achieved by encapsulation. The data sent by the host computer are encapsulated into an IP datagram. The IP header identifies the address of the receiving host computer. The IP datagram and header are further encapsulated into the specific protocol of the transit network. For example, a transit network could be an X.25 network or an Ethernet LAN.

After the transit network has delivered the traffic to an IP gateway, its control information is stripped away. The gateway then uses the destination address in the datagram header to determine where to route the traffic. Typically, it then passes the datagram to a subnetwork by invoking a subnetwork access protocol (for example, Ethernet on a LAN, or X.25 on a WAN). This protocol is used to encapsulate the datagram header and user data into the headers and trailers that are used by the subnetwork. This process is repeated at each gateway, and eventually, the datagram arrives at the final destination where it is delivered to the receiving station.

IP addresses.   TCP/IP networks use a 32-bit address to identify a host computer and the network to which the host is attached. The structure of the IP address is:

IP Address = Network Address + Host Address.

IP addresses are classified by their formats (see Figure B–8). Four formats are permitted: class A, class B, class C, or class D. The first bits of the address specify the format of the remainder of the address field in relation to the network and host subfields. The host address is also called the local address (also called the REST field).

The *class A* addresses provide for networks that have a large number of hosts. The host ID field is 24-bits. Therefore, 2^{24} hosts can be identified. Seven bits are devoted to the network ID, which supports an identification scheme for as many as 127 networks (bit values of 1 to 127).

Class B addresses are used for networks of intermediate size. Fourteen bits are assigned for the network ID, and 16-bits are assigned for the host ID. *Class C* networks contain fewer than 256 hosts (2^8). Twenty-one bits are assigned to the network ID. Finally, *class D* addresses are reserved for multicasting, which is a form of broadcasting but within a limited area.

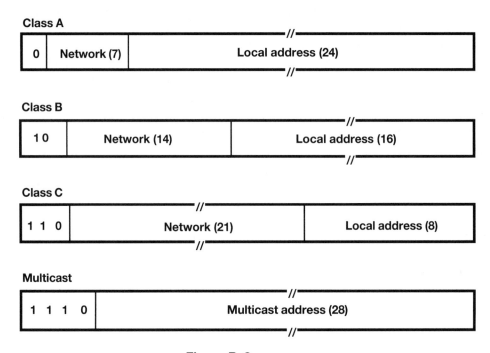

Figure B–8
IP address formats.

TCP Operations

The Internet Protocol (IP) is not designed to recover from certain problems, nor does it guarantee the delivery of traffic. IP is designed to discard datagrams that are outdated, have exceeded the number of permissible transit hops in an internet, etc. UDP also is a connectionless protocol and, as such, discards traffic under similar conditions.

However, certain user applications require assurance that all datagrams have been delivered safely to the destination. Furthermore, the transmitting user may need to know that the traffic has been delivered at the receiving host. The mechanisms to achieve these important services reside in TCP.

The job of TCP may be quite complex. It must be able to satisfy a wide range of applications requirements, and equally important, it must be able to accommodate to a dynamic environment within an internet.

TCP establishes and manages sessions (logical associations) between its local users and these users' remote communicating partners. This means that TCP must maintain an awareness of the user's activities in order to support the user's data transfer through the internet.

TCP resides in the transport layer of the conventional seven-layer model. It is situated above IP and below the upper layers. Figure B–7 also illustrates that TCP is not loaded into the gateway. It is designed to reside in the host computer or in a machine that is tasked with end-to-end integrity of the transfer of user data. In practice, TCP is usually placed in the user host machine.

TCP is designed to run over the IP. Since IP is a connectionless network, the tasks of reliability, flow control, sequencing, opens, and closes are given to TCP. Although TCP and IP are tied together so closely that they are used in the same context, "TCP/IP," TCP can also support other connectionless protocols, such as UDP.

Many of the TCP functions (such as flow control, reliability, sequencing, etc.) could be handled within an application program. But it makes little sense to code these functions into each application. Moreover, applications programmers are usually not versed in error-detection and flow control operations. The preferred approach is to develop generalized software that provides community functions applicable to a wide range of applications, and then invoke these programs from the application software. This allows the application programmer to concentrate on solving the application problem and it isolates the programmer from the nuances and problems of networks.

TCP is a *connection-oriented protocol*. TCP maintains status information about each user datastream flowing into and out of the TCP module. TCP is responsible for the *reliable transfer* of each of the bytes passed to it from an upper layer.

A sequence number is assigned to each byte transmitted. The receiving TCP module uses a checksum routine to check the data for damage that may have occurred during the transmission process. If the data are acceptable, TCP returns a positive acknowledgment (ACK) to the sending TCP module.

TCP receives the data from an upper layer protocol (ULP) in a *stream-oriented* fashion. This operation is in contrast to many protocols in the industry. Stream-oriented protocols are designed to send individual characters and *not* blocks, frames, datagrams, etc.

TCP also checks for duplicate data. In the event the sending TCP transmits duplicate data, the receiving TCP discards the redundant data.

TCP also supports the concept of a *push* function. This operation is used when an application wants to make certain that all the data that it has passed to the lower layer TCP has been transmitted. In so doing, it governs TCP's buffer management.

TCP uses an inclusive acknowledgment scheme. The acknowledgment number acknowledges all bytes up to and including the acknowledgment number less one.

The receiver's TCP module is also able to *flow control* the sender's data, which is a very useful tool to prevent buffer overrun and a possible saturation of the receiving machine.

TCP also has a very useful facility for *multiplexing* multiple user sessions within a single host computer onto the ULPs. This is accomplished through some rather simple naming conventions for ports and sockets in the TCP and IP modules. TCP provides *full duplex transmission* between two TCP entities.

TCP also provides the user with the capability to specify levels of *security* and *precedence* (priority level) for the connection. Even though these features are not implemented on all TCP products, they are defined in the TCP standard.

TCP provides a *graceful close* to a virtual circuit (the logical connection between the two users). A graceful close ensures that all traffic has been acknowledged before the virtual circuit is removed.

TCP ports. A TCP upper layer user in a host machine is identified by a *port* address. The port address is concatenated with the IP internet address to form a *socket*. This address must be unique throughout the internet and a pair of sockets uniquely identifies each endpoint connection. As examples:

Sending socket = Source IP address + Source port number
Receiving socket = Destination IP address + Destination port number

Although the mapping of ports to higher-layer processes can be handled as an internal matter in a host, the Internet publishes numbers for frequently used higher-level processes.

Even though TCP establishes numbers for frequently used ports, the numbers and values above 255 are available for private use. The remainder of the values for the assigned port numbers have the low-order 8-bits set to zero. The remainder of these bits are available to any organization to use as they choose. Be aware that the numbers 0 through 255 are reserved and they should be avoided.

Management Information Bases (MIBs)

INTRODUCTION

Purpose of a MIB

The management information base (MIB) is one of the most important parts of a network management system. The MIB identifies the network elements (managed objects) that are to be managed. It also contains an unambiguous name that is to be associated with each managed object.

From the conceptual viewpoint, MIBs are really quite simple. Yet, if they are not implemented properly, the network management protocols (such as CMIP or SNMP) are of little use. These network management protocols rely on the MIB to define the managed objects in the network.

The MIB represents the managed objects. Conceptually, the MIB is a virtual information store. In reality, it is a database that contains information about the managed objects. For example, a MIB can contain information about the number of packets that have been sent and received across an X.25 interface; it can contain statistics on the number of connections that exist on a TCP port, etc.

Objects are defined by the use of a special language called Abstract Syntax Notation One (ASN.1). Each object must have a name, a syntax, and an ASN.1 encoding.

The MIB defines the syntax (Integer, Boolean, etc.) of the information carried with the network management protocols. It also contains information that describes each network management user's ability to access elements of the MIB. For example, user A might have read-only capabilities to a MIB while another user may have read/write capabilities.

Network management protocols (with few exceptions) do not operate directly on the managed object, they operate on the MIB. In turn, the MIB is the reflection of the managed object. How this reflection is conveyed is a proprietary decision. The important aspect of the MIB is that it defines (a) the elements that are managed, (b) how the user accesses them, and (c) how they can be reported.

EXAMPLES OF MIB OBJECTS AND OTHER ENTRIES

This section shows several examples of MIBs and the managed objects that reside in the MIB.

- A MIB provides a unique id of each managed object (a managed network element).
  Example: "ifType" for type of interface at a physical port

- A MIB can store read-only values that are standardized across a network, enterprise.
  Example: The value 6 always identifies an Ethernet interface
  Example: The value 19 always identifies an E1 interface

- Thus, a network management message could contain an object name and a reserved value.
  Example: ifType = 6, which means the message is reporting on an Ethernet interface

- A MIB defines how users may access network management data.
  Examples: READ ONLY, WRITE ONLY, etc.

- A MIB stores statistical (performance) information pertaining to managed objects.
  Example: "ipInReceives = 4000", which means an IP router has received 4000 datagrams

Example: "snmpInTooBigs = 32", which means 32 SNMP messages have been received with an error status of tooBig

- A MIB may describe the contents/structure of fields of the network management message.

Example:

Type (T)   Always defined (e.g., Integer, ASCII, etc.)

Length (L)   Sometimes defined (e.g., one byte for Ethernet ifType, but variable for ipInReceives)

Value (V)   Sometimes defined (e.g., 6 for Ethernet ifType, but variable for ipInReceives)

- Network management components must have compatible MIBs for full internetworking to occur.

Emerging Communications Technologies Worksheet

Technology name: _____

New technology? _____

Targeted applications? _____

Topology dependent? _____

Media dependent? _____

LAN/WAN based? _____

Competes with: _____

Cell/Frame based? _____

Connection management? _____

Flow control (explicit/implicit)? _____

ACKs and NAKs? _____

Traffic discard option? _____

Bandwidth on demand? _____

Addressing/identification scheme? _____

References

In addition to the formal standards for the systems described in this seminar, these references should prove useful to the seminar delegate. Many of them were used for the development of this workbook.

[AHMA93] Amhad, R., & Halsall, F. (1993). Interconnecting high-speed LANs and backbones, *IEEE Network*.

[ARMT93] Armitage, G.J., and Adams, K.M. (1993). Packet reassembly during cell loss, *IEEE Network*.

[ATM92a] ATM Forum. (June 1, 1992). *ATM user-network interface specification, Version 2.0.*

[ATM93a] ATM Forum. (August 5, 1993). *ATM user-network interface specification, Version 2.4,* (to be released as Version 3.0).

[ATT89a] Observations of error characteristics of fiber optic transmission systems, CCITT SGXVIII, San Diego, CA, January, 1989.

[BELL90a] (May, 1993). Generic requirements for frame relay PVC exchange service, TR-TSV-001369, Issue 1.

[BLAC93] Black, U. (1993). *Data link protocols.* Prentice Hall.

[BNR92a] Bell Northern Research. (1992). Global systems for mobile communications, Telesis, *94*.

[CDPD93] Cellular Digital Packet Data System Specification. (July 19, 1993 [Release 1.0]). Published by a consortium: Ameritech Mobile Communications, Inc., Bell Atlantic Mobile Systems, Contel Cellular, Inc., GTE Mobile Communications, Inc., McCaw Cellular Communications, Inc., NYNEX Mobile Communications, Inc., Pactel Cellular, Southwestern Bell Mobile Systems.

[CHER92] Cherukuri, R., (August 26, 1992). Voice over frame relay networks. A technical paper issued as Frame Relay Forum, FRF 92.33.

[CHEU92] Cheung, N.K. (1992). The infrastructure of gigabit computer Networks. *IEEE Communications Magazine*, April.

[DAVI91] Davidson, R.P., & Muller, N.J. (1991). *The guide to SONET*, Telecom Library, Inc.

[DELL92] Dell Computer, Intel, and University of Pennsylvania, A study compiled by Marty Baumann, *USA Today,* date not available.

[dePr91] dePrycker, M. (1992a). *Asynchronous transfer mode.* Ellis Harwood Ltd.

[dePR92] de Prycker, M. (1992b). ATM in Belgian Trial. *Communications International,* June.

[ECKB92] Eckberg, A. E. (1992). B-ISDN/ATM traffic and congestion control. *IEEE Network*, September.

[FORU92] Frame Relay Forum Technical Committee. (1992). Frame relay network-to-network interface, phase 1 implementation agreement, Document Number FRF 92.08R1–Draft 1.4, May 7, 1992.

[GASM93] Gasman, L. (1993). ATM CPE—Who is providing what? *Business Communications Review*, October.

[GOKE93] Goke, L.R., & Lipovski, G.J. (1973). Banyan networks for partitioning multiprocessor systems. First Annual Symposium on Computer Architecture.

[GRIL93] Grillo, D., MacNamee, R.J.G., & Rashidzadeh, B. (1993). Towards third generation mobile systems: A European possible transition path. *Computer Networks and ISDN Systems*, 25(8).

[GRON92] Gronert, E. (1992). MANs make their mark in Germany. *Data Communications International*, May.

[HALL92] Hall, M. (ed.). (1992). LAN-based ATM products ready to roll out. *LAN Technology*, September.

[HAND91] Handel, R., Huber, MN (1991). *Integrated broadband networks: An introduction to ATM-based networks.* Addison-Wesley.

[HEWL91] Hewlett-Packard, Inc. (1991). Introduction to SONET, A tutorial. Author.

[HEYW93] Heywood, P. (1993). PTTs gear up to offer high-speed services. *Data Communications,* August.

[HUNT92] Hunter, P. (1992). What price progress? *Communications International,* June.(need vol#, pages)

[ITU93a] ITU-TS. (1993). ITU-TS draft recommendation Q93.B, B-ISDN user-network interface layer 3 specification for basic call/bearer control. May, 1993.

[JOHN91] Johnson, J.T. (1991). Frame relay mux meets cell relay switch, *Data Communications*, October.

[JOHN92] Johnson, J.T. (1992). Getting access to ATM. *Data Communications LAN Interconnect*, September 21.

[LEE89] Lee, W.C.Y. (1989). *Mobile cellular telecommunications systems.* McGraw-Hill.

[LEE93] Lee, B.G., Kang, M., & Lee, J. (1993). *Broadband telecommunications technology.* Artech House.

[LISO91] Lisowski, B. (1991). Frame relay: What it is and how it Works. *A Guide to Frame Relay, Supplement to Business Communications Review*, October.

[LYLE92] Lyles, J.B., & Swinehart, D.C. (1992). The emerging gigabit environment and the role of the local ATM. *IEEE Communications Magazine*, April.

[MCQU91] McQuillan, J.M. (1991). Cell relay switching, *Data Communications,* September.

[MINO93] Minoli D., (June, 1993). Proposed Cell Relay Bearer Service Stage 1 Description, T1S1.1/93-136 (Revision 1), ANSI Committee T1 (T1S1.1).

[NOLL91] Nolle, T. (1991). Frame relay: Standards advance, *Business Communications Review Supplement,* October.

[PERL85] Perlman, R. (1985). An algorithm for distributed computation of spanning tree in an extended LAN. *Computer Communications Review*, *15*(4) September.

[ROSE92] Rosenberry, W., Kenney, D., & Fisher, G. (1992). *Understanding DCE.* O'Reilly & Associates.

[SALA92] Salamone, S. (1992). Sizing up the most critical issues. *Network World.*

[STEW92] Steward, S.P. (1992). The world report 92. *Cellular Business,* May.

[WADA89] Wada, M. (June 1989). "Selective Recovery of Video Packet Loss Using Error Concelement," IEEE Journal of Selected Areas in Communications.

[WALL91] Wallace, B. (1991). Citicorp goes SONET. *Network World*, November.

[WERK92] Wernik, M., Aboul-Magd, O., & Gilber, H. (1992). Traffic management for B-ISDN services. *IEEE Network*, September.

[WEST92] Westgate, J. (1992). *OSI Management.* NCC Blackwell.

[WILL92] Williamson, J. (1992). GSM bids for global recognition in a crowded cellular world. *Telephony*, April.

[YOKO93] Yokotani, T., Sato, H., & Nakatsuka, S. (1993). A study on a performance improvement algorithm in DQDB MAN. *Computer Networks and ISDN Systems, 25*(10).

Abbreviations

AAL: ATM adaptation layer
AC: Access control
ACF: Access control field
ADM: Add-drop multiplexer
ADPCM: Adaptive differential pulse code modulation
AFI: Authority format identifier
AMPS: Advanced mobile phone system
ANSI: American National Standards Institute
APS: Automatic protection switching
ASN.1: Abstract Notation One
AT: Abatement threshold
ATM: Asynchronous transfer mode
AU: Access unit
AU: Administration unit
AUG: Administration unit group
B bit: Busy bit
B-ICI: Broadband intercarrier interface
B_c: Committed burst rate
BCD: Binary coded decimal
B_e: Excess burst rate
BECN: Backward explicit congestion notification
BER: Basic encoding rules
BIP-8: Bit interleaved parity
BISDN: Broadband ISDN
BLER: Block error rate
BOCs: Bell Operating Companies
BOM: Beginning of message
BRI: Basic rate interface
BSC: Base station controller
BSHR: Bidirectional SHR
BT: Burst tolerance
BTS: Base transceiver station

C-plane: Control plane
C/R: Command/response
C/SAR: Convergence services and segmentation and reassembly services
C: Containers
CBR: Constant bit rate
CDPD: Cellular digital data packet system specification
CDV: Cell delay variation
CEPID: Connection endpoint identifier
CES: Circuit emulation switching
CIB: CRC 32 indication bit
CIR: Committed information rate
CLLM: Consolidated link layer management
CLNP: Connectionless network layer protocol
CLNS: Connectionless network service
CMIP: Common Management Information Protocol
CNM: Customer network management
COM: Continuation of message
CP: Common part
CPE: Customer premises equipment
CRC: Cyclic redundancy check
CS: Convergence sublayer
CSU: Channel service unit
CUGs: Closed user groups
DA: Destination address
DAC: Dual attachment concentrator
DAS: Dual attachment station
DCS: Digital cross connect
DE: Discard eligibility
DFI: Domain specific part identifier
DLCI: Data link connection identifier
DMPDU: Derived MAC PDU

417

DNS: Domain name system
DQDB: Distributed queue dual bus
DSP Domain specific part (DSP), 283
DSU: Data service unit
DTE: Data terminal equipment
E/O: Electrical/optical converter
EA: Address extension
EC: European Commission
ECSA: Exchange Carriers Standards
    Association
EIR: Equipment identity register
EIR: Excess information rate
EOM: End of message
ES: End systems
ESI: End system identifier
ETSI: European Telecommunications
    Standards Institute
F-ES: Fixed end systems
FC: Frame control
FCC: Federal Communications Commission
FCS: Frame check sequence
FDDI: Fiber distributed data interface
FDM: Frequency division multiplexing
FEC: Forward error correction
FECN: Forward explicit congestion notification
FIFO: First in-first out
FISU: Fill-in signal unit
FNS: FDDI network service
FPS: Fast packet switching
FR-SSCS: Frame relay service specific conver-
    gence sublayer
FRF: Frame Relay Forum
FRS: Frame relay service
FS: Full status
GCRA: Generic cell-rate algorithm
GFC: Generic flow control
GIO: Generic interface of operation
GSM: Groupe Speciale Mobile/Glogal System
    for Mobile Communications
HCS: Header check sequence
HDLC: High level data link control
HDTV: High-definition TV
HLPI: High level protocol identifier
HLR: Home location register
HRC: Hybrid ring control
I field: Information field
ICIP: ICI protocol
IDI: Initial domain identifier
IDN: Integrated digital network
IDP: Initial domain part
ILMI: Interim local management interface
IMAC: Isochronous MAC
IMPDU: Initial MAC protocol data unit
INSI: Intra-network switching interface
IP: Internet Protocol
IS: Intermediate system
ISDN: Integrated services digital network
ISSI: Inter-switching system interface
IT: Information type

ITU: International Telecommunication Union
IXC: Interexchange carriers
LANs: Local area networks
LAPB: Link access procedure, balanced
LAPD: Link access procedure, D channel
LAPM: Link access procedure for modems
LATA: Local access and transport area (LATA)
LCN: Logical channel number
LD: Laser diode
LECs: Local exchange carriers
LED: Light emitting diode
LLC: Logical link control
LME: Layer management entity
LMI: Local management interface
LSB: Least significant bit
LSSU: Link status signal unit
MAC SDU: MAC service data unit
MAC: Media access control
MAN: Metropolitan area network
MAP: Mobile application part
MBS: Maximum burst size
MCF: MAC convergence function
MD-IS: Mobile data intermediate system
MHF: Mobile home function
MIB: Management information base
MID: Message identifier
MNLP: Mobile network location protocol
MS: Mobile station
MSB: Most significant bit
MSC: Mobile switching center
MTBSO: Mean time between service outages
MTP: Message transfer part
MTSO: Mobile telephone switching office
MTTR: Mean time to restore
MTTSR: Mean time to service restoral
NCC: Network control centers
NCI: Network control information
NEI: Network entity identifier
NLPID: Network level protocol identifier
NMP: Network management protocol
NNI: Network-to-network interface
NSAP: Network service access points
OAM&P: Operations, administration, mainte-
    nance, and provisioning
OAM: Operation, administration, and mainte-
    nance
OC-1: Optical carrier-level 1 signal
OS/NE: Operations system/network element
OS: Operating System
OSPF: Open shortest path first
OT: Onset threshold
OUI: Organizational unique identifier
P: Preamble
PA: Pre-arbitrated
PARC: Xerox Palo Alto Research Center
PC: Payload CRC
PCM: Pulse code modulation
PCR: Peak cell rate
PCS: Personal computer systems

PDN: Public data network
PDU: Protocol data units
PHY: Physical layer
PI: Protocol identification
PID: Protocol identifier
PL: Payload length
PLCP: Physical layer convergence protocol
PLP: Packet layer procedures
PMD: Physical layer medium dependent interface
PPS: Path protection switching
PSR: Previous slot read
PTTs: Postal Telephone and Telegraph Ministries
PVC: Permanent virtual circuits
QA: Queued arbitrated
QD: Queuing delay
QOS: Quality of service
QPSX: Queued packed synchronous exchange
R: Reserved
RBOCs: Regional Bell Operating Companies
RD: Routing domain identifier
RER: Residual error rate
RN: Routing number
S: Status
SA: Source address
SAAL: Signaling ATM adaptation layer
SAP: Service access points
SAPI: Service access point identifier
SAR: Segmentation and reassembly sublayer
SAS: Signal attachment station
SCP: Service control points
SD: Starting delimiter
SD: Switching delay
SDDI: Shielded twisted pair specifications
SDH: Synchronous digital hierarchy
SDT: Structured data transfer
SDU: Service data unit
SE: Status enquiry
SEFS: Severely erred framing seconds
SHR: Self-healing ring
SIP: SMDS interface protocol
SIR: Sustained information rate
SMDS: Switched multi-megabit data service
SMT: Station management function
SN: Sequence number
SNAP: Subnetwork access protocol
SNI: Subscriber-to-network interface
SNMP: Simple Network Management Protocol
SNP: Sequence number protection
SONET: Synchronous optical network
SPE: Synchronous payload envelope
SPVC: Semi-permanent virtual circuit

SS7: Signaling system #7
SS: Signaling system
SSCF: Service specific coordination function
SSCOP: Service specific connection-oriented part
SSCP: Signaling connection control part
SSM: Single segment message
SSP: Service specific part
SSP: Service switching points
ST: Segment type
ST: Slot type field
STDM: Statistical time division multiplexer
STM: Synchronous transport module
STP: Signaling transfer point
STS-n: Synchronous transport signal
STS: Synchronous transport signal
SVC: Switched virtual circuit
TA: Terminal adapter
TACS: Total access communications system
TAT: Theoretical arrival time
T_c: Time interval
TCP: Transmission control protocol
TDM: Time division multiplexing
TE1, TE2: Terminal types 1 and 2
TEI: Terminal endpoint identifier
THT: Token holding timer
TID: Terminal identifier
TMM: Transmission monitoring machine
TTRT: Target token rotation time
TU: Tributary units
TUG: TU groups
UDP: User datagram protocol
UDT: Unstructured data transfer
UI: Unnumbered information
ULPs: Upper layer protocols
UMTS: Universal Mobile Telecommunications System
UNI: User-to-network interface
UPC: Usage parameter control
UPT#: Universal personal telecommunications number
USHR: Unidirectional self-healing ring
UT: User terminal
VBR: Variable bit rate
VC: Virtual containers
VCI: Virtual channel connection
VLR: Visitor location register
VPC: Virtual path connection
VPI: Virtual path identifier
VPN: Virtual private networks
WANs: Wide area networks
XA SMDS: Exchange access SMDS
XID: Exchange identifier

Index

CARROLL COLLEGE LIBRARY

2 5052 00621966 3

WITHDRAWN
CARROLL UNIVERSITY LIBRARY

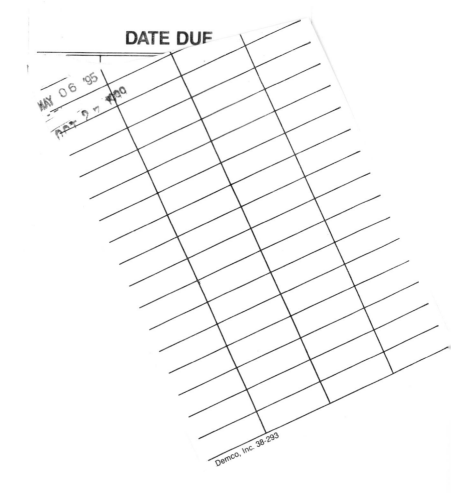

DATE DUE

MAY 06 '95

Demco, Inc. 38-293